CIRCUIT DESIGN for CMOS VLSI

by

John P. Uyemura
Georgia Institute of Technology

Kluwer Academic Publishers
Boston/Dordrecht/London

Distributors for North America:
Kluwer Academic Publishers
101 Philip Drive
Assinippi Park
Norwell, Massachusetts 02061 USA

Distributors for all other countries:
Kluwer Academic Publishers Group
Distribution Centre
Post Office Box 322
3300 AH Dordrecht, THE NETHERLANDS

Library of Congress Cataloging-in-Publication Data

Uyemura, John P. (John Paul), 1952-
Circuit design for CMOS VLSI / by John P. Uyemura.
p. cm.
Includes index.
ISBN 0-7923-9184-5 (acid free paper)
1. Integrated circuits--Very large scale integration. 2. Metal oxide semiconductors, Complementary. 3. Electronic circuit design.
I. Title.
TK7874.U93 1992
621.39'5--dc20 91-33124
CIP

Printed on acid-free paper.

Printed in the United States of America

Foreword

During the last decade, CMOS has become increasingly attractive as a basic integrated circuit technology due to its low power (at moderate frequencies), good scalability, and rail-to-rail operation. There are now a variety of CMOS circuit styles, some based on static complementary conductance properties, but others borrowing from earlier NMOS techniques and the advantages of using clocking disciplines for precharge-evaluate sequencing. In this comprehensive book, the reader is led systematically through the entire range of CMOS circuit design. Starting with the individual MOSFET, basic circuit building blocks are described, leading to a broad view of both combinatorial and sequential circuits. Once these circuits are considered in the light of CMOS process technologies, important topics in circuit performance are considered, including characteristics of interconnect, gate delay, device sizing, and I/O buffering. Basic circuits are then composed to form macro elements such as multipliers, where the reader acquires a unified view of architectural performance through parallelism, and circuit performance through careful attention to circuit-level and layout design optimization. Topics in analog circuit design reflect the growing tendency for both analog and digital circuit forms to be combined on the same chip, and a careful treatment of BiCMOS forms introduces the reader to the combination of both FET and bipolar technologies on the same chip to provide improved performance.

Both designers new to CMOS as well as those with considerable experience will find a rich selection of topics in this up-to-date book, written in a style that is thoughtful, detailed, and broad from the system perspective. Readers can have confidence that designs based on these well-taught principles will produce circuits of outstanding contemporary performance for increasingly demanding applications.

Jonathan Allen
Consulting Editor

Dedication

This book is dedicated to the ladies in my life:

Melba Valerie Uyemura
(my wife)

Ruby Shizue Uyemura
(my mother)

Valerie Elizabeth Hanako Uyemura
and
Rebecca Christine Shizue Uyemura
(my daughters)

for their love and encouragement.

植　村

Contents

Preface

The field of CMOS integrated circuits has reached a level of maturity where it is now a mainstream technology for high-density digital system designs. This book deals with circuit design in an integrated CMOS environment. Emphasis is placed on understanding the operation, performance, and design of basic digital circuits such as logic gates, latches, and adders. The topical outline and the level of presentation has been gauged such that the book should be of use to both students and practicing engineers.

CMOS admits to a wide variety of design variations. While static logic is the most common, powerful dynamic system design styles have been developed and applied to improve performance and packing density. When selecting the material for the book, the topics were examined for unique or interesting circuit/systems techniques, useful application examples, and the occurrence of a particular subject in the open literature. Once the topics were chosen for inclusion in the book, decisions on the length and depth of the presentation had to be made. In general, the discussions center around the basic operational and design concepts, with circuit examples to reinforce the material. References are provided at the end of each chapter if more details are needed.

No specific knowledge of CMOS circuits is assumed; however, some background in electronics and integrated circuit fabrication and layout is required to understand some of the more advanced sections. A general treatment of digital ICs at the level of **Analysis and Design of Digital Integrated Circuits** (Hodges and Jackson, McGraw-Hill) is sufficient. Georgia Tech offers a 2-quarter sequence in digital MOS IC design. The first course uses the textbook **Fundamentals of MOS Digital Integrated Circuits** (Uyemura, Addison-Wesley), and centers around the basics of circuits, devices, and chip design. The second course has evolved to the point where it is exclusively devoted to CMOS using this book and the current literature.

This book attempts to present a field which has evolved from a highly specialized technique to the dominant silicon technology over a span of twenty-plus years. It is difficult to give complete discussions of all of the topics which are covered. Hopefully, the knowledge gained from reading the book will be worth the investment of time. I apologize for all errors and omissions in the writing; even with a CMOS-based microcomputer, these problems seem to persist!

Acknowledgments

I would like to thank the reviewers who spent many hours reading the original manuscript. Professor Charles Zukowski (Columbia University) did an amazingly thorough job analyzing the approach, content, and writing. His detailed analyses was a great help when preparing the final version. Professor M. Annaratone (ETH, Zurich; DEC) deserves a special note of thanks. The original proposal for this project was an updating of his earlier title **Digital CMOS Circuit Design** (Kluwer, 1986). However, as the writing progressed, the book took on distinct characteristics of its own due to the dynamic evolution of the field. In spite of the obvious divergence from the original plan, Professor Annaratone graciously read and provided thoughtful comments on the entire manuscript. Professor Reginald J. Perry (FSU/FAMU) also caught many errors in the original manuscript. I would also like to thank the many readers who have sent in corrections to the first printing.

Mr. Carl Harris of Kluwer has constantly given encouragement and support to this project. I have watched Carl take Kluwer from a small start-up operation to one which is a leading publisher in the field of advanced electrical and computer engineering. I congratulate him on his success, and am happy that this book can be added to his growing list of titles.

I would like to thank my wife Melba for her love and support during yet another long writing project. Although lost hours can never be regained, she still remains enthusiastic about my work. My parents, Reverend George and Ruby Uyemura, have always supported me in everything I have ever attempted. On this, the occasion of their 50th wedding anniversary, I again extend my love and thanks. Finally, I would like to thank my little Valerie for her help during the writing. She always keeps me in line by making it clear when it is time to stop work and start playing with her and *Bear*. Watching her grow and learn has taught me what is really important in life. This has become even more meaningful for me now that she has a new sister Christine as her best friend.

J. P. Uyemura
February, 1993
Atlanta, GA

CIRCUIT DESIGN FOR CMOS VLSI

Chapter 1

Introduction to CMOS

Integrated electronic circuits which are based on complementary metal-oxide-semiconductor (**CMOS**) technologies are firmly established in modern electronics. CMOS provides the important characteristics needed for high-density logic designs. Moreover, with recent developments in the field of BiCMOS, it is anticipated that we have a technology which will provide a transition to the next century.

In this chapter we will examine CMOS from a bird's eye vantage point in an effort to place the subject into a reasonable perspective. Then we will be in a position to discuss the content and structure of the book.

1.1 Why Study CMOS?

CMOS circuits have many advantages when compared with other integrated circuit technologies. The ones which are most often quoted are

- Low power dissipation
- High integration density

but there are other reasons which make silicon CMOS the technology of choice for many applications. Let's examine some of the more important aspects of the technology.

Power Dissipation

Static CMOS circuits have the nice characteristic that power supply current is only required when a gate is undergoing a switching event. If the inputs are at stable logic 0 or logic 1 values, then the static power dissipation P_{static}

is limited to that generated by leakage currents. When compared with nMOS or bipolar circuits, this is a distinct advantage since it leads to reduced heating.

Although the static power dissipation in a CMOS circuit is very small, the circuit exhibits dynamic dissipation whenever the inputs are changed. The power consumption of a switched network is proportional to the switching frequency:

$$P_{dynamic} \propto f. \tag{1.1}$$

The total power dissipation P of the circuit is then written as

$$P = P_{static} + P_{dynamic}. \tag{1.2}$$

At low frequencies, both contributions are small. However, the dynamic power dissipation increases with switching frequency, and can dominate in high-speed circuits. At the system level, the total power dissipation depends on both the frequency of the switching and the number of gates which are undergoing a transition.

High Logic Integration Density

CMOS circuits can be structured in various ways. Standard static CMOS gates require two transistors for every input. Thus, a 3-input NAND or NOR gate will use 6 MOSFETs. An equivalent nMOS-only design uses 4 transistors, so we can conclude that CMOS does not lead to a smaller device count. Instead, the size of the device becomes the important consideration. CMOS circuit design allows us freedom to adjust the size according to the desired transient response specifications. This results in a tradeoff between speed versus area.

Advances in silicon processing technology have been instrumental in propelling CMOS to the forefront of integrated circuits. Linewidths as small as one-half of a micron are possible using optical lithography; x-ray and electron-beam patterning allows 0.1 [μm] resolution. Logic density is increased as the device size decreases, allowing CMOS to achieve greater integration densities than possible with a bipolar technology. These features have helped propel CMOS to a leading technology.

Logic Swings

CMOS gates allow **rail-to-rail** output logic voltage swings. For example, if a 5 volt power supply is used, then the output voltage range is (approximately) [0,5] volts. As a comparison, a bipolar TTL (transistor-transistor logic) gate only gives outputs in the range [0.3,3.6] volt with the same power supply. Rail-to-rail swings provide better noise immunity and aid in designing reliable logic circuits.

Symmetrical Transient Response

Transient gate response is critical for synchronizing the flow of logic signals. A CMOS circuit can be designed to give symmetrical output switching times by adjusting the size of the transistors. This means that the time needed for the output to switch from a logic 0 to a logic 1 can be made equal to the time needed to switch from a logic 1 to a logic 0. Symmetrical response patterns helps to simplify timing in a large system design.

Dynamic Circuit Designs

MOS-based circuits intrinsically possess capacitive nodes, i.e., the nodes are dominated by parasitic capacitance. The value of the capacitance directly affects the switching times in static logic gates, with large values giving long switching times. Although large capacitances tend to slow down the switching speed of static logic circuits, it is possible to use these nodes for dynamic charge storage. The charge then acts as the logic parameter and can quickly be transferred using transistor networks. The best known example of this type of circuit is probably a DRAM (dynamic random-access memory) cell. However, dynamic CMOS has been developed to a much higher level of sophistication, with entire system design styles possible. Dynamic logic circuits can generally provide higher logic densities and also have other advantages in timing and speed.

Bipolar Integrated Circuits

Bipolar emitter-coupled logic (ECL) is the fastest silicon logic available. It has been the basis of mainframe computer design in the past, and is still holding strong as we move towards the 21st century. The main reason why it has not taken over the microprocessor market is due to much higher power dissipation levels and subsequent heating. Bipolar technology is still an important approach to integrated logic circuits, but it fills a different niche than CMOS. Merging bipolar and CMOS into BiCMOS tends to provide the best aspects of both worlds. A chapter on BiCMOS has thus been included in the book to deal with this evolving technology.

Gallium Arsenide?

One question which arises when benchmarking various circuit technologies is

> *What about gallium arsenide? Doesn't it have the potential to "blow silicon out of the water"?*

Well, the answer is a loud, resounding **Yes!**. Gallium arsenide (GaAs) does indeed have the potential to create much faster switching circuits than those possible in silicon. The main reason for this is that the electron mobility μ_n is much larger in GaAs, indicating that we can make transistors react to higher frequencies.

Assuming that we know the answer to the question, then we must ask

> *Why study silicon CMOS?*

to at least justify this book, if not our careers. There are actually several reasons why silicon CMOS is so important, even when compared to evolving GaAs (and other) technologies. Some of these are as follows.

Materials Costs: After engineering costs have been amortized, the cost of materials becomes a limiting factor in pricing an integrated circuit. Gallium arsenide does not occur naturally, and must be grown; this increases the cost of production regardless of other factors. Silicon, on the other hand, is as plentiful as dirt. Thus, the costs of materials is minimal.

Technology Know-How: Silicon is the best studied element on earth. Billions of dollars have been invested to understand and use silicon for semiconductor device applications. Although we have also amassed a tremendous amount of knowledge about GaAs, it does not yet compare to our silicon technology. In addition, the problems of treating compound semiconductors are in a different league, increasing the problem further.

Applications: There are applications where the speed advantage offered by GaAs is needed. However, the vast majority of logic chips do not need to process data at frequencies in the gigabit/sec range due to other problems. For these circuits, silicon is still the designer's choice.

The above comments illustrate that the growth in popularity of CMOS circuits is due to many factors. CMOS constitutes the primary tool for tackling very large-scale integration (VLSI) system design problems, and for exploration into the regime of ultra large-scale integration (ULSI). We might go so far as to say that the current generation of microprocessors may not have been possible without CMOS technology.

1.2 Basic Concepts

Metal-oxide-semiconductor field-effect transistors (MOSFETs) are excellent voltage-controlled electronic switches for constructing logic circuits. The unique complementary nature of CMOS arises from the use of opposite polarity MOSFETs in designing circuits. The MOSFET types are **n-channel** (nMOS) and **p-channel** (pMOS) transistors, and will be represented by the schematic symbols illustrated in Figure 1.1. Those on the left are the complete symbols and explicitly show the four device electrodes. Simplified symbols such as those shown on the right are commonly used in practice. An nMOS transistor conducts current using negatively-charged electrons, while a pMOS transistor relies on positively-charged holes for current flow. The opposite polarities give the devices complementary characteristics which can be used to create a unique circuit family.

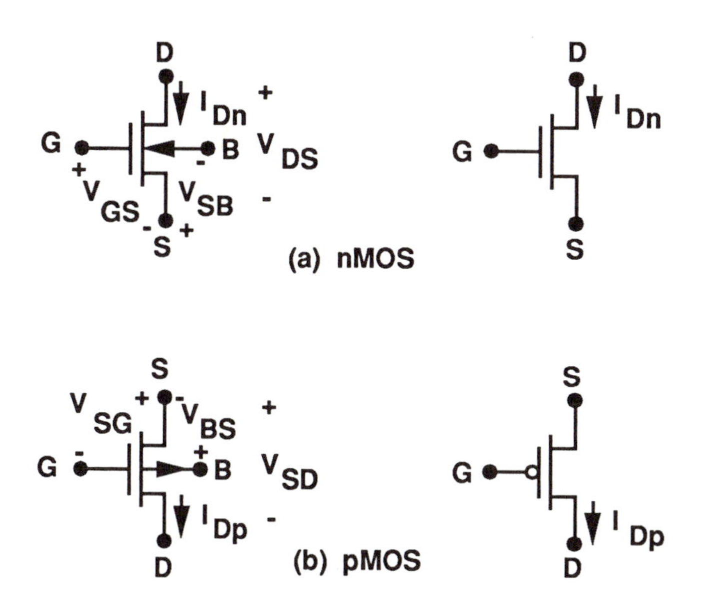

Figure 1.1: nMOS and pMOS Transistor Symbols

MOSFETs have four terminals: the **gate**, **drain**, **source**, and **bulk** (or, **substrate**). The **drain current** I_D flows between the drain and source, and is the main current flow path through a MOSFET. The primary device voltages used to control the current flow are V_{GS} and V_{DS}. The source-bulk voltage V_{SB} affects the turn-on characteristics of the transistor, which in turn affects the value of I_D. As shown by the simplified symbols in Figure 1.1, the bulk electrode (which specifies the polarity) is often omitted for simplicity[1]; to distinguish between nMOS and pMOS transistors, we use an "inversion bubble" at the gate of a pMOSFET as shown.

Applying a voltage to the gate electrode establishes a conduction path between the drain and source. An n-channel MOSFET conducts with a positive gate-source voltage $V_{GS} > 0$ applied; positive current is defined as flowing into the drain and out of the source. A p-channel MOSFET is exactly opposite: a negative gate-source voltage $V_{GS} < 0$ is required to induce current flow. In a pMOSFET, positive current enters the source and leaves the drain. The nMOS and pMOS transistors are electrical complements of each another. This means that (a) the currents flow in opposite directions and (b) the voltages are reversed. CMOS circuits are often designed to exploit this characteristic by using an nMOS/pMOS pair of transistors as a complementary switch. In standard CMOS circuits, only one transistor in the pair conducts current at a time, which can be used to give the low power property discussed above.

1.2.1 Switch Logic

Ideal MOSFETs can be modelled as simple voltage-controlled switches as shown in Figure 1.2. Conduction from drain to source is controlled by the gate voltage. A closed switch corresponds to a conducting transistor, while an open switch indicates that the applied gate-source voltage is not sufficient to induce current flow through the device. In an nMOS transistor, a high gate voltage gives $G = 1$ which closes the switch. A pMOSFET is exactly opposite: a low voltage with $G = 0$ is needed to close the switch. MOSFET switches are the most basic logic element. The operation of complex digital logic circuits is often modelled using this simplistic model.

Digital MOS circuits perform binary logic by associating logic 0 and logic 1 states with voltage levels. Assume that the power supply voltage is V_{DD}. In an ideal **positive logic** system[2],

$$\begin{aligned} \text{Logic 0} &= 0 \quad [\text{V}] \\ \text{Logic 1} &= V_{DD} \end{aligned} \tag{1.3}$$

[1] In the simplified symbols, the bulk electrode of an nMOS transistor is assumed to be grounded, while the bulk electrode of a pMOS transistor is assumed to be connected to the positive power supply.

[2] We will use positive logic throughout the book.

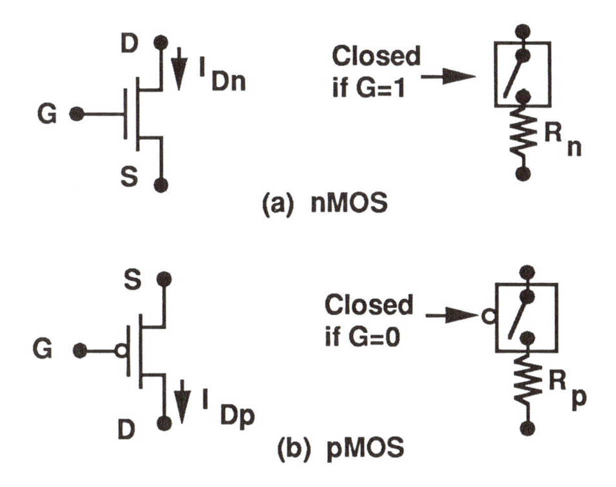

Figure 1.2: nMOS and pMOS Switch Models

defines the relation between Boolean variables and circuit voltages. Logic transferral is achieved by using MOSFETs as switches to provide conduction paths from a node to either V_{DD} or ground. MOS nodes are dominated by parasitic capacitance, and many basic switching properties can be described by using simple RC switching networks.

Figure 1.3 illustrates how we can construct an inverter using MOSFET switches. Functionally, the circuit is designed to give a high voltage at the output with a low voltage applied to the input, and vice-versa. In this circuit, the input voltage V_{in} is connected to both the nMOS and pMOS gates. To charge the output load capacitance C_L from 0 volts to V_{DD}, the pMOS transistor must be conducting. This is obtained by setting V_{in} to a low value. The output voltage is then given by

$$V_{out}(t) = V_{DD}[1 - e^{-(t/\tau_p)}] \tag{1.4}$$

where $\tau_p = R_p C_L$ is the charging time constant. R_p represents the pMOS drain-source resistance when the device is active (i.e., in a conducting state). To discharge the output node, the n-channel MOSFET must be on to provide a conducting path to ground. With V_{in} at a high voltage, this is described by

$$V_{out}(t) = V_{DD}e^{-(t/\tau_n)} \quad (1.5)$$

where $\tau_n = R_nC_L$ is the discharge time constant and R_n the nMOS drain-source resistance.

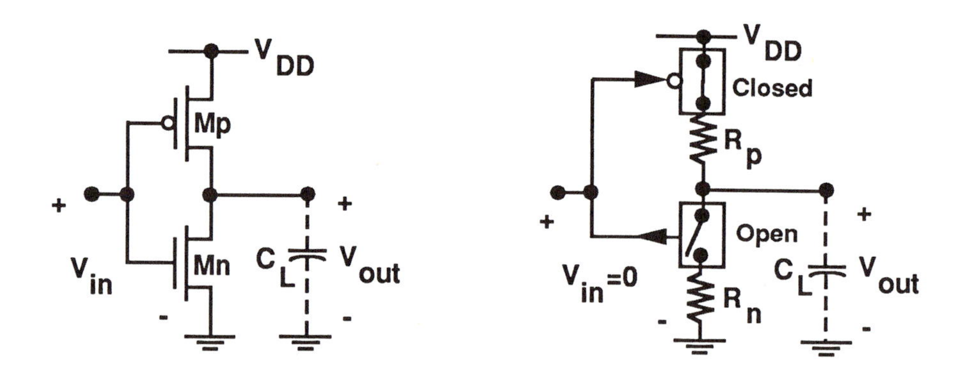

Figure 1.3: Inverter Switching Network

The simple example above shows how a CMOS inverter circuit can be implemented. Equations (1.4) and (1.5) show that the maximum switching frequency f_{max} is limited by the MOSFET resistances R_n and R_p, and the value of the load capacitance C_L. All of these parameters are critical in high-speed chip design and depend on the processing parameters and the layout geometry.

Another important property of the circuit is the amount of dissipated power. If a stable input voltage is applied to the circuit, the complementary arrangement of the transistors insures that only one device is conducting at a time. Power supply current is required only when the input is being switched. The values of the average current I_{av} and the dissipated power $P_{av} = I_{av}V_{DD}$ are thus dependent on the switching frequency f. The power consumption and switching speed characteristics of a digital logic circuit are contained in a single parameter called the **power-delay product** (PDP), which is discussed in Chapter 3.

We can expand the inverter circuit and create higher-order logic gates. A 2-input NAND gate is shown in Figure 1.4. Two MOSFETs have been

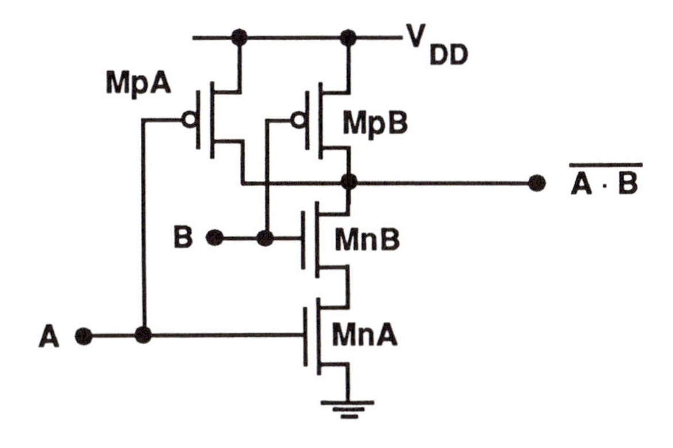

Figure 1.4: CMOS NAND Gate

added to the inverter to create the new function. The nMOS and pMOS segments complement each another such that the nMOS transistors are in series while the pMOS transistors are in parallel. To pull the output to ground, both MnA and MnB must be conducting, i.e., we need $A = 1, B = 1$. Otherwise, at least one pMOSFET provides a conduction path to the power supply, giving $V_{out} = V_{DD}$. Other gates, such as the NOR function, can be obtained by rearranging the transistor switching networks.

Switch logic can be used to analyze the CMOS array shown in Figure 1.5. Applying the switch model to a complicated logical network provides

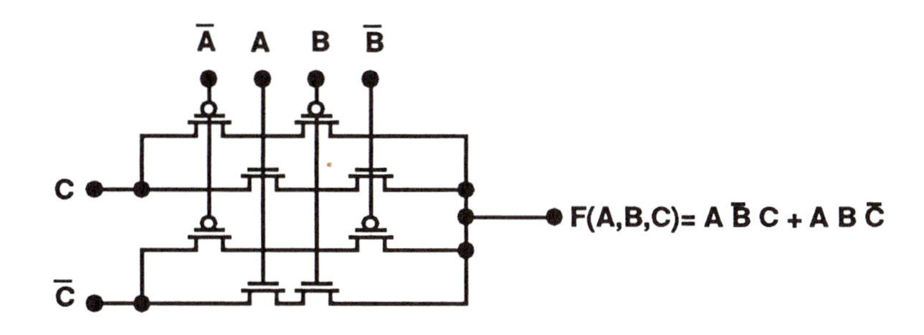

Figure 1.5: Basic Switch Logic

first-order verification of the functionality. However, other problems such as timing delays, the effects of parasitics, and inclusion of leakage currents require a more careful analysis of the circuit.

Several types of differential switch logic circuits can be created by using complementary transistor arrays to drive a flip-flop. An example of a static **Cascode-Voltage Switch Logic** (CVSL) circuit is shown in Figure 1.6. A pair of complementary nMOS switching arrays are used to perform differential logic. The outputs Q and $\overline{Q}$ are taken from a pMOS latching circuit which is made up of Mp1 and Mp2. The main objective of this type of design is to obtain fast switching (using the differential circuit) while maintaining a structured transistor array to facilitate layout. Another similar

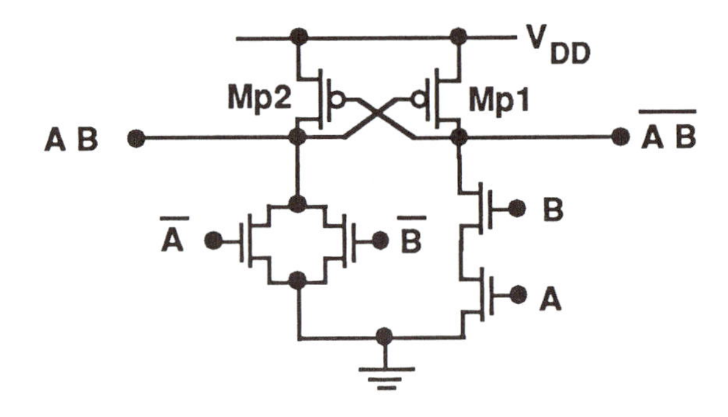

Figure 1.6: CVSL Circuit

approach, called **Complementary Pass-transistor Logic** (CPL) uses n-channel MOSFET arrays to realize basic gate functions. Figure 1.7 illustrates a basic AND/NAND gate in CPL. The outputs can be connected to a pMOS latch or to a CMOS inverter which increases the flexibility of the logic.

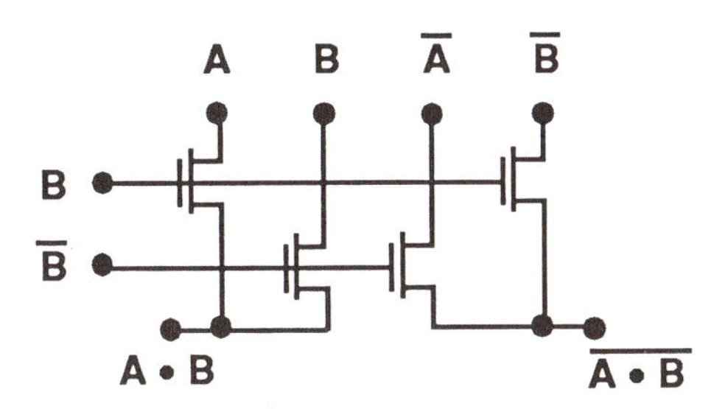

Figure 1.7: CPL AND/NAND Array

1.2.2 Logic Transmission

One problem which is critical to large system design is synchronizing logic flow in an organized way. System clocks are used to direct the data and control signals so that everything is at the right place at the right time. The simplest clocking scheme is to use a signal $\phi(t)$ as shown in Figure 1.8; CMOS works well with complementary signals, so we have also included $\overline{\phi}(t)$ in the drawing. The clock period T [sec] is the length of one cycle, while the frequency $f = (1/T)$ [Hz] is the **clock rate**.

The main idea of a system clock is to open and close data paths so that we can synchronize the transfer of information among the various sub-units in the system. In a CMOS circuit, we can introduce a bidirectional **transmission gate** (TG) as shown in Figure 1.9. Logically this may be viewed as a voltage- controlled switch. If $\phi = 1$, then the switch is closed and data moves from input to output. Setting $\phi = 0$ is equivalent to opening the switch, which blocks data movement. Transmission gates are built by placing an nMOS and a pMOS transistor in parallel; both polarities are used because this gives the best electrical transmission characteristics.

Data flow can be controlled using a simple approach. The system clock is distributed and combined with control signals using a multiplexer; the resulting control variables are used to open and close transmission gates. Logic blocks between the TGs operate on the data before transferring it on to the next stage. As an example, consider the concept of a **data bus** which carries data around on a chip. Figure 1.10 provides a general illustration. Many sub-units need access to the data; for example in a microcontroller,

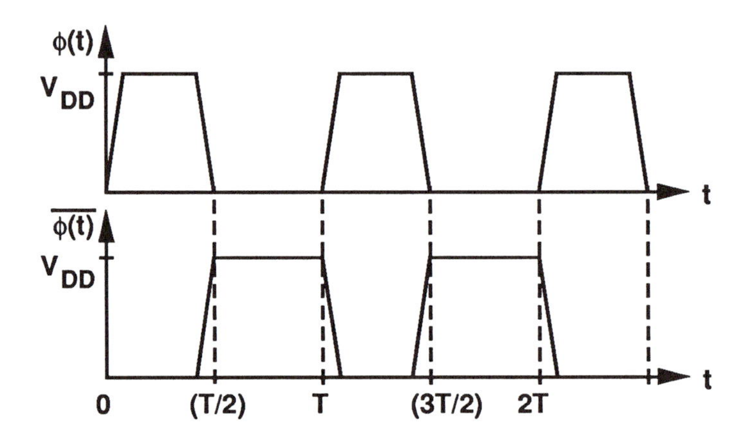

Figure 1.8: System Clocking

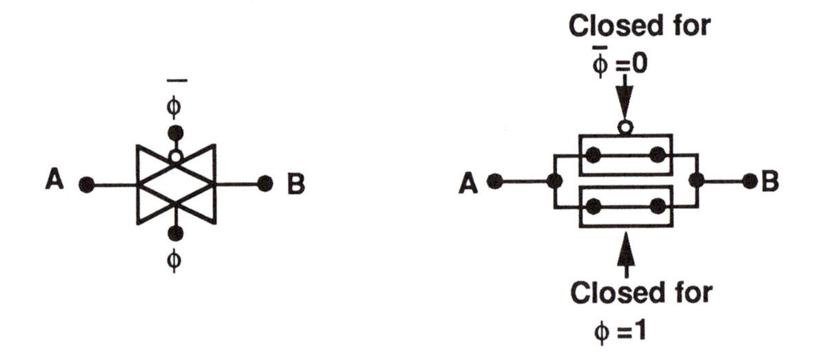

Figure 1.9: CMOS Transmission Gates

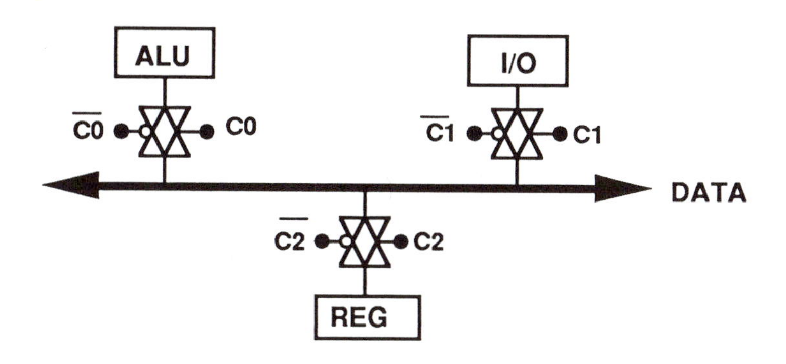

Figure 1.10: Controlled Data Transfer

the storage registers, the ALU (arithmetic logic unit), and the input/output circuits all use the data in one way or another. To control the operations, we transfer the data to the proper sub-units using control signals; these are shown as $(C0, C1, C2)$ in the drawing. Transmission gates can be used to bridge the data bus and the logic sections; both timing and combinational logic establish the data path. As an example, suppose we want to input data into the ALU when a signal IN=1 and the clock is high. Defining the control signal

$$C0 = \mathrm{IN} \cdot \phi \tag{1.6}$$

achieves the desired result.

This example illustrates a logic structuring which can be used for arbitrarily complex systems. The simplicity of the basic circuitry allowed by CMOS is immediately obvious.

1.2.3 Data Storage

Data storage can be accomplished using the circuit shown in Figure 1.11. The load control signal LD (and its complement $\overline{\mathrm{LD}}$) is used to admit new data when LD=1, or to hold the existing state when LD=0. The operation is easy to see. A load condition (LD,$\overline{\mathrm{LD}}$)=(1,0) places TG1 in a transmitting state. The input DATA is thus admitted to the cell which consists of series connected inverters; during this time TG2 is off. Once the data has been loaded, it is held in the cell by the condition (LD,$\overline{\mathrm{LD}}$)=(0,1). This control state blocks transmission of new data by placing TG1 in an

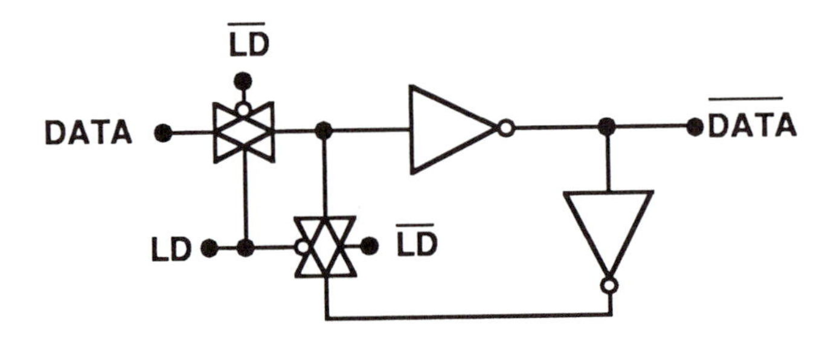

Figure 1.11: Simple Storage Register

off state. Transmission gate TG2 is on, completing the series connected inverter circuit. If we replace LD by the composite TG control signal

$$C = \mathrm{LD} \cdot \phi, \tag{1.7}$$

register loading becomes synchronized with the clock signal ϕ.

1.2.4 Dynamic CMOS

Digital logic circuits are based on the notion of charging and discharging capacitors. Output transients are calculated by analyzing the amount of time needed to transfer charge on to and off of output nodes. MOSFET gates can be viewed as capacitors whose charge state determines whether the transistor is on or off. In a dynamic CMOS circuit, logic 0 and logic 1 states are defined by the amount of charge held on a capacitive node. To perform a logical operation, we dynamically control the transfer of charge among parasitic capacitances and MOSFETs. This approach generally results in faster and smaller logic circuits.

Switching transistors are used to control charge flow to and from a node. It is possible to set a charge state on a capacitor, and then isolate it from the rest of the network. When isolated, this **dynamic charge storage node** acts as a simple voltage memory device[3] for the data bit.

The basic concepts can be explained using the nMOS dynamic RAM (DRAM) cell shown in Figure 1.12. A single pass transistor Mn connects the storage capacitor C_{store} to the data bit line D. Access to the cell is gained by means of the cell select signal CS. To **write** data to the cell, CS is set to 1, which turns on Mn. Charge transfer from the bit line to

[3]Recall the physical explanation of the integral $v(t) = v(0) + (1/C)\int_0^t i(\tau)d\tau$ when describing the behavior of a capacitor

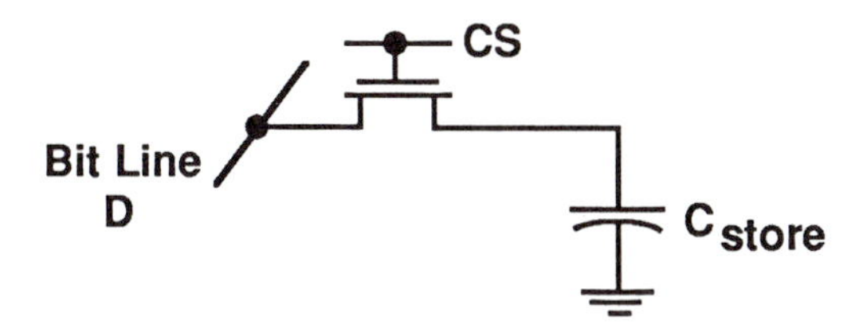

Figure 1.12: nMOS DRAM Cell

C_{store} sets the voltage. **Reading** is just the opposite; charge is transferred from C_{store} to the bit line. Due to the small amount of charge involved (usually a few femtocoulombs), a sensitive detection circuit called a **sense amplifier** is used to determine the logical value.

When $CS = 0$, the cell is in a **hold** state. Ideally, the storage capacitor holds the charge indefinitely. In a realistic chip environment, leakage current paths exist, allowing small current flows which alter the amount of charge stored on the node. This limits the amount of time that a storage node can hold a logic state. For example, DRAM cells typically can hold a charge state for a maximum of a few milliseconds. Leakage problems require that we perform a **refresh** operation where the contents of the cell are read and then rewritten. The basic principles used to create the memory cell can be extended to generalized logic circuits.

Dynamic circuit techniques can also be used to construct digital logic gates; an example of a dynamic CMOS circuit is shown in Figure 1.13. Transistors Mp1 and Mn1 are respectively termed the **precharge** and **evaluate** devices. Both are controlled by the clock signal ϕ in a complementary manner. When the clock is in a state $\phi = 0$, Mp1 is active and C_{out} charges to a voltage V_{DD}. This *precharge event* provides the charge needed to drive the next stage. *Logic evaluation* takes place when $\phi = 1$ and Mn1 is driven into an active mode. C_{out} will discharge if the logic block provides a conducting path from C_{out} to Mn1; otherwise, the output remains at a high voltage. The event is termed a **conditional discharge** for obvious reasons. Although some problems arise due to leakage paths and other charge-altering events, the circuit exhibits excellent switching characteristics. It may also be used as the basis for more advanced design techniques.

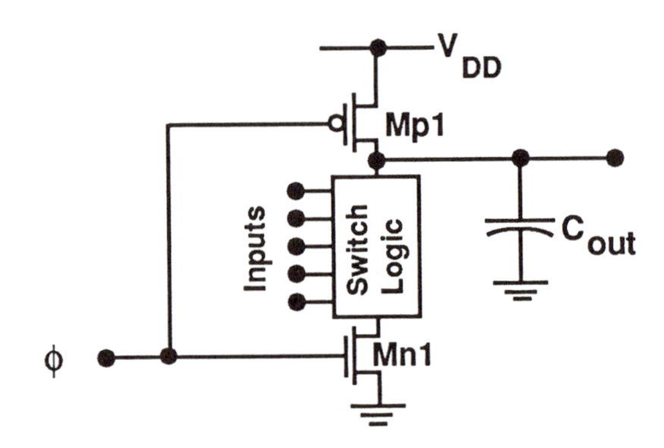

Figure 1.13: Dynamic CMOS Circuit

1.2.5 CMOS System Design

Individual circuits combine to form large-scale integrated CMOS systems. Obviously, the characteristics of the building blocks set the overall system performance, so that constructing an integrated chip design requires care at each level. Ideally, the two extreme cases of system/chip design are

- **Top-down design:** System specifications ("top") are set forth and guide the design "down" towards the circuit level;
- **Bottom-up design:** Circuits are designed first, and then connected "upwards" according to the required system structure.

However, in a mature field such as CMOS logic design, one rarely encounters either extreme. Many projects turn in to a *tug-of-war* among performance criteria.

What do we aim for in a complete CMOS system design? The answer depends on the particular application, but the following items recur quite frequently:

- High-density logic,
- Fast switching time,
- Low power dissipation.

The system performance is related to that of individual circuits through considerations such as

- Floor plan,
- Interconnect strategies,
- Input/output coupling,
- Clock distribution,
- Interfacing,

and others.

The overall problem can be seen by studying the floorplan in Figure 1.14. Each section has a specific function (or, set of functions) which are implemented using basic CMOS circuits. Interconnects among the various functional units complete the system layout. The chip performance depends on all aspects of the design including the individual circuits, the placement of the sub-units, and the characteristics of the interconnects. Although the goals and problems described here are identical to those for any electronic system, CMOS solutions tend to be unique. A wide variety of compatible circuit design styles have been developed allowing a mixing as needed to achieve the performance specifications.

1.3 Plan of the Book

This book deals with digital CMOS logic circuits from the viewpoint of a circuit designer. Emphasis is directed towards analyzing the circuit properties as functions of the fundamental semiconductor properties, and then applying the results to higher level systems.

The presentation has been structured to permit access to specific topics in CMOS circuits and design. Although system considerations are introduced at strategic points in the discussion, the treatment does not go too deep into large-scale system design, as that warrants a separate book. Instead, the current work emphasizes the properties of relatively small CMOS circuits which form the basis for more complex systems. Coverage and level have been adjusted to address the needs of both circuit and system designers.

The outline below summarizes the content, approach, and philosophy of the book.

The **physics and modeling of MOSFETs** is covered in **Chapter 2**. Emphasis is placed on the relationship between the processing parameters and device operation. Various analytic techniques are discussed, and

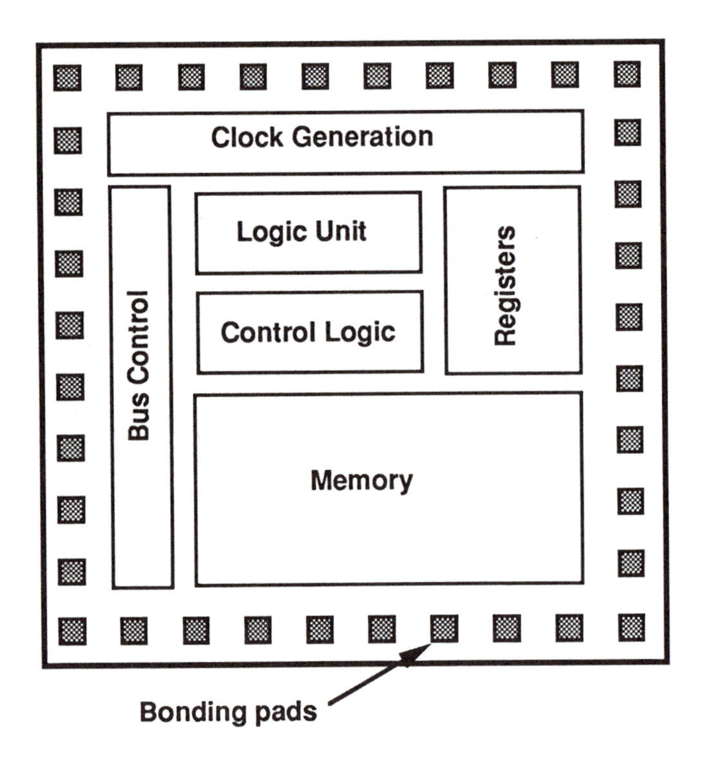

Figure 1.14: Chip Floorplan Example

SPICE modelling is presented. Topics have been chosen according to their importance to using of MOSFETs in designing circuits. Some advanced discussions on scaling theory and small-device modelling have been included for completeness, but are not required for understanding the rest of the book.

CMOS circuit design begins with **Chapter 3** which is devoted entirely to the **static inverter**. The discussion provides the reader with the basic foundations of CMOS circuits which are used for the remainder of the book. Important analyses include finding DC logic voltages, capacitances, and transient time intervals. Notational conventions are also established in this chapter. The concepts are extended to **static logic circuits** in **Chapter 4**. Static NAND and NOR gates are analyzed and compared in detail. Procedures for constructing random combinational logic circuits are summarized and applied to illustrate basic circuits such as adders. The chapter includes the design and analysis of "nMOS-like" and "pMOS-like" (also known as pseudo-nMOS and pseudo-pMOS) circuits. An introduction to bistable circuits (flip-flops, Schmitt triggers) completes the chapter.

Chapter 5 is concerned with **alternate types of CMOS logic circuits**. Most of these are based on switching networks which implement logic functions using structured transistor arrays. The CMOS transmission gate (TG) provides good logic transmission characteristics, so it is analyzed and compared with alternate approaches. TG-intensive circuit design is introduced by means of examples such as logic gates, multiplexers, and latches. Using complementary logic arrays to drive a bistable circuit gives the concept of differential switch logic gates. The basic differential families of cascode-voltage switch logic (CVSL), complementary pass-transistor logic (CPL), and differential split-level (DSL) logic are discussed.

The material in **Chapter 6** completes the discussion of basic logic circuits in CMOS by examining the interface between schematics and realistic **chip design**. After a review of process flows and isolation techniques, design rules are studied to set the stage for the discussion. The analysis is then directed towards the properties and modeling of interconnects from both the physics and circuit viewpoints. Voltage transmission from a CMOS circuit is discussed using both lumped-element and transmission line models. Crosstalk, latchup, and other important problems are also covered in this chapter.

Synchronous CMOS logic circuits are presented in **Chapter 7**. The discussion starts with basic clocking of static cascades and examines problems of charge leakage and charge sharing. Dynamic gates are constructed and analyzed. These are used as a basis for developing the advanced system design styles of Domino Logic (DL) [including Multiple-Output Domino Logic (MODL) and Latched-Domino (LDom) variations] and NORA circuits. Items such as logic design and clock limitations are covered for each technique.

Chapter 8 is devoted to presenting some **important CMOS circuits** which are either very useful to know or form the basis for more advanced designs. Topics include optimizing the performance static logic chains, clock drivers and clock distribution techniques, input circuits, off-chip drivers, and others[4].

Although digital circuits are the main subject of the book, basic **analog CMOS** is discussed in **Chapter 9**. The treatment was written as a introduction to linear CMOS for digital chip designers with a standard electronics background, and is not meant to be comprehensive. Coverage is limited to current sources, voltage dividers, and simple amplifiers.

The merging of bipolar and CMOS into **BiCMOS** is presented in **Chapter 10**; this represents the extension of silicon technology into the next century. Although BiCMOS is relatively new, circuit design techniques have been studied to the point where they can be presented in a coherent manner. After reviewing the properties of bipolar transistors, Chapter 10 examines the basic BiCMOS circuits such as logic gates and level-converters from the viewpoint of a circuit designer.

1.4 References

A few general references on digital integrated circuits including CMOS and GaAs techniques are listed below. Each chapter in the book has a detailed reference list to aid in studying the material.

[1] H. Hazedar, **Digital Microelectronics**, Addison-Wesley, Reading, MA, 1991.

[2] D.A. Hodges and H.G. Jackson, **Analysis and Design of Digital Integrated Circuits**, 2nd ed., McGraw-Hill, 1990.

[3] N. Kanopoulos, **Gallium Arsenide Digital Integrated Circuits**, Prentice-Hall, Englewood Cliffs, NJ, 1989.

[4] O. Wing, **Gallium Arsenide Digital Circuits**, Kluwer Academic Publishers, Boston, 1990.

[4] In other words, this chapter is a "catch-all".

Chapter 2

MOSFET Characteristics

Current flow through a metal-oxide-semiconductor field-effect transistor (MOSFET) is understood by analyzing the response of the charge carriers in the semiconductor to applied electric fields. This chapter summarizes the physics of MOS transistors needed to understand circuit models.

Figure 2.1 shows a typical n-channel MOSFET in a LOCOS[1] technology which will be used for the analysis. The central region of the device consists

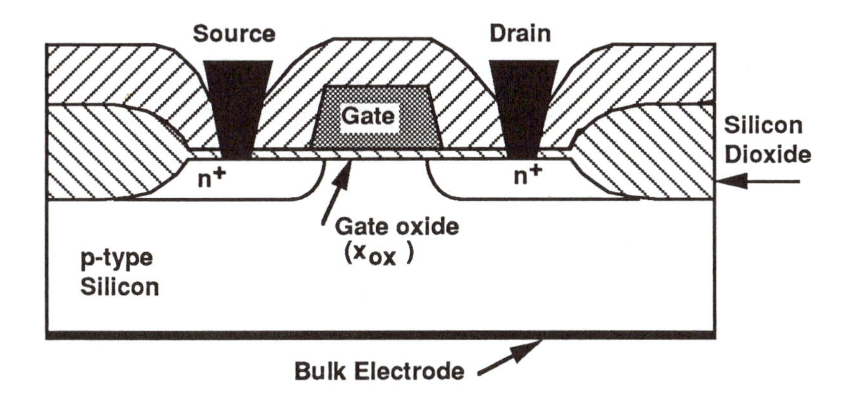

Figure 2.1: LOCOS MOSFET

of an MOS subsystem made up of the gate, the insulating gate oxide layer, and the p-type silicon. The $I-V$ characteristics of the transistor result from

[1] The acronym "LOCOS" stands for the LOCal Oxidation of Silicon, a processing technique that is described in Section 6.1.1

the physics of the MOS system when coupled to the drain and source n^+ regions. Current flow from drain to source is controlled by the gate-source voltage V_{GS}, the drain-source voltage V_{DS}, and the source-bulk voltage V_{SB}.

The current flow characteristics of a MOSFET are referenced to the **threshold voltage** V_T of the device. An **enhancement-mode** (E-mode) n-channel MOSFET is defined to have $V_{Tn} > 0$. The threshold voltage V_{Tn} is especially important to circuit design. In an ideal MOSFET, setting the gate-source voltage to a value $V_{GS} < V_{Tn}$ places the transistor into **cutoff** with $I_D = 0$. Increasing the gate-source voltage to a value $V_{GS} > V_{Tn}$ allows the transistor to conduct current I_D; this defines the **active** mode of operation. In other words, the value of V_{GS} relative to V_{Tn} determines if the transistor is ON (active) or OFF (no current flowing). The actual value of the current depends on the voltages applied to the device.

2.1 Threshold Voltage

Conduction from the drain to the source in a MOSFET is possible because the central MOS structure has the characteristics of a simple capacitor. The oxide capacitance per unit area is given by

$$C_{ox} = \frac{\epsilon_{ox}}{x_{ox}} \quad [\mathrm{F/cm^2}], \tag{2.1}$$

where x_{ox} is the oxide thickness in [cm] underneath the gate, and $\epsilon_{ox} \simeq (3.9)\epsilon_o$ [F/cm] for silicon dioxide. In current technologies, x_{ox} is less than about 400 Å$=$ 0.04 [μm]. Since the value of the free-space permittivity is $\epsilon_o = 8.854 \times 10^{-14}$ [F/cm], C_{ox} is typically on the order of 10^{-8} [F/cm^2].

The charge carrier population at the semiconductor surface is controlled by the gate voltage V_{GS}. Consider an enhancement-mode device. If a positive voltage is applied to the gate electrode, negative charge is induced in the semiconductor region underneath the oxide. This is due to the penetration of the electric field into the silicon, and is termed the **field-effect**: the charge densities are controlled by the external voltage through the electric field. For small values of V_{GS}, a depletion region provides sufficient negative space charge to support the electric field. However, if V_{GS} exceeds the threshold voltage V_{Tn} of the structure, a thin electron **inversion layer** is created at the silicon surface. This allows drift current to flow between the source and drain when V_{DS} is applied. Figure 2.2 shows the formation of the conducting channel.

The value of the threshold voltage is set by the electrical properties of the MOS system. Internal device parameters such as doping densities, oxide thickness, and ion implant dose are established during the processing, and establish the basic value denoted by V_{T0n}. The "0" subscript is used to denote the **zero body-bias** case where $V_{SB} = 0$.

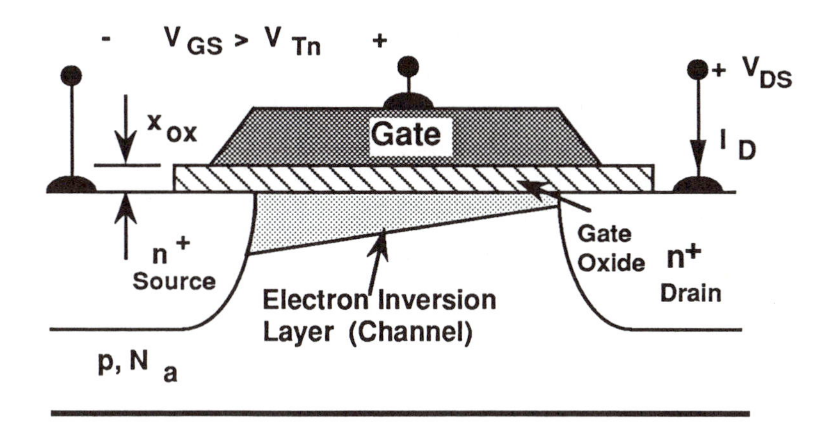

Figure 2.2: MOSFET Conducting Channel

The threshold voltage may be calculated from [13]

$$V_{T0n} = V_{FB} + \phi_S + \frac{1}{C_{ox}}\sqrt{2q\epsilon_{Si}N_a\phi_S} \pm \frac{qD_I}{C_{ox}}, \tag{2.2}$$

where V_{FB} is the **flatband voltage**, ϕ_S is the **surface potential**, $\epsilon_{Si} \simeq (11.8)\epsilon_o$ is the silicon permittivity, N_a is the acceptor doping in the substrate, and D_I the **dose** [cm^{-2}] of the **threshold adjustment ion implant**. The flatband voltage accounts for differences between the gate and substrate materials, in addition to any charges in the oxide [16]. The surface potential term represents the surface voltage needed to internally create the electron layer, while the square root term gives the voltage needed to support the bulk depletion charge. An ion implantation step is provided to adjust V_{T0} to a desired value, and is accounted for in the last term.

Flatband Voltage

The flatband voltage is found from

$$V_{FB} = \Phi_{GS} - \frac{Q_f}{C_{ox}} - \frac{1}{C_{ox}}\int_0^{x_{ox}} \frac{x'}{x_{ox}}\rho_{ox}(x')dx' \tag{2.3}$$

with Φ_{GS} the work function difference between the gate (G) and substrate (S), Q_f the fixed surface charge density in units of [C/cm^2], and ρ_{ox} the trapped oxide charge density in units of [C/cm^3]. Current processing techniques generally reduce the effect of the trapped charge to negligible levels by performing the oxidation in a chlorinated atmosphere. The fixed charge Q_f, on the other hand, is due in part to the change in composition from silicon to silicon dioxide, and cannot be eliminated [2].

The value of Φ_{GS} depends on the gate material as discussed below.

(a) Metal Gate: If a metal with work function Φ_M is used for the gate,

$$\Phi_{GS} = \Phi_M - \Phi_S, \tag{2.4}$$

where Φ_S is the work function of the semiconductor as determined by the position of the Fermi level [3]. For a p-type substrate with an acceptor doping of N_a,

$$\Phi_S = \chi_S + \frac{E_g}{2q} + \left(\frac{kT}{q}\right)\ln\left(\frac{N_a}{n_i}\right) \tag{2.5}$$

where $\chi_S \simeq 4.15$ [V] is the electron affinity in silicon and E_g is the energy gap value in units of joules. Numerically, $(E_g/2q) \simeq (1.12/2)$ [V] at room temperature.

(b) Polysilicon Gate: Gates fabricated out of poly can be doped either n-type (with $N_{d,poly}$) or p-type (with $N_{a,poly}$). The value of Φ_{GS} can be approximated by using either

$$\Phi_{GS} \simeq -(\frac{kT}{q})\ln\left(\frac{N_a N_{d,poly}}{n_i^2}\right), \tag{2.6}$$

or,

$$\Phi_{GS} \simeq (\frac{kT}{q})\ln\left(\frac{N_{a,poly}}{N_a}\right), \tag{2.7}$$

depending on the gate doping. In both equations, N_a is the background acceptor substrate doping. The factor (kT/q) is the thermal voltage, and n_i is the intrinsic concentration. At room temperature ($T = 300°$ [K]) these are given by $(kT/q) \simeq 0.026$ [V] and $n_i \simeq 1.45 \times 10^{10}$ [cm^{-3}].

Surface Potential

ϕ_S is the silicon surface potential. At the onset of strong inversion,

$$\phi_S \simeq 2|\phi_F|, \tag{2.8}$$

where

$$|\phi_F| \simeq \left(\frac{kT}{q}\right)\ln\left(\frac{N_a}{n_i}\right) \tag{2.9}$$

is the **bulk Fermi potential** in a p-type substrate. Note that the value of the surface potential required to produce inversion varies with the doping N_a.

Threshold Adjustment Ion Implant

The working value of V_{T0n} is set by an ion implantation step in the processing. Ion implanted dopings are specified by the implant dose D_I [cm^{-2}]. Implanting acceptor ions into the substrate induces a positive shift in V_{T0} so that the "+" sign is applicable. The opposite holds for a donor ion implant; in this case, the threshold voltage is made more negative and the "-" must be used. Using the term $\pm(qD_I/C_{ox})$ models the effect of the implanted ions as a charge sheet and ignores the actual distribution into the substrate. A more accurate analysis can be performed using a simulator such as SUPREM.

EXAMPLE 2.1: Threshold Voltage Calculation

A MOSFET is characterized by $x_{ox} = 400$ [Å] and $N_a = 10^{15}$ [cm^{-3}]. An n-type poly gate is used with $N_{d,poly} = 10^{19}$ [cm^{-3}]. The fixed oxide charge is $Q_f \simeq q(10^{11})$ [C/cm^2] and dominates the oxide charge. The acceptor ion implant dose is $D_I = 5.20 \times 10^{11}$ [cm^{-2}]. To determine the threshold voltage, first calculate

$$\begin{aligned} C_{ox} &= [(3.9)(8.854 \times 10^{-14})/(0.04 \times 10^{-4})] \\ &\simeq 8.63 \times 10^{-8} \quad [\text{F/cm}^2]. \end{aligned}$$

The flatband voltage is computed to be

$$\begin{aligned} V_{FB} &= -(0.026)\ln[(10^{15}10^{19})/(1.45 \times 10^{10})^2)] \\ &\qquad -[(1.6 \times 10^{-19})(10^{11})/(8.63 \times 10^{-8})] \\ &\simeq -0.819 - 0.185 \\ &\simeq -1.004 \quad [\text{V}], \end{aligned}$$

and the surface potential is

$$\begin{aligned} \phi_S &\simeq 2(0.026)\ln[(10^{15})/(1.45 \times 10^{10})] \\ &\simeq 0.579 \quad [\text{V}]. \end{aligned}$$

The bulk charge term contributes a value of

$$\frac{\sqrt{2(1.6 \times 10^{-19})(11.8)(8.854 \times 10^{-14})(10^{15})(.579)}}{(8.63 \times 10^{-8})} \simeq 0.161 \quad [\text{V}].$$

Finally, the ion implantation step increases the value of V_{T0} by an amount

$$\frac{(1.6 \times 10^{-19})(5.20 \times 10^{11})}{(8.63 \times 10^{-8})} \simeq +0.964 \ [\mathrm{V}].$$

Combining terms gives the threshold voltage as

$$\begin{aligned} V_{T0n} &\simeq -1.004 + 0.579 + 0.161 + 0.964 \\ &\simeq +0.700 \ [\mathrm{V}] \qquad \square. \end{aligned}$$

The ion implant step is required to produce an n-channel enhancement-mode MOSFET with $V_{T0n} > 0$ in this example.

2.1.1 Body-Bias

The threshold voltage of a MOSFET is altered when a source-bulk voltage V_{SB} is applied as shown in Figure 2.3. This **body-bias** effect in-

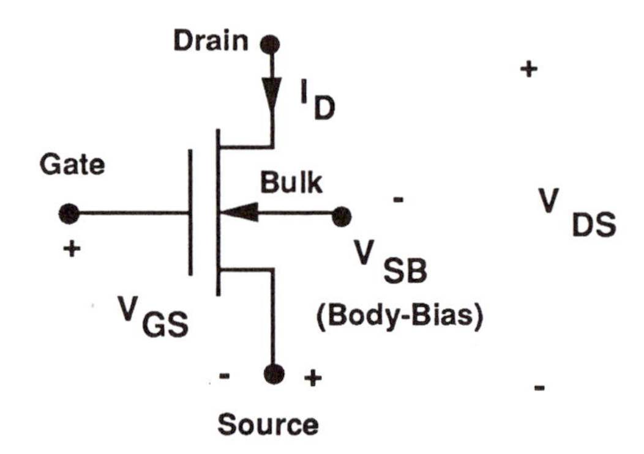

Figure 2.3: Body-Bias Voltage

creases the threshold voltage V_{Tn} because V_{SB} adds reverse-bias across the n-channel/p-substrate boundary which increases the bulk depletion charge. Applying body-bias increases the third term of equation (2.2) to

$$\frac{1}{C_{ox}}\sqrt{2q\epsilon_{Si}N_a(\phi_S + V_{SB})}. \tag{2.10}$$

Comparing this with equation (2.2) for V_{T0n} shows that the increase ΔV_{Tn} in threshold voltage due to body-bias is given by

$$\Delta V_{Tn} = \gamma_n(\sqrt{\phi_S + V_{SB}} - \sqrt{\phi_S}), \tag{2.11}$$

where

$$\gamma_n = \frac{1}{C_{ox}}\sqrt{2q\epsilon_{Si}N_a} \quad [\mathrm{V}^{1/2}] \tag{2.12}$$

is the **body-bias factor**. In the general case we write

$$V_{Tn} = V_{T0n} + \gamma_n(\sqrt{2|\phi_F| + V_{SB}} - \sqrt{2|\phi_F|}), \tag{2.13}$$

with V_{T0n} the zero body-bias value.

EXAMPLE 2.2: Body-Bias Effect

The process parameters in Example 2.1 give a body-bias factor of

$$\begin{aligned} \gamma_n &= \sqrt{2(1.6 \times 10^{-19})(11.8)(8.854 \times 10^{-14})(10^{15})}/(8.63 \times 10^{-8}) \\ &\simeq 0.212 \quad [\mathrm{V}^{1/2}] \end{aligned}$$

so that

$$V_{Tn} \simeq 0.700 + 0.212(\sqrt{0.579 + V_{SB}} - \sqrt{0.579}).$$

Figure 2.4 shows V_{Tn} as a function of V_{SB} □.

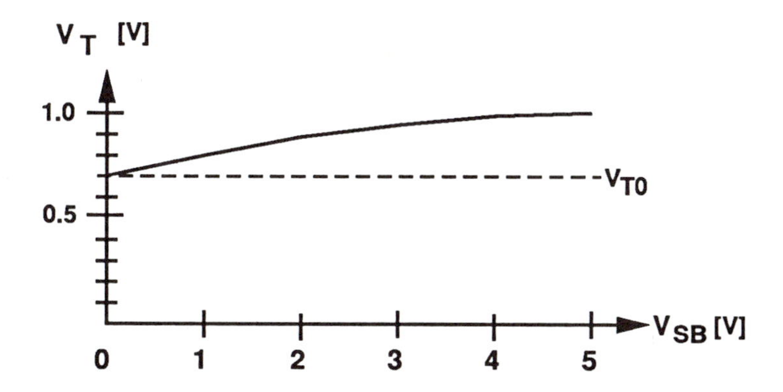

Figure 2.4: Body-Bias Effect

2.2 Current-Voltage Characteristics

The MOSFET $I - V$ characteristics can be extracted by modelling the electron inversion layer created when $V_{GS} > V_{Tn}$. The layer itself forms the conduction channel from drain to source, which allows us to compute I_{Dn} as a function of V_{GS} and V_{DS}. Modelling can be performed at various levels with the general tradeoff being complexity versus accuracy.

The basic analytic models are obtained using charge control arguments within the **gradual-channel approximation**. Consider the device cross-section shown in Figure 2.5(a). To induce current flow, two conditions are needed. First, $V_{GS} > V_{Tn}$ is required to create the channel. Second, a drain-to-source voltage V_{DS} must be applied to produce the channel electric field $\mathcal{E}$. This field forces electrons from the source to the drain, thereby giving current flow I_D in the opposite direction.

The electron inversion charge Q_n [C/cm^2] in the channel is given by a standard capacitor relation of the form

$$Q_n(y) = -C_{ox}[V_{GS} - V_{Tn} - V(y)], \tag{2.14}$$

where $V(y)$ represents the voltage in the channel due to V_{DS}. The voltage dV across a differential segment dy of the channel is

$$dV = \frac{I_D dy}{-\mu_n Q_n W}, \tag{2.15}$$

with μ_n as the electron surface mobility in units of [cm^2/V-sec]. Rearranging and integrating y from $y = 0$ to $y = L$ gives the generic integral

$$I_D = k'_n\left(\frac{W}{L}\right)\int_0^{V_{DS}}\left(V_{GS} - V_{Tn} - V\right)dV. \tag{2.16}$$

We have introduced the nMOS **process transconductance** k'_n by

$$k'_n = \mu_n C_{ox} \quad [\mathrm{A/V^2}]. \tag{2.17}$$

The device geometry is specified by the channel width W and the channel length L; the **aspect ratio** (W/L) is the important geometrical factor which determines the current. Since the aspect ratio is set by the device layout, it is the easiest parameter to control for circuit design. The **device transconductance**

$$\beta_n = k'_n\left(\frac{W}{L}\right) \tag{2.18}$$

is used to characterize a specific device. Table 2.1 provides a listing of the important MOSFET parameters. The basic device equations obtained from the gradual-channel analysis are discussed below.

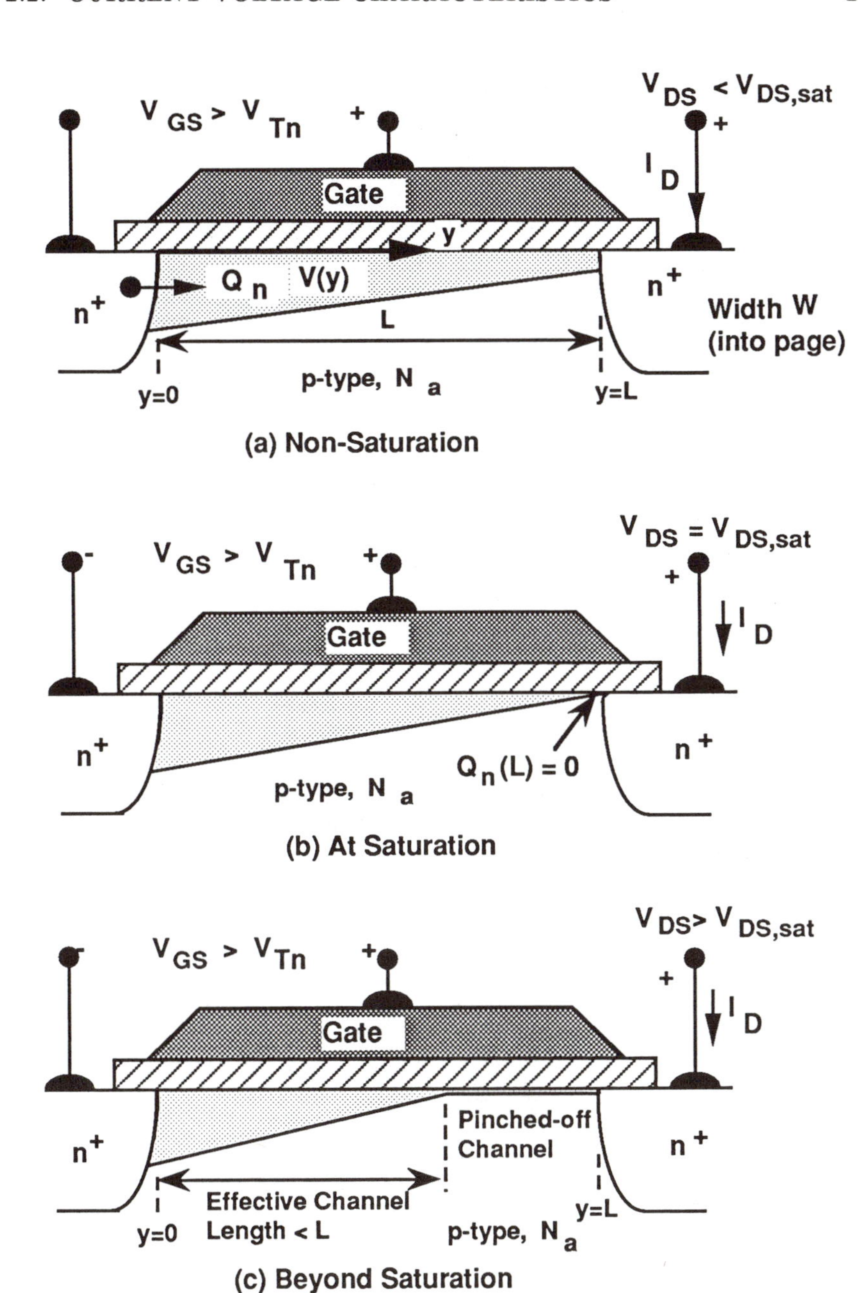

Figure 2.5: Conducting Channel in a MOSFET

Symbol	Parameter	Units
V_T	Threshold voltage	V
V_{T0}	Zero-body bias threshold voltage	V
V_{GS}	Gate-source voltage	V
V_{DS}	Drain-source voltage	V
V_{SB}	Source-bulk voltage	V
I_D	Drain current	A
$k' = \mu C_{ox}$	Process transconductance	A/V^2
μ	Mobility	cm^2/V-sec
C_{ox}	Oxide capacitance	F/cm^2
L	Channel length	cm
W	Channel width	cm
(W/L)	Aspect ratio	
$\beta = k'(W/L)$	Device transconductance	A/V^2
γ	Body bias factor	$V^{1/2}$
λ	Channel modulation parameter	V^{-1}

Table 2.1: MOSFET Parameters

2.2.1 Square-Law Model

The simplest description of current flow through a MOSFET is obtained by assuming that V_{Tn} is a constant in the channel. The integral (2.16) may then be evaluated to give

$$I_{Dn} = \frac{\beta_n}{2}[2(V_{GS} - V_{Tn})V_{DS} - V_{DS}^2], \tag{2.19}$$

which describes **non-saturated** current flow. Given a gate-source voltage V_{GS}, this predicts a non-linear increase in current flow with increasing V_{DS}. The peak current occurs at the **saturation voltage**

$$V_{DS,sat} = (V_{GS} - V_{Tn}) \tag{2.20}$$

such that eqn. (2.19) is valid for $V_{DS} \leq V_{DS,sat}$.

The physical significance of the saturation voltage is shown in Figure 2.5(b). When $V_{DS} = V_{DS,sat}$, the channel is "pinched off" at the drain side of the transistor. This can be verified mathematically by noting that the saturation condition corresponds to a channel voltage of $V(y = L) = V_{DS,sat}$, so that the inversion charge in eqn. (2.14) evaluates to $Q_n(y = L) = 0$. At this point the channel is viewed as being "compressed" to its minimum thickness.

When the drain-source voltage is increased to $V_{DS} \geq V_{DS,sat}$, the device conducts in the **saturated** mode where the current flow has only a weak dependence on the drain-source voltage. As V_{DS} increases, the effective length of the channel decreases as shown in Fig. 2.5(c); this phenomenon is termed **channel-length modulation**. Since the drain current is proportional to $(1/L)$, channel-length modulation tends to increase the saturated current flow.

The saturated current is approximated by using the maximum value of the non-saturated current with an effective channel length. The simplest form for the saturated current is given by

$$I_{Dn} = \frac{\beta_n}{2}(V_{GS} - V_{Tn})^2[1 + \lambda(V_{DS} - V_{DS,sat})], \tag{2.21}$$

where λ $[\text{V}]^{-1}$ is the **channel-length modulation parameter**. Channel-length modulation effects can be important in analog networks. However, we will usually take $\lambda \simeq 0$ as a reasonable approximation when performing hand calculations on a digital circuit.

Figure 2.6 illustrates the family of curves generated by the square law model. Each curve corresponds to a different value of V_{GS}. The border between saturation and non-saturation is predicted by equation (2.19), and is approximately parabolic. Also note that the saturation voltage depends on the applied gate-source voltage. Another useful plot is the **transfer**

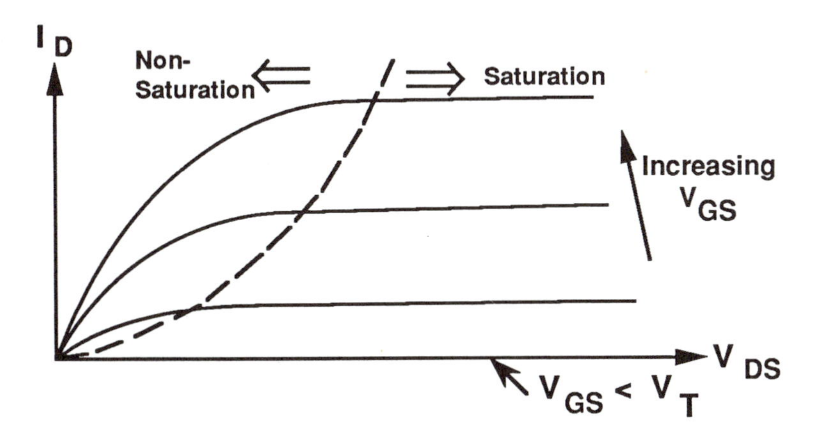

Figure 2.6: MOSFET I-V Characteristic

curve shown in Figure 2.7 which gives $\sqrt{I_{Dn}}$ as a function of V_{GS} for a saturated MOSFET. Thresholding at $V_{GS} = V_{Tn}$ is evident from the figure. A saturated MOSFET is particularly useful for measuring the threshold

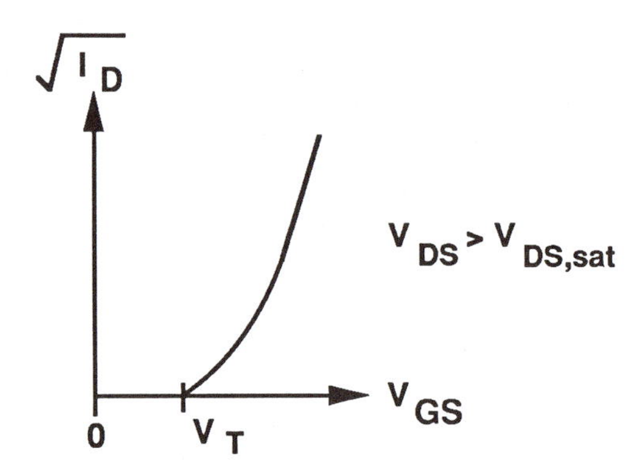

Figure 2.7: MOSFET Transfer Curve

voltage of transistors.

Square-law MOSFET models are usually chosen for circuit analysis due to their simplicity. Since this approach ignores some fundamental device physics, errors are automatically introduced into the analysis. This is not a problem so long as the equations are only used for general calculations. Crucial results must always be checked using computer simulations.

2.2.2 Bulk-Charge Model

A more accurate equation set is obtained by noting that the channel voltage V is underneath the oxide and increases the effective bias on the MOS system. This increases the threshold voltage bulk charge term in equation (2.2) to

$$\frac{1}{C_{ox}}\sqrt{2q\epsilon_{Si}N_a(\phi_S + V)}, \tag{2.22}$$

where we assume for simplicity that $V_{SB} = 0$. Since V_{Tn} is now a function of the channel voltage V, integrating equation (2.16) gives

$$\begin{aligned} I_{Dn} &= \beta_n\Big[(V_{GS} - V_{FB} - 2|\phi_F| \mp \frac{qD_I}{C_{ox}})V_{DS} - \frac{1}{2}V_{DS}^2 \\ &\quad -\frac{2}{3}\gamma[(2|\phi_F| + V_{DS})^{3/2} - (2|\phi_F|)^{3/2}]\Big] \end{aligned} \tag{2.23}$$

as the non-saturated drain current. V_{T0n} is still termed "the" threshold voltage, and physically represents the gate voltage needed induce surface inversion at the source end of the MOSFET. The device enters saturation at a drain-source voltage of $V_{DS,sat}$ corresponding to the value where I_{Dn} is a maximum. Explicitly,

$$\begin{aligned} V_{DS,sat} &= V_{GS} - V_{FB} - 2|\phi_F| \pm \frac{qD_I}{C_{ox}} \\ &\quad - \frac{q\epsilon_{Si}N_a}{C_{ox}^2}\left[\sqrt{1 + \frac{2C_{ox}^2}{q\epsilon_{si}N_a}(V_{GS} - V_{FB})} - 1\right] \end{aligned} \tag{2.24}$$

in this model. The value of I_{Dn} evaluated at this voltage is the first order approximation to the saturation current.

The bulk-charge model is more accurate than that predicted by the simpler square-law approximation. However, the increased complexity can offset the desire for precision, particularly when performing calculator-based estimates. Because of this reason, square-law equations are the most common for manual circuit analysis. A more complete comparison between the two models can be found in the literature [13,18].

Simplified Bulk-Charge Model

A simpler model which retains some of the accuracy of the bulk charge analysis can be obtained by performing a Taylor series expansion on the voltage terms in the bulk charge equation. Including body-bias due to V_{SB} and keeping only the first order term gives

$$\frac{1}{C_{ox}}\sqrt{2q\epsilon_{Si}N_a(2|\phi_F| + V + V_{SB})} \simeq \gamma\sqrt{2|\phi_F| + V_{SB}} + \delta\ V, \tag{2.25}$$

where

$$\delta = \frac{\gamma}{2\sqrt{2|\phi_F| + V_{SB}}} \tag{2.26}$$

is the slope. Integrating the current integral in equation (2.16) now gives

$$I_{Dn} = \frac{\beta_n}{2}[2(V_{GS} - V_{Tn})V_{DS} - (1+\delta)V_{DS}^2]. \tag{2.27}$$

This expression is similar to the square-law model, but has a factor of $(1+\delta)$ in the V_{DS}^2 term, reducing the current to a more correct value. The threshold voltage V_{Tn} is again interpreted as the value needed to invert the surface at the source end of the channel. The saturation voltage is now given by

$$V_{DS,sat} = \frac{(V_{GS} - V_{Tn})}{(1+\delta)} \tag{2.28}$$

so that the saturated current is

$$I_{Dn} = \frac{\beta_n}{2(1+\delta)}(V_{GS} - V_{Tn})^2[1 + \lambda(V_{DS} - V_{DS,sat})]. \tag{2.29}$$

This equation set retains the simplicity of the square-law model allowing it to be used in analytic treatments.

The accuracy of these analytic models is limited by the fact that the gradual-channel approximation is a 1-dimensional approximation to the 3-dimensional MOSFET geometry. While computer device simulations provide the key to understanding the intricacies of the transistor operational modes, circuit analysis and design can be based on very simple equations of current flow. We will follow this philosophy here and base most of our circuit discussions on the basic square-law model.

2.3 p-Channel MOSFETs

Perhaps the most important circuit design aspect of CMOS is the use of both n-channel and p-channel transistors in complementary arrangements. Adhering to this technique guarantees the desired properties of minimum DC power dissipation and rail-to-rail output logic swings.

The operational physics of a p-channel MOSFET (pMOS) is the exact complement of that used to describe the operation of an n-channel device. This means that all positive and negative quantities are reversed, and n-type and p-type silicon regions are interchanged. Instead of duplicating the discussion already presented for n-channel MOSFETs, we will just summarize the important equations. Square-law models will be used throughout.

A p-channel MOSFET must be constructed in an n-type background region. If a p-substrate is used as a starting point, then an n-well must be provided at all pMOS locations. Figure 2.8 illustrates a pMOS transistor in this type of process; note that the pMOS bulk (n-well) is connected to the highest voltage in the system (typically the power supply rail). Conduction is achieved by making the gate sufficiently negative to attract minority carrier holes to the silicon surface. Current flow is thus controlled by V_{SGp} and V_{SDp}, and I_{Dp} flows out of the drain electrode. Enhancement-mode p-channel MOSFETs have negative zero body-bias threshold voltages: $V_{T0p} < 0$. It is convenient to use absolute values such that $V_{Tp} = -|V_{Tp}|$; the current flow equations then look very similar to those for nMOS transistors.

Cutoff occurs when $V_{SGp} < |V_{Tp}|$, indicating that the source-gate voltage is insufficient to support the formation of an inversion layer. As expected, cutoff is characterized by $I_{Dp} \simeq 0$, and only leakage currents exist in the device.

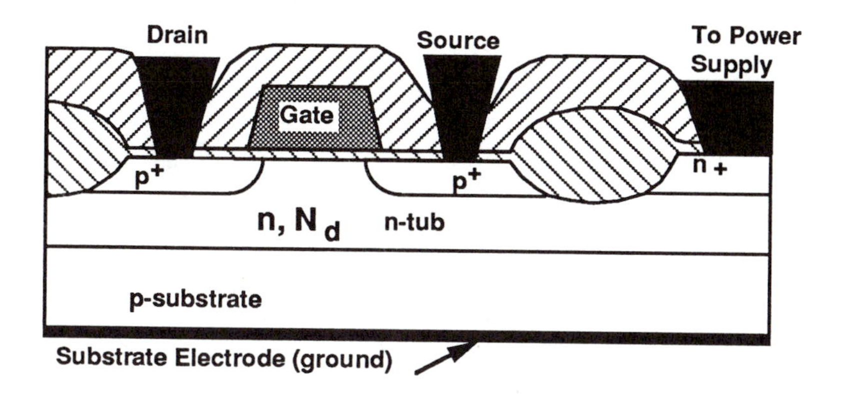

Figure 2.8: p-Channel MOSFET

Active operation requires a source-gate voltage of $V_{SGp} > |V_{Tp}|$. Saturation occurs at the point $V_{SDp,sat} = (V_{SGp} - |V_{Tp}|)$. If $V_{SDp} \leq V_{SDp,sat}$, the device is non-saturated with a current of

$$I_{Dp} = \frac{\beta_p}{2}[2(V_{SGp} - |V_{Tp}|)V_{SDp} - V_{SDp}^2]. \tag{2.30}$$

In this equation, $\beta_p = k'_p(W/L)$ is the pMOS device transconductance, with μ_p the hole mobility. Saturated current flow occurs when $|V_{SDp}| \geq |V_{SDp,sat}|$ such that

$$I_{Dp} = \frac{\beta_p}{2}(V_{SGp} - |V_{Tp}|)^2[1 + \lambda_p(V_{SDp} - V_{SDp,sat})]. \tag{2.31}$$

Body-bias effects are described by writing the threshold voltage in the form

$$V_{Tp} = V_{T0p} - \gamma_p(\sqrt{2\phi_{Fn} + V_{BSp}} - \sqrt{2\phi_{Fn}}). \tag{2.32}$$

Since $V_{T0p} < 0$, body-bias makes V_{Tp} more negative, i.e., applying a bulk-source voltage V_{BSp} increases $|V_{Tp}|$.

The complementary aspects of nMOS and pMOS transistors are exploited in many CMOS circuit design techniques. However, it should be noted that since electrons move faster than holes (which are vacant electron states), the electron mobility is always larger than the hole mobility, i.e., $\mu_n > \mu_p$ (for equal background doping) [16]. This implies that, in general,

$$k'_n > k'_p, \tag{2.33}$$

with $k'_n \sim (2.5)k'_p$ being typical. The difference between conduction levels due to different mobilities can significantly influence the circuit design choices. For example, this implies that nMOS transistors should be chosen over pMOS devices if the device speed is critical. Tradeoffs of this type will be discussed in great details in the remaining chapters of the book.

2.4 MOSFET Capacitances

MOSFETs exhibit a number of parasitic capacitances which must be accounted for in circuit design. The simplest capacitance model is illustrated in Figure 2.9; the elements of the model are summarized in Table 2.2. The

Symbol	Name
C_{GS}	Gate-to-Source Capacitance
C_{GD}	Gate-to-Drain Capacitance
C_{GB}	Gate-to-Bulk Capacitance
C_{SB}	Source-to-Bulk Capacitance
C_{DB}	Drain to Bulk Capacitance

Table 2.2: MOSFET Capacitances

capacitances originate from the central MOS gate structure, the characteristics of the channel charge, and the pn junction depletion regions. MOS contributions are constants, but the channel and depletion capacitances are both nonlinear and vary with the applied voltage.

Parasitic capacitances give the fundamental limitation on the switching speed. Although computer simulations are usually required for an accurate analysis, the analytic estimates below suffice for a first-order calculations. MOS and depletion capacitance properties will be discussed first, and then applied to the device model.

2.4.1 MOS-Based Capacitances

The basic MOS capacitance is due to the physical separation of the gate conductor and the semiconductor by the gate oxide, which has a thickness x_{ox}. We characterize the capacitance per unit area by writing

$$C_{ox} = \frac{\epsilon_{ox}}{x_{ox}} \quad [\mathrm{F/cm^2}]. \tag{2.34}$$

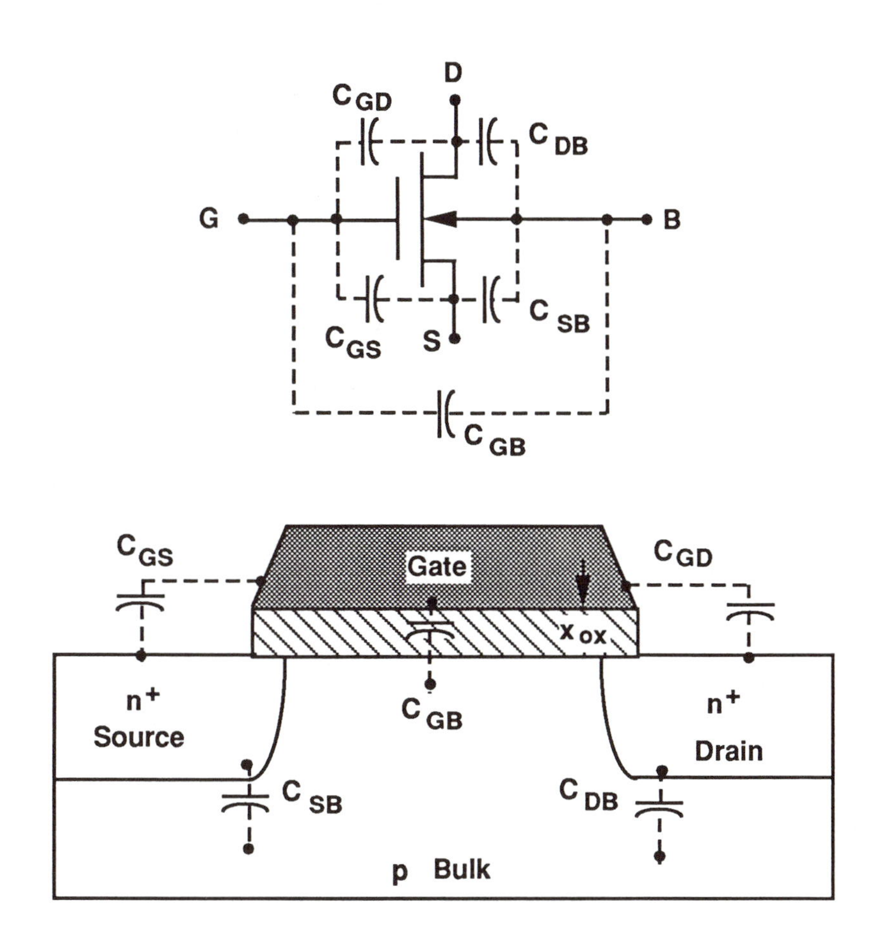

Figure 2.9: MOSFET Capacitance Model

The oxide capacitance is an intrinsic part of the transistor structure and the value of C_{ox} is important to the MOSFET current equations through k'. This simple equation ignores the presence of fringing electric fields, which may be significant in the device. However, neglecting fringing for simplicity gives the **gate capacitance** as

$$C_g = C_{ox} W L, \tag{2.35}$$

where L is the channel length and W is the channel width. Although C_g is not explicitly shown in the MOSFET capacitance model, it is useful for calculating the gate-related transistor capacitances.

An important parasitic is the **overlap capacitance** illustrated in Figure 2.10. The overlap distance L_o is measured beyond the normal p-type channel region on both the source and drain sides. It is used to insure an operational device where the electron inversion layer can electrically connect the drain and source n^+ regions. The overlap capacitance per unit gate width is

$$C_{o\ell} = C_{ox} L_o \quad [\text{F/cm}], \tag{2.36}$$

such that

$$C_o = C_{o\ell} W \quad [\text{F}] \tag{2.37}$$

gives the total overlap capacitance. Overlap exists on both the drain and source sides of the transistor and should be included in all calculations.

There is also a gate-bulk overlap capacitance due to the **gate overhang** from the active region to the surrounding field oxide. Denoting the thickness of the field oxide by X_{FOX}, the gate-bulk overlap capacitance is

$$C_{o,GB} = C_{FOX}(L + 2L_o) w_o \tag{2.38}$$

where

$$C_{FOX} = \frac{\epsilon_{ox}}{X_{FOX}} \tag{2.39}$$

is the field oxide capacitance per unit area and w_o is the overhang distance beyond the drain/source regions. In realistic device geometries, gate-bulk capacitance may be a complicated non-parallel plate structure. In this case, the above formulas should be used with care.

2.4.2 Depletion Capacitance

Depletion capacitance originates from the ionized dopants in the vicinity the junction. Assuming a step (or, abrupt) doping profile with doping N_a and N_d on the p-side and n-side, respectively, the **zero-bias capacitance** per unit area is given by

$$C_{j0} = \frac{\epsilon_{Si}}{x_d}, \tag{2.40}$$

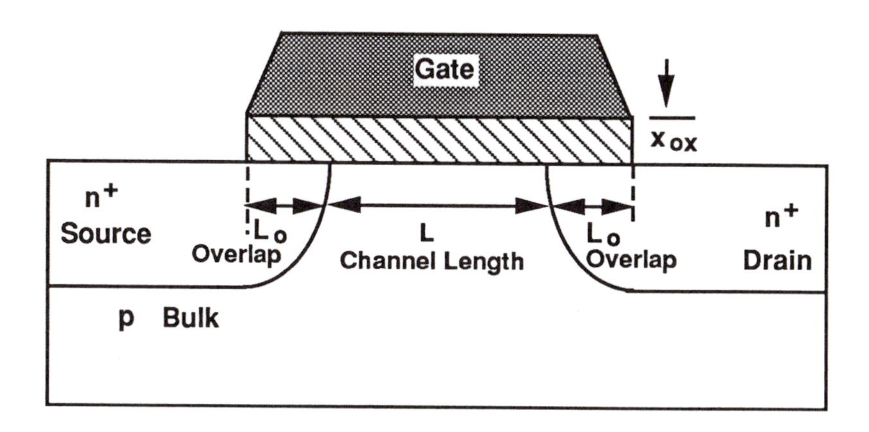

(a) Side View

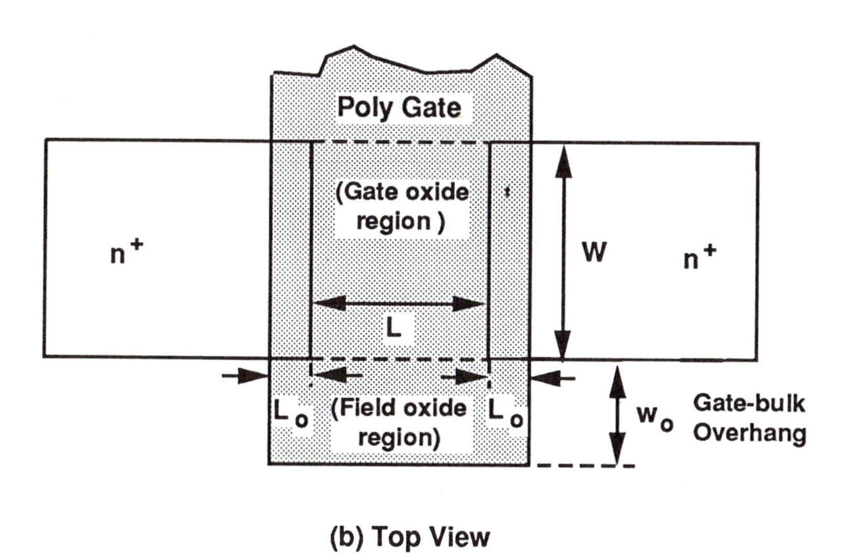

(b) Top View

Figure 2.10: MOSFET Geometry for Capacitance Estimates

where

$$x_{d0} = \sqrt{\frac{2\epsilon_{Si} V_{bi}}{q}\left(\frac{1}{N_d} + \frac{1}{N_a}\right)} \tag{2.41}$$

is the **zero-bias depletion width**. In this equation,

$$V_{bi} = \left(\frac{kT}{q}\right) \ln\left(\frac{N_d N_a}{n_i^2}\right) \tag{2.42}$$

denotes the **built-in voltage** which is set by the processing. Depletion widths increase with an applied reverse-bias voltage V_R according to

$$x_d = \sqrt{\frac{2\epsilon_{Si}(V_{bi} + V_R)}{q}\left(\frac{1}{N_d} + \frac{1}{N_a}\right)}. \tag{2.43}$$

The junction capacitance C_j is then a nonlinear function of V_R with

$$\begin{aligned} C_j(V_r) &= \frac{\epsilon_{Si}}{x_d(V_R)} \\ &= \frac{C_{j0}}{\sqrt{1 + (V_R/V_{bi})}}, \end{aligned} \tag{2.44}$$

which shows that $dC_j/dV_r < 0$, i.e., the junction capacitance decreases as the reverse-bias voltage increases. The general dependence is illustrated in Figure 2.11. The actual capacitance in farads is obtained by multiplying C_j by the area A of the junction. This gives

$$C = C_j A \tag{2.45}$$

which may be calculated for each depletion region.

MOSFET depletion capacitances are found at the drain and source regions. A typical n^+p nMOS region is illustrated in Figure 2.12. For this device, the **sidewall** acceptor doping $N_{a,sw}$ is usually larger than the **bottom** doping N_a. The difference is due to field implants which are used for device isolation, or increased surface doping levels from a ion implantation step which adjusts the threshold voltage. Denoting the zero-bias sidewall capacitance per unit area by C_{j0sw}, we define

$$C_{jsw} = C_{j0sw} x_j \quad [\mathrm{F/cm}] \tag{2.46}$$

as the **sidewall capacitance per unit perimeter**; x_j is the junction depth of the n^+ well. The total sidewall capacitance is given by

$$C_{sw} = C_{jsw} \ell \; [\mathrm{F}] \tag{2.47}$$

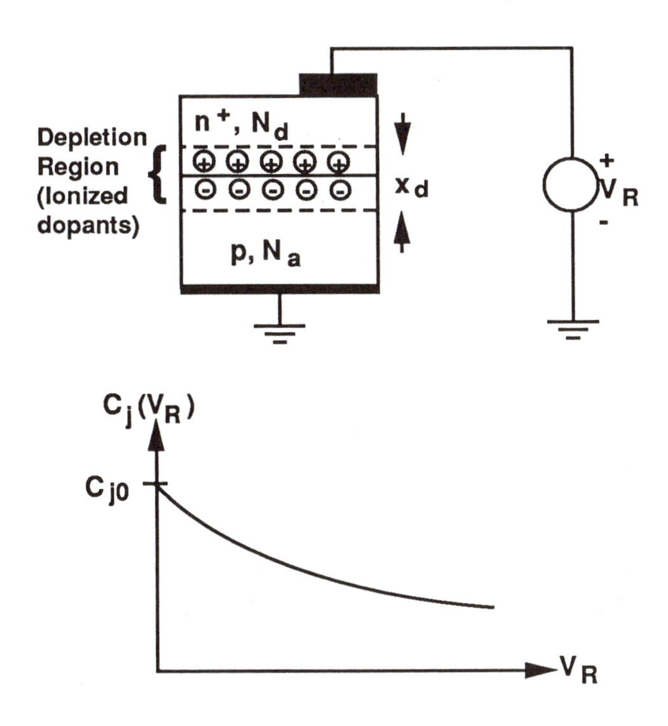

Figure 2.11: Depletion Capacitance

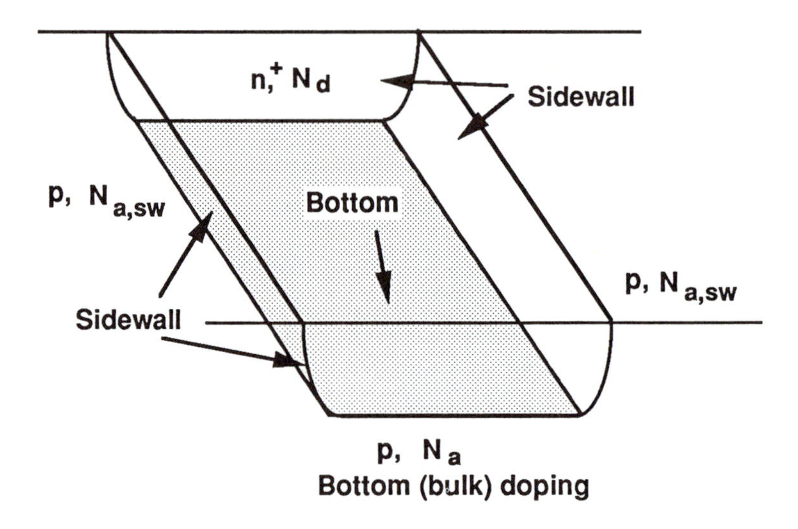

Figure 2.12: MOSFET n^+ Drain/Source Region

where ℓ is the **total perimeter length** around the n^+ region in units of centimeters. The value of ℓ is obtained from the layout geometry for each transistor.

Linearly-graded junction models are sometimes more accurate for describing a real process than is possible using a simple step-profile. In this case, the voltage-dependent capacitance assumes the form

$$C_j(V_r) = \frac{C_{j0g}}{[1 + (V_r/V_{bi})]^{1/3}} \quad [\text{F/cm}^2], \tag{2.48}$$

where C_{j0g} is the zero-bias capacitance per unit area. Use of a linearly-graded junction is usually restricted to computer simulations due to the increased complexity of the equations.

SPICE provides for various doping profiles via the grading parameter m. Step junctions have m=0.5 corresponding to the square root dependence, while a linearly graded junction is described by a cubed root dependence with m=0.33. Other values of m usually correspond to empirical results, and are commonly used in circuit simulation.

2.4.3 Channel Capacitances

MOSFET channel capacitances couple the gate electrode to the bulk conducting region. The important values are for C_{gs} (gate-source), C_{gd} (gate-drain), and C_{gb} (gate-bulk), where the lower-case subscripts are used to indicate coupling to the channel[2]. Since the channel properties vary with the applied voltage conditions, the parasitics also change with bias. A simple model for hand-calculations can be developed from the channel charge expression used to calculate the $I-V$ equations. The results are summarized in Table 2.3. Although the values are only approximate, they are

Operational Mode	**Average Values**
Non-Saturation	$C_{gs} = (1/2)C_g$ $C_{gd} = (1/2)C_g$ $C_{gb} = 0$
Saturation	$C_{gs} = (2/3)C_g$ $C_{gd} = 0$ $C_{gb} = 0$
Cutoff	$C_{gs} = 0$ $C_{gd} = 0$ $C_{gb} = C_g$

Table 2.3: Average Gate-Channel Capacitance Values

sufficient for initial design estimates. Worst-case values can be used to obtain order-of-magnitude transient time intervals; a full computer simulation is generally required for verification.

2.4.4 Device Model

When describing the overall properties of a MOSFET, it is convenient to introduce capacitances which are made up of different contributions. The most important gate-related parameters which were shown in Fig. 2.9 are

$$\begin{aligned} C_{GS} &= C_o + C_{gs}, \\ C_{GD} &= C_o + C_{gd}, \\ C_G &= 2C_o + C_g, \end{aligned} \tag{2.49}$$

[2] Upper-case subscripts, such as in C_{GS}, are used for the device capacitances in Figure 2.9

where the individual elements have already been discussed. Note that both C_{GS} and C_{GD} are nonlinear.

To use the MOSFET capacitance model for analyzing a circuit, we interpret the gate capacitance $C_G = C_{ox}W(L+2L_o)$ as the total **MOSFET input capacitance** in our calculations. The nonlinear gate-source and gate-drain capacitances C_{GS} and C_{GD} are replaced by constant LTI capacitors with the worst-case values of

$$\begin{aligned} C_{GS} &= C_o + \frac{2}{3}C_g, \\ C_{GD} &= C_o + \frac{1}{2}C_g. \end{aligned} \tag{2.50}$$

Basing our calculations on worst-case values tends to yields conservative performance estimates.

The drain-bulk and source-bulk capacitances C_{DB} and C_{SB} are more complicated. Consider the geometry shown in Figure 2.13. The n^+ region is characterized by a donor doping N_d, while the p-region underneath has an acceptor doping N_a. Assuming a step doping profile, this gives a zero-bias depletion capacitance per unit area of C_{j0} such that

$$C_j(V_R) = \frac{C_{j0}}{\sqrt{1+(V_R/V_{bi})}} \tag{2.51}$$

gives the junction capacitance for arbitrary reverse voltages V_R. In this expression,

$$V_{bi} = (\frac{kT}{q})\ln(\frac{N_dN_a}{n_i^2}) \tag{2.52}$$

is the built-in voltage. The problem with using this formula is that the capacitance changes with the reverse voltage. Analog circuits operate around bias points, so that V_R is set at a quiescent value then varied with the signal. Digital circuits, on the other hand, experience voltage changes over the entire power supply range. To include depletion capacitance contributions, we direct our interest to modelling the nonlinear capacitance using a simpler linear time-invariant (LTI) element. This can be accomplished by averaging over a range of voltages.

The depletion capacitance for an arbitrary doping profile can be written as

$$C_d(V_R) = C_{j0}(1+\frac{V_R}{V_{bi}})^{-m} \tag{2.53}$$

with m the grading constant. The average capacitance over a voltage range (V_1, V_2) is given by

$$C_{av} = \frac{1}{(V_2-V_1)}\int_{V_1}^{V_2} C_d(V_R)dV_R. \tag{2.54}$$

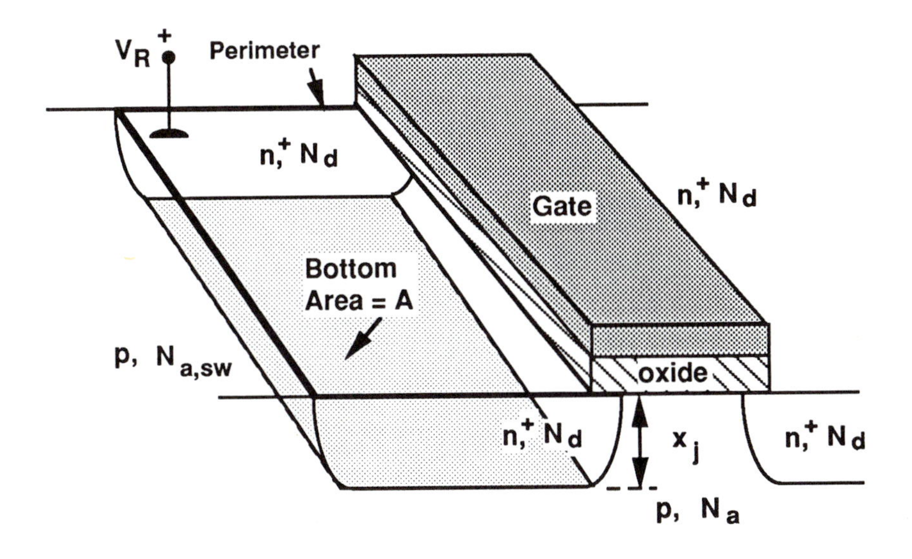

Figure 2.13: Depletion Region Geometry

Integrating gives

$$C_{av} = K_m(V_1, V_2)C_{j0}, \tag{2.55}$$

where

$$K_m(V_1, V_2) = \frac{V_{bi}}{(-m+1)(V_2 - V_1)}\left[(1+\frac{V_2}{V_{bi}})^{-m+1} - (1+\frac{V_1}{V_{bi}})^{-m+1}\right] \tag{2.56}$$

is less than unity and is a factor used to describe the averaging. For a step profile junction with $m = 1/2$,

$$K_{1/2}(V_1, V_2) = \frac{2V_{bi}}{(V_2 - V_1)}\left[\sqrt{1+\frac{V_2}{V_{bi}}} - \sqrt{1+\frac{V_1}{V_{bi}}}\right]; \tag{2.57}$$

this is usually a reasonable approximation for the bottom pn junction. The average capacitance of the bottom is then estimated by

$$C_{bottom} = K_{1/2}(V_1, V_2)C_{j0}A, \tag{2.58}$$

where A is the area of the bottom junction.

The sidewall regions tend to be heavier doped due to threshold voltage ion implants or field implants. In the n^+p junction in Figure 2.13, the sidewall acceptor doping is shown as $N_{a,sw} > N_a$. Sidewall contributions must be accounted for in realistic models. We thus introduce the sidewall built-in potential

$$V_{bi,sw} = (\frac{kT}{q}) \ln (\frac{N_d N_{a,sw}}{n_i^2}) \quad (2.59)$$

and the sidewall capacitance

$$C_{sw}(V_R) = \frac{C_{jsw}\ell}{[1 + (V_R/V_{bi,sw})]^{1/3}} \text{ [F]}, \quad (2.60)$$

where we have chosen a grading constant of $m = (1/3)$ as being typical. In this equation, ℓ is the perimeter length around the sidewall, and

$$C_{jsw} = C_{j0sw} x_j \text{ [F/cm]} \quad (2.61)$$

is the sidewall capacitance per unit length as determined by the zero-bias value C_{j0sw} [F/cm^2] and the junction depth x_j. The average sidewall contribution is given by

$$C_{sw} = K_{1/3}(V_1, V_2) C_{jsw}\ell \text{ [F]}, \quad (2.62)$$

where

$$K_{1/3}(V_1, V_2) = \frac{3V_{bi}}{2(V_2 - V_1)} \left[(1 + \frac{V_2}{V_{bi,sw}})^{2/3} - (1 + \frac{V_1}{V_{bi,sw}})^{2/3} \right] \quad (2.63)$$

is the appropriate averaging function for a linearly graded junction.

The MOSFET depletion capacitances consist of both bottom and sidewall terms. Summing gives

$$\begin{aligned} C_{DB} &= K_{1/2}(V_1, V_2) C_{j0} A_D + K_{1/3}(V_1, V_2) C_{jsw} \ell_D, \\ C_{SB} &= K_{1/2}(V_1, V_2) C_{j0} A_S + K_{1/3}(V_1, V_2) C_{jsw} \ell_S \end{aligned} \quad (2.64)$$

where A_D and A_S are the bottom areas for the drain and source, respectively, and ℓ_D and ℓ_S are the respective sidewall perimeter lengths. It is important to note that the parasitic capacitances are very sensitive to layout. Also, it should be mentioned that the sidewall capacitance often dominates the total depletion capacitance, and should not be ignored for the sake of simplifying the calculations.

The analytic approximations described here are reasonable first estimates of the depletion capacitance contributions. Since a ten percent error (or more) introduced by the averaging is quite common, these equations must be used with caution. The situation is really not as bad as it may seem, since process variations often mask out the errors in calculating capacitance values. Furthermore, critical designs are always checked with simulation results.

2.5 Junction Leakage Currents

Ideally, a reverse-biased pn junction acts as a block against current flow. Realistic junctions admit small leakage currents which may become important when small charges are involved. Dynamic logic is particularly susceptible to junction leakage problems as discussed in Chapter 6.

Consider a step-profile pn junction with a voltage V applied as illustrated in Figure 2.14. By convention, $V > 0$ corresponds to a forward bias while $V < 0$ is a reverse bias. In general, the junction current is described by

$$I = I_o(e^{V/\phi_T} - 1) + I_{dep}, \tag{2.65}$$

where I_o is the saturation current, $\phi_T = (kT/q)$ is the thermal voltage, and I_{dep} is due to recombination-generation currents in the depletion region. In

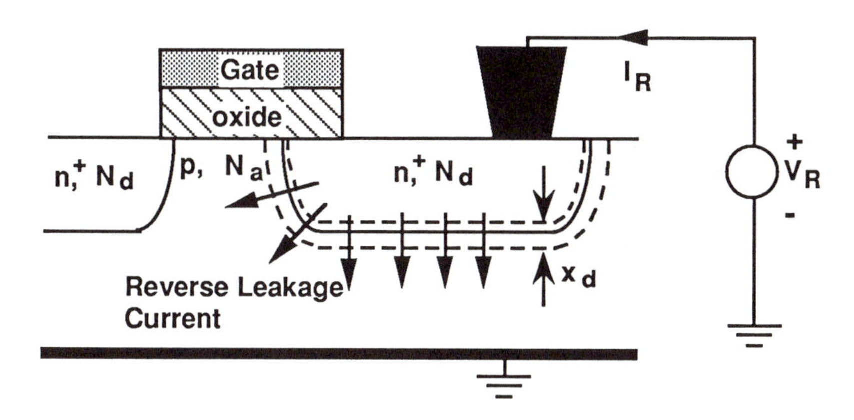

Figure 2.14: Reverse-biased pn Junction

reverse bias with $V = -V_R$ ($V_R > 0$), the reverse leakage current is

$$I_R = I_o + I_{gen}; \tag{2.66}$$

I_{gen} is the generation current due to trap states in the depletion region. Analyzing the problem gives [20]

$$I_o = qAn_i^2\left(\frac{D_n}{L_n N_a} + \frac{D_p}{L_p N_d}\right) \tag{2.67}$$

where A is the junction area, D_n and D_p are the minority carrier diffusion coefficients, and L_n and L_p are the minority carrier diffusion lengths. The generation current is approximated by

$$I_{gen} = \frac{qAn_i x_d}{2\tau_o} e^{V_R/\phi_T}, \tag{2.68}$$

where x_d is the depletion width and τ_o is the effective carrier lifetime. In silicon, $I_o << I_{gen}$, so that we approximate reverse leakage by

$$I_R \simeq I_{gen}. \tag{2.69}$$

Since the thickness of the depletion region varies with reverse voltage according to

$$x_d(V_R) = \sqrt{\frac{2\epsilon_{Si}(V_{bi} + V_R)}{q}(\frac{1}{N_a} + \frac{1}{N_d})}, \tag{2.70}$$

the reverse leakage current increases with reverse voltage.

Bulk leakage currents in MOSFETs occur whenever a drain or source region is at a different potential than the bulk. Figure 2.15 illustrates the leakage path in a simple inverter. A nonzero value of V_{SB} or V_{DB} induces the flow of leakage current which may be important to operation of the circuit. Temperature affects the level of reverse current flow; in general, I_R increases with temperature T.

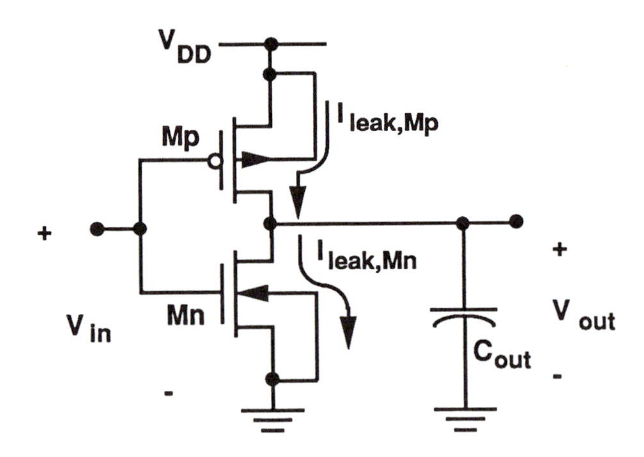

Figure 2.15: Leakage Currents in a CMOS Inverter

2.6 Parasitic Resistances

Parasitic resistances are always present in real devices. MOSFETs are no exception. Two types are particularly important.

2.6.1 Drain and Source Resistance

All semiconductors exhibit resistance which affects the currents and voltages. Accurate modelling of the drain and source n^+ regions introduces the parasitic elements R_D and R_S which are strongly dependent on the geometry. To characterize these values, we introduce the concept of a **sheet resistance**

$$R_s = \frac{\rho}{x_j} \; [\Omega], \tag{2.71}$$

where ρ is the resistivity of the layer in units of [Ω-cm] and x_j the junction depth. The sheet resistance has strict units of [Ω], but is often modified to [Ω/□] (read as "ohms per square") since it represents the edge-to-edge resistance of a square surface, e.g., $w \times w$. The drain and source resistance are then given by

$$\begin{aligned} R_D &= R_s n_D \\ R_S &= R_s n_S, \end{aligned} \tag{2.72}$$

where n_D and n_S are the equivalent number of drain and source squares in the direction of current flow as set by the device layout. Note that the value of R_s is different for n^+ and p^+ regions. Figure 2.16 illustrates the concept applied to a device layout.

Sheet resistance values are set by the drain and source doping levels which are typically greater than 10^{19} [cm^{-3}], and also by the junction depth x_j. The resulting values of R_D and R_S are usually negligible when compared to the transistor resistance between the drain and source of a MOSFET, but may dominate over the parasitic resistance of the line[3].

2.6.2 Contact Resistance

Metal-semiconductor interfaces exhibit **contact resistance** due to the nature of the boundary. In silicon, interfacing a metal to a p-type region gives a good ohmic contact, while the boundary between a metal and an n^+ region operates as a tunnel ohmic connection. Both provide reasonably

[3]The properties of the interconnect are discussed in Chapter 6

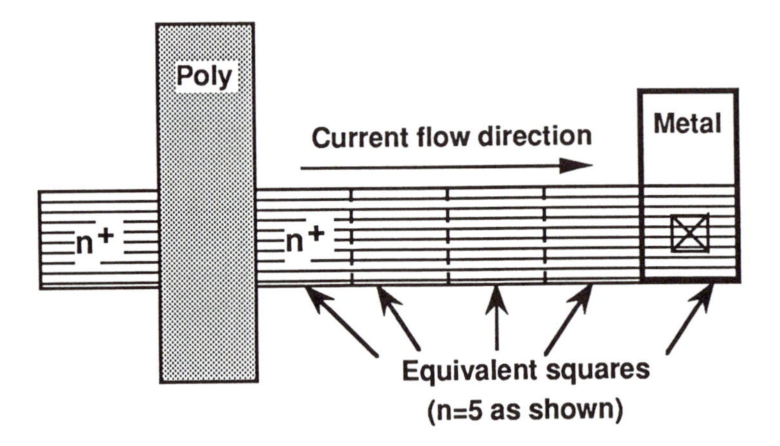

Figure 2.16: Drain and Source Resistance

small contact resistance values that are characterized by

$$\begin{aligned} R_{C-p} &= R_{p+}A_P \\ R_{C-n} &= R_{n+}A_N \end{aligned} \tag{2.73}$$

where R_{p+} (R_{n+}) is the contact resistance [Ω/cm^2] for a metal-to-p^+ (n^+) region, and A_P (A_N) is the area of the contact.

Contacts may be limited to a single standard size to improve reliability and yield. Large areas are accessed using multiple contacts as illustrated by the example in Figure 2.17. These are usually approximated as parallel resistors so that if R_C is the resistance of a single contact, N connections gives

$$R_{Total} \approx \frac{R_C}{N} \tag{2.74}$$

for the total resistance. This neglects the physical mechanism of current crowding, but is usually sufficient for estimating circuit performance.

2.7 Non-Rectangular MOSFET Gates

Minimizing circuit real estate often results in gate geometries which are not simple rectangles. CAD layout software is classified into two types: Manhattan, which only allows right angle (90o) turns, and non-Manhattan, where other angles [such as (45°)] are possible. When the shape of the gate

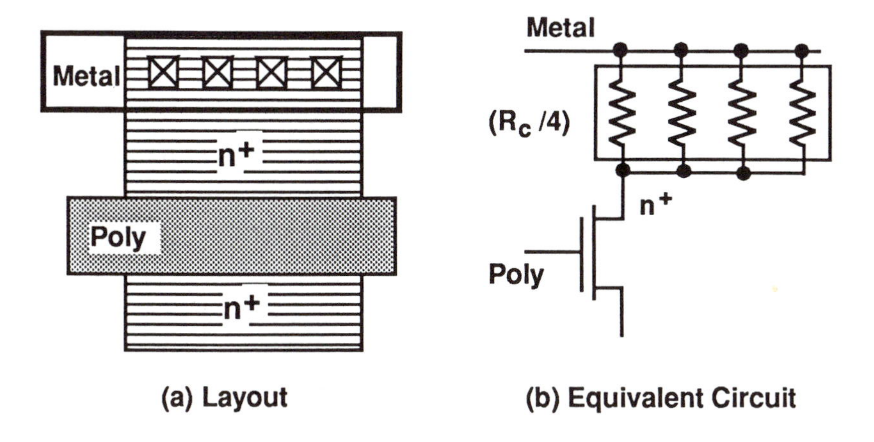

Figure 2.17: Multiple Contacts

becomes anything other than a rectangle, the equivalent value of the aspect ratio (W/L) must be determined to compute the current flow..

Calculating the effective aspect ratio of a complex gate geometry is based on the current flow pattern between the source and drain. This can be obtained using a 2-dimensional electrostatic analysis such as a flux plot or a conformal map. In practice, only approximate values are used. This is due in part to the fact that the exact value of (W/L) can be difficult to compute; moreover, processing variations counteract the need for a more accurate analysis.

Figure 2.18 provides the effective aspect ratios for two non-rectangular gate geometries. Alternately, we may use average values W_{av} and L_{av} to estimate a reasonable $(W/L)_{av}$ for use in circuit simulations. While averaging of this type may appear to be a gross oversimplification, it usually suffices for first estimates.

2.8 Mobility Variations

The surface mobility μ_n in the transconductance parameter k' is a sensitive function of fabrication and operating voltages. Moreover, physical phenomena such as velocity overshoot become important in small devices. In this section we will examine the mobility in more detail to understand how it changes the $I - V$ curves.

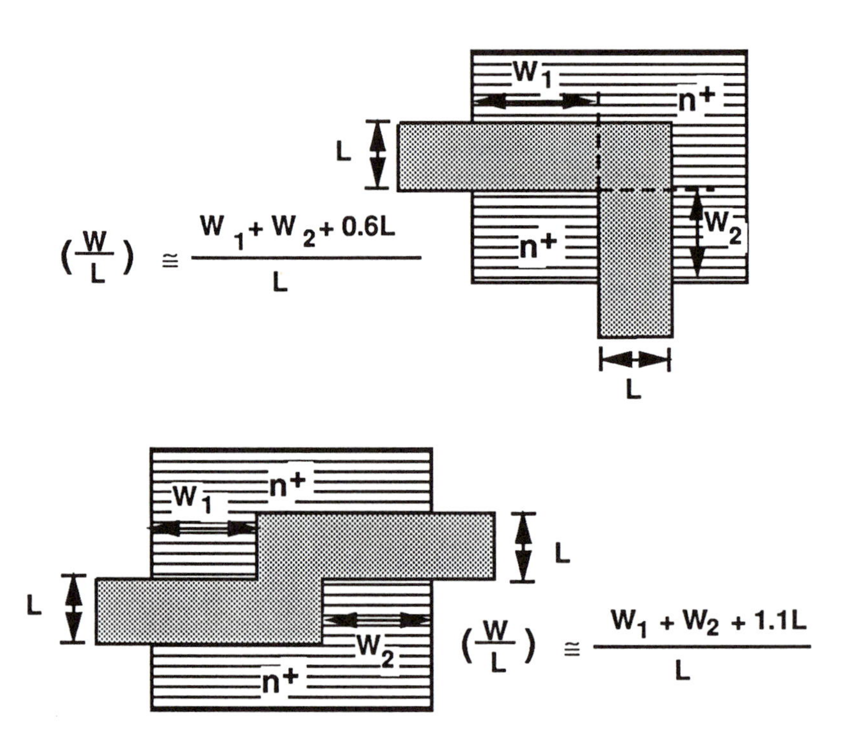

Figure 2.18: Aspect Ratios for Non-rectangular MOSFETs

2.8.1 Velocity-Saturation Effects

The two MOSFET models discussed in this chapter are based on low-electric field charge transport where the electron drift velocity v_n is proportional to the electric field such that

$$v_n = \mu_n \mathcal{E} \tag{2.75}$$

defines the linear mobility. In a MOSFET, the field intensity can easily exceed 10^3 [V/cm], driving the transport into the nonlinear regime[4]. A simple model for the velocity is obtained by writing

$$v_n = \frac{\mu_n \mathcal{E}}{1 + (\mu_n / v_s)\mathcal{E}} \tag{2.76}$$

where $v_s \simeq 10^7$ [cm/sec] is the **saturation velocity** for electrons in silicon. Defining the **critical electric field** $\mathcal{E}_c$ by

$$\mathcal{E}_c = \frac{v_s}{\mu_n} \tag{2.77}$$

yields a field-dependent velocity expression in the form

$$v_n = \frac{v_s(\mathcal{E}/\mathcal{E}_c)}{1 + (\mathcal{E}/\mathcal{E}_c)}. \tag{2.78}$$

This can be used to obtain a nonlinear mobility $\mu_n(\mathcal{E})$ with $v_n = \mu_n(\mathcal{E})\mathcal{E}$.

Velocity-saturation results in a non-saturated drain current expression of the form

$$I_D = \beta \frac{[2(V_{GS} - V_{TO})V_{DS} - V_{DS}^2]}{[1 + (\mu / L v_s) V_{DS}]}. \tag{2.79}$$

The saturation voltage is calculated by finding the maximum current where $(\partial I_D / \partial V_{DS}) = 0$. This gives

$$V_{DS,sat} = \frac{L v_s}{\mu} \left[\sqrt{1 + \frac{2\mu}{L v_s}(V_{GS} - V_{TO})} - 1 \right] \tag{2.80}$$

such that the saturated current is

$$I_D = \frac{1}{2} \beta V_{DS,sat}^2. \tag{2.81}$$

For small channel lengths L,

$$I_D \simeq W C_{ox} v_s (V_{GS} - V_T), \tag{2.82}$$

indicating that the saturated current is an approximate linear function of the gate-source voltage V_{GS}. Although the model is highly simplified, this functional dependence is observed in short-channel MOSFETs.

[4] This is discussed in more detail in Section 2.14 in the context of hot electrons

2.8.2 Gate Voltage Reduction

The surface motion of electrons is affected by the gate voltage applied to the device. A simple model which describes this phenomena is

$$\mu = \frac{\mu_n}{1 + \theta(V_{GS} - V_T)} \tag{2.83}$$

where μ_n is the "normal" mobility and θ [V^{-1}] is the **mobility modulation factor**. This model changes the device transconductance to

$$\beta = \frac{k_n'(W/L)_n}{1 + \theta(V_{GS} - V_T)} \tag{2.84}$$

where $k_n' = \mu_n C_{ox}$. The additional term in the denominator shows that the conduction is decreased by this effect.

An empirical model used in SPICE LEVEL=2 models mobility reduction by

$$\beta = [k_n'(W/L)] \left[\frac{\epsilon_{Si}}{\epsilon_{ox}} \frac{U_c x_{ox}}{(V_{GS} - V_T - U_t V_{DS})} \right]^{U_e} \tag{2.85}$$

where $0 \leq U_t \leq 0.5$ models the effect of the drain-source voltage, U_c is the **critical gate field**, and $U_e \sim 0.1$ gives an exponential influence.

2.9 Subthreshold Current

We have assumed that the MOSFET is in cutoff with $I_D = 0$ whenever $V_{GS} < V_T$. Although the transistor does indeed work quite well as a switch, the cross-sectional geometry of the device indicates the presence of a small **subthreshold current** I_{sub} which flows even when $V_{GS} < V_T$. Leakage of this type is particularly important in dynamic networks where small amounts of stored charge are used to define logic 0 and logic 1 states.

Subthreshold current can be understood by examining the MOSFET cross-section shown in Figure 2.19. Tracing the surface semiconductor polarity from source to drain shows an n^+pn^+ pattern which is similar to an npn bipolar junction transistor. The important voltages are V_{GS}, V_{DB}, and V_{SB}. Since V_{DB} and V_{SB} are normally positive, both pn junctions are reverse biased. This allows us to construct the subthreshold current in the form

$$I_{sub} = I_0(e^{-V_{DB}/\phi_T} - e^{-V_{SB}/\phi_T}), \tag{2.86}$$

where $\phi_T = (kT/q)$ is the thermal voltage. One familiar with the Ebers-Moll equations will immediately see the resemblance[5]. An analysis of the

[5] if not, then turn to Chapter 10 on BiCMOS circuits!

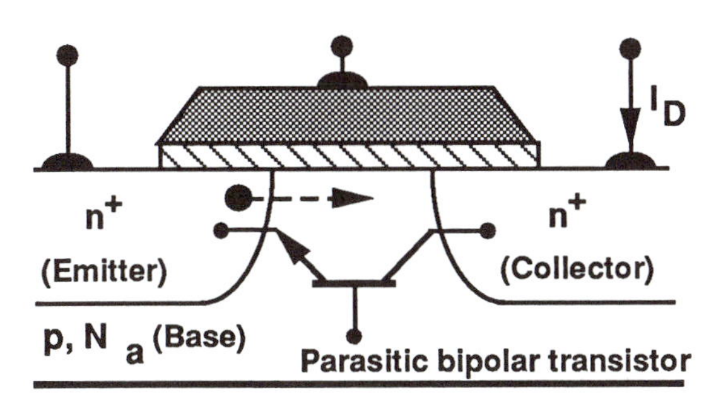

Figure 2.19: Origin of Subthreshold Current

problem gives

$$I_0 = \frac{\mu_n C_{ox} \gamma \phi_T^2}{2\sqrt{\phi_W}} \left(\frac{W}{L}\right) e^{(\phi_W - 2|\phi_F|)/\phi_T}, \tag{2.87}$$

where

$$\phi_W = \left[-\frac{\gamma}{2} + \left(\frac{\gamma^2}{4} + V_{GB} - V_{FB} \right)^{1/2} \right]^2 \tag{2.88}$$

is the surface potential ϕ_S when the device is biased into **weak inversion**.

Subthreshold current can be important in critical charge-storage circuits. A typical plot for I_D including subthreshold behavior is shown in Figure 2.20. Other analytic formulations can found in the literature with the same general results [11].

2.10 Temperature Dependence

Temperature variations in the $I - V$ behavior occur because of two main factors. First, the surface mobility varies with temperature according to the approximate behavior

$$\mu(T) \simeq \mu|_{T=300^\circ} \left(\frac{300}{T} \right)^2, \tag{2.89}$$

where $\mu|_{T=300^\circ}$ is the room temperature mobility. This dependence makes $k' = \mu C_{ox}$ a decreasing function of temperature T.

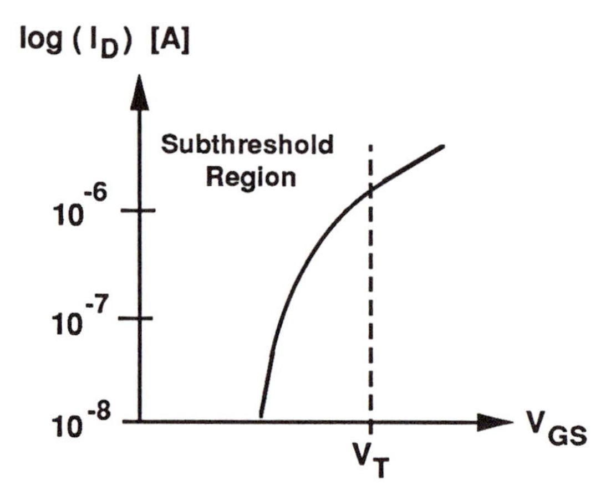

Figure 2.20: Subthreshold Current

The second variation is in the threshold voltage which exhibits a temperature dependence due to the bulk Fermi potential

$$|\phi_F| = \left(\frac{kT}{q}\right) \ln\left[\frac{N_a}{n_i(T)}\right]. \tag{2.90}$$

This shows a direct dependence on T, and also a variation due to the fact that the intrinsic density $n_i(T)$ increases with temperature. Differentiating V_{T0} gives directly that

$$\frac{dV_{T0}}{dT} = \frac{d|\phi_F|}{dT}\left[2 + \frac{\gamma}{2|\phi_F|}\right]. \tag{2.91}$$

Evaluating the derivative shows that $(dV_{T0}/dT) < 0$, so that V_{T0} decreases with increasing temperature. A simple model for this dependence is given by

$$V_{T0} = V_{T0,300^\circ} - a(T - 300^\circ), \tag{2.92}$$

where a is in the approximate range $(0.5, 5)$ [mV/°K]. The value of a tends to increase with substrate doping and oxide thickness.

Drain current variations can be expressed using $k'(T)$ and $V_T(T)$ in the basic equations. For example, the saturated MOSFET current is

$$I_D(T) = \frac{\beta(T)}{2}[V_{GS} - V_T(T)]^2[1 + \lambda V_{DS}]. \tag{2.93}$$

The basic temperature dependence of a MOSFET is illustrated in Figure 2.21.

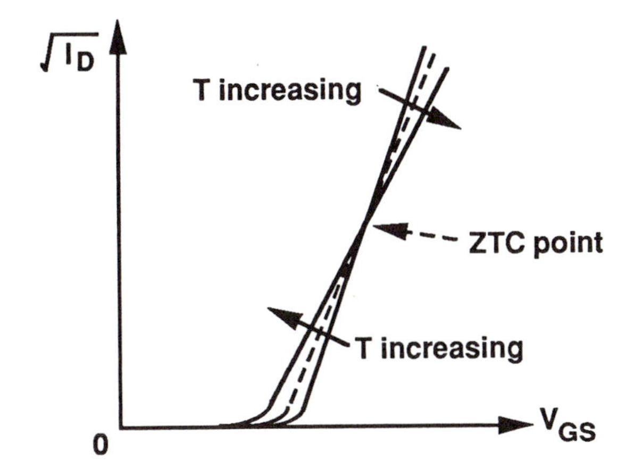

Figure 2.21: Temperature Variations in a Saturated MOSFET

Field-effect transistors such as the MOSFET exhibit an interesting characteristic: they possess a **zero-temperature coefficient** (ZTC) point where the $I - V$ curves are invariant with respect to temperature. The ZTC point can be seen in Fig. 2.21. If the device is biased at the ZTC current-voltage point, then the variation in $\mu(T)$ is balanced by the variation in $V_T(T)$. For low currents, the threshold voltage term dominates, increasing the current flow. At high current levels, mobility reduction dominates and decreases the current.

The basic temperature dependence of MOSFETs may be approximated by using the formulas above. However, when investigating temperature variations at the circuit level, computer simulations are indispensable.

2.11 Scaling Theory

Reducing device dimensions allows higher density logic integration and faster speeds. This philosophy has been the primary motivating factor for developing micron-level lithography. One price paid for this advancement is an increase in the complexity of the device physics. Effects which are negligible in "large" MOSFETs become extremely important when the transistor dimensions are reduced.

Scaling theory provides a systematic approach to modelling the situation where the size of a device is reduced while maintaining the *form* of the $I-V$ equations. Consider a MOSFET with channel length L and channel width

W. To "shrink" the real estate, we introduce a **scaling factor** $S > 1$ such that

$$\begin{aligned} L' &= \frac{L}{S} \\ W' &= \frac{W}{S} \end{aligned} \tag{2.94}$$

gives the new device dimensions. Note that the aspect ratio remains unchanged: $(W/L) = (W'/L')$. Figure 2.22 illustrates the concept. To be consistent with the surface scaling, we should also reduce the vertical dimensions by the same amount. This implies that

$$\begin{aligned} x'_{ox} &= \frac{x_{ox}}{S} \\ x'_j &= \frac{x'_j}{S}; \end{aligned} \tag{2.95}$$

achieving these values may not be possible due to limits in the fabrication. However, assuming that vertical scaling is a viable goal, this shows that the important oxide capacitance is scaled up according to

$$C'_{ox} = SC_{ox}. \tag{2.96}$$

The electric field is thus stronger in the smaller device. A deeper examination of the scaling problem shows that the doping densities must also be increased by a factor S to preserve the form of the Poisson equation $\nabla^2\phi = -\rho/\epsilon$ which describes the charge-field relation in the scaled device. Geometrical scaling is only part of the overall issue. Voltage levels are also important in determining the device equations. Two main possibilities exist in structured scaling.

2.11.1 Full-Voltage Scaling

Full-voltage scaling assumes that the power supply voltage is reduced according to

$$V'_{DD} = \frac{V_{DD}}{S}. \tag{2.97}$$

The MOSFET terminal voltages are also lowered by the same amount.

Now consider the "large" MOSFET equation

$$I_D = \frac{\beta}{2}[2(V_{GS} - V_T)V_{DS} - V_{DS}^2] \tag{2.98}$$

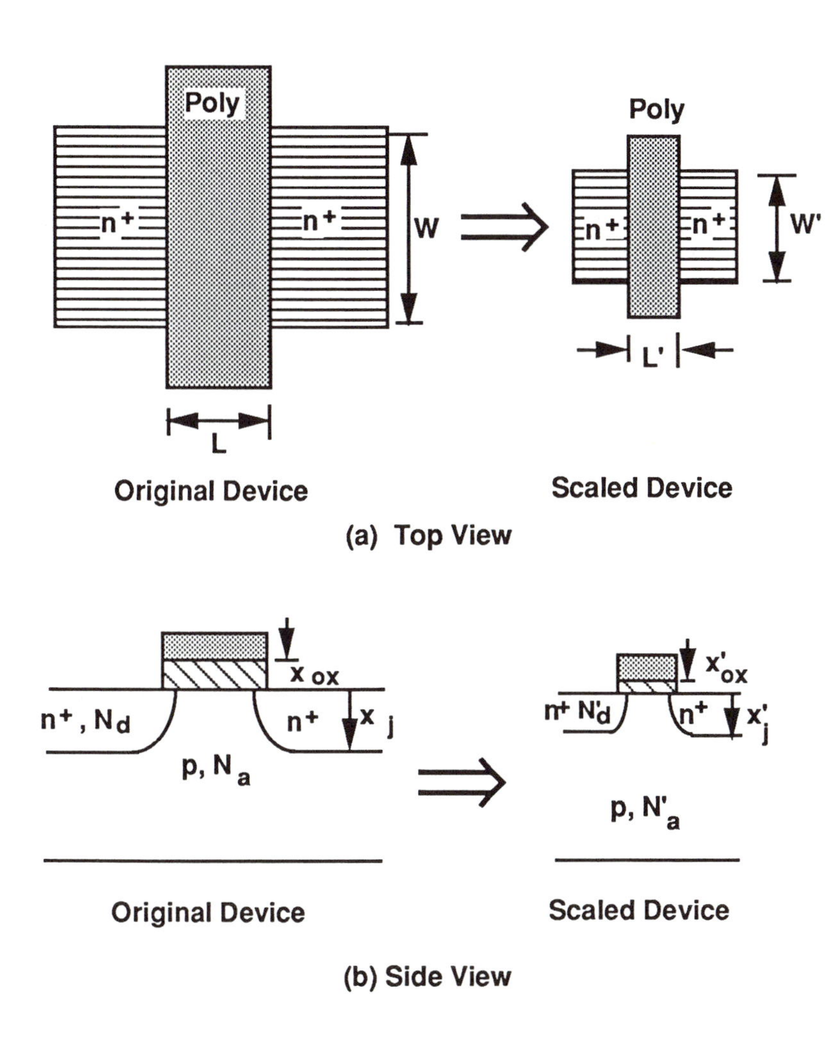

Figure 2.22: MOSFET Scaling

as subjected to the scaling rules. The device transconductance scales according to

$$\begin{aligned} \beta' &= \mu' C'_{ox} \\ &\simeq S\beta \end{aligned} \tag{2.99}$$

where the approximation enters because μ' is slightly less than μ due to increased mobility scattering in the heavier doped substrate. Applying this to the drain current then gives

$$\begin{aligned} I'_D &= \frac{\beta'}{2}[2(V'_{GS} - V'_T)V'_{DS} - (V'_{DS})^2] \\ &= \frac{I_D}{S}, \end{aligned} \tag{2.100}$$

where it has been assumed that $V'_T = (V_T/S)$ can be set in the processing. The scaled device has the same $I - V$ equation form as the original, which was the goal of the transformation. The saturated current scales according to

$$\begin{aligned} I'_D &= \frac{\beta'}{2}(V'_{GS} - V'_T)^2 \\ &= \frac{I_D}{S}, \end{aligned} \tag{2.101}$$

if we ignore channel-length modulation.

An important characteristic of full-voltage scaling is power reduction. In the original large device

$$P = I_D V_{DS} \tag{2.102}$$

so that the scaled device is described by

$$P' = \frac{P}{S^2}. \tag{2.103}$$

Note that the power dissipated per unit area is the same, but that the scaled device geometry allows a higher packing density.

2.11.2 Constant-Voltage Scaling

Full-voltage scaling is not always practical, particularly when the designer is faced with power supply standards. In constant-voltage scaling, the voltages are the same in both the original and the scaled MOSFET. The drain current then scales according to

$$\begin{aligned} I'_D &= \frac{\beta'}{2}[2(V'_{GS} - V'_T)V'_{DS} - (V'_{DS})^2] \\ &= SI_D \end{aligned} \tag{2.104}$$

indicating increased current flow. The saturated current also scales upward in the same manner, so that the power dissipation is increased to

$$P' = SP \tag{2.105}$$

for this case.

2.11.3 Second-Order Scaling Effects

The first-order scaling discussed above deals with direct transformations of the MOSFET dimensions, doping levels, voltages, and currents. Second order effects arise when the interdependence of device characteristics on these parameters is examined.

The mobility provides an example of a second order effect. In general, μ is a function of background doping N such that

$$\frac{d\mu}{dN} < 0, \tag{2.106}$$

i.e., the mobility decreases with increased doping. Since scaling theory uses an upward scaling to $N' = SN$, increased impurity scattering gives that

$$\mu'(N') < \mu(N). \tag{2.107}$$

When this is included in the scaling equations, the transconductance no longer exhibits linear scaling.

The effects of ion implants change when the vertical dimensions are scaled. An example of this is the term (qD_I/C_{ox}) in V_{T0} [see eq. (2.2)]. This assumes that the implanted layer is thin enough to be modeled as a sheet. However, if the drain and source junction depths are scaled according to $x'_j = x_j/S$, then the ion implant penetration depth (i.e., the projected range) may be comparable with x'_j. This then dictates the need for a more accurate analysis.

2.12 Short-Channel Effects

Reducing a MOSFET channel to less than about 3 [μm] invalidates many of the geometrical assumptions included either directly or indirectly in the gradual channel analysis. Although it would be desirable to find equations for the small transistors, the two- and three-dimensional effects are too complicated to expect closed form results. However, it is possible to alter some of the *large* device parameters, or to introduce new physical effects, in an effort to gain better design perspectives [1]. This section will introduce some of the more important small-geometry changes.

2.12.1 Short-Channel Definition

The definition of a "short-channel" device varies with the treatment. If we only account for the threshold voltage reduction in the next section, then a short-channel MOSFET may be defined as one where

$$L \sim x_j, \tag{2.108}$$

where x_j is the junction depth of the drain (or source) region.

An alternate definition is obtained by calculating an empirical channel parameter

$$L_{min} = 0.4[x_j \; x_{ox}(y_S + y_D)^2]^{1/3} \tag{2.109}$$

where x_j is the junction depth in microns [μm], x_{ox} is the oxide thickness in Angstroms [Å], and $(y_S + y_D)$ is the sum of source and drain pn junction depletion depths in units of microns [μm]. L_{min} constitutes the smallest L for which long-channel analysis is valid. If $L < L_{min}$, short-channel effects must be included.

2.12.2 Threshold Voltage Reduction

The MOSFET threshold voltage V_{T0} was obtained by assuming that all of the depletion charge beneath the gate originates from the MOS field effects. This ignores the fact that the depletion regions around the source and drain n^+ areas are due to normal pn junction effects. Consequently, the actual value of V_{T0} is **smaller** than that found in the simple analysis.

Figure 2.23 illustrates the basic problem. Assuming the simple depletion charge geometry shown gives a short-channel effect (SCE) threshold voltage reduction of [1]

$$(\Delta V_{T0})_{SCE} = -\gamma\left(\frac{x_j}{L}\right)\left(\sqrt{1 + \frac{2x_{dm}}{x_j}} - 1\right), \tag{2.110}$$

such that

$$(V_{T0})_{SCE} = V_{T0} - (\Delta V_{T0})_{SCE}, \tag{2.111}$$

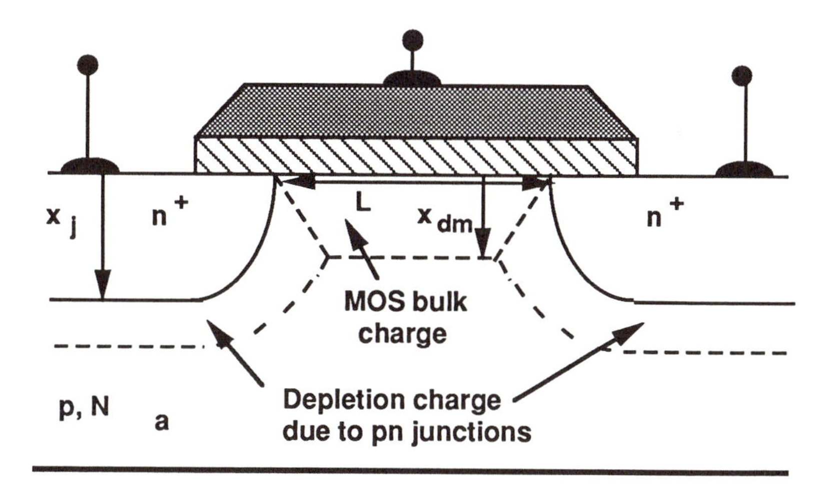

Figure 2.23: Short-Channel Effect

where V_{T0} is the "long-channel" threshold voltage. The depth parameter x_{dm} is the maximum thickness of the MOS-induced depletion region underneath the gate oxide with

$$x_{dm} = \sqrt{\frac{2\epsilon_{Si}(2|\phi_F|)}{qN_a}}. \qquad (2.112)$$

The variation in V_T as a function of channel length is shown in Figure 2.24.

The reduction in threshold voltage predicted by this expression depends on the technique used to separate MOS depletion from pn-junction depletion. In reality, this is a 2-dimensional effect which cannot be described using a simple expression. Process specifications usually provide working values for threshold voltages which can be used to design circuits.

2.12.3 Short-Channel MOSFET Model

Various approaches to modeling short channel MOSFETs have appeared in the literature. A unified analytic model by Toh, Ko, and Meyer [17] provides a basis for hand calculations which accounts for the dominant small transistor effects. We will summarize the equation set here to illustrate how the various effects are included.

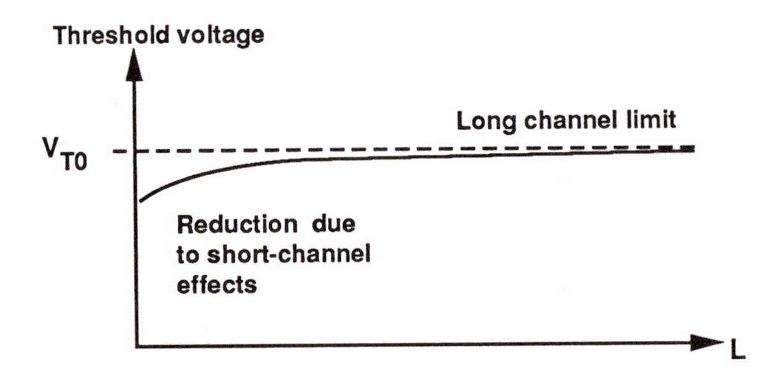

Figure 2.24: Short-Channel Threshold

Saturation is described by using

$$V_{DS,sat} = (1 - K)(V_{GS} - V_T), \tag{2.113}$$

where

$$K = \frac{1}{1 + (\mathcal{E}_c L_e)/(V_{GS} - V_T)}. \tag{2.114}$$

In this expression, $\mathcal{E}_c$ is the critical electric field

$$\mathcal{E}_c = \frac{2v_s}{\mu} \tag{2.115}$$

and v_s the saturation velocity. The electron velocity v_n is calculated from by

$$\begin{aligned} v_n &= \frac{\mu\mathcal{E}}{1 + (\mathcal{E}/\mathcal{E}_c)} \quad (\mathcal{E} \leq \mathcal{E}_c) \\ &= v_s \qquad (\mathcal{E} \geq \mathcal{E}_c). \end{aligned} \tag{2.116}$$

Also, L_e is the electrical channel length, i.e.,

$$L_e = L - x_d, \tag{2.117}$$

with x_d the depletion width at the drain junction. The mobility μ is a critical input parameter, and has a large affect on the accuracy of the model.

The non-saturated transistor current for $V_{DS} \leq V_{DS,sat}$ is described by

$$I_D = \frac{\beta_e}{1 + (V_{DS}/\mathcal{E}_c L_e)}[(V_{GS} - V_T)V_{DS} - \frac{1}{2}V_{DS}^2], \tag{2.118}$$

where

$$\beta_e = \mu C_{ox}\left(\frac{W}{L_e}\right) \tag{2.119}$$

is the effective device transconductance. Saturated current flow occurs with $V_{DS} \geq V_{DS,sat}$. The present model describes this by

$$I_D = v_{sat} C_{ox} W (V_{GS} - V_T - V_{DS,sat}), \tag{2.120}$$

which agrees with our earlier discussion.

This type of model provides corrections to the simpler long-channel MOSFET equations. Increased accuracy is gained at the expense of more complicated calculations. Although advanced device models are usually too complicated to use in analyzing large circuits, they are quite useful for predicting specific circuit benchmarks, e.g., peak current flow levels.

2.13 Narrow-Width Threshold Voltage

Narrow-width MOSFETs exhibit threshold voltages which are larger than that predicted with the gradual channel approximation. The amount of increase $(\Delta V_{T0})_{NWE}$ of the threshold voltage is due to bulk charge outside of the gate region which is ignored in the simpler analysis. Figure 2.25 illustrates the basic problem: fringing fields deplete the silicon beyond the gate region. To model this, we assume that the depletion edge is circular with a radius x_{dm} as shown. The additional bulk charge gives a positive threshold voltage shift by an amount

$$(\Delta V_{T0})_{NWE} = +\frac{\pi}{2}\frac{\gamma x_{dm}}{W} \tag{2.121}$$

which must be added to the long-channel value V_{T0} such that

$$(V_{T0})_{NWE} = V_{T0} + (\Delta V_{T0})_{NWE}. \tag{2.122}$$

If body-bias is present, then we must modify γ by

$$\gamma \rightarrow \sqrt{2q\epsilon_{Si}N_a(2|\phi_F| + V_{SB})} \tag{2.123}$$

since the depletion charge is increased accordingly.

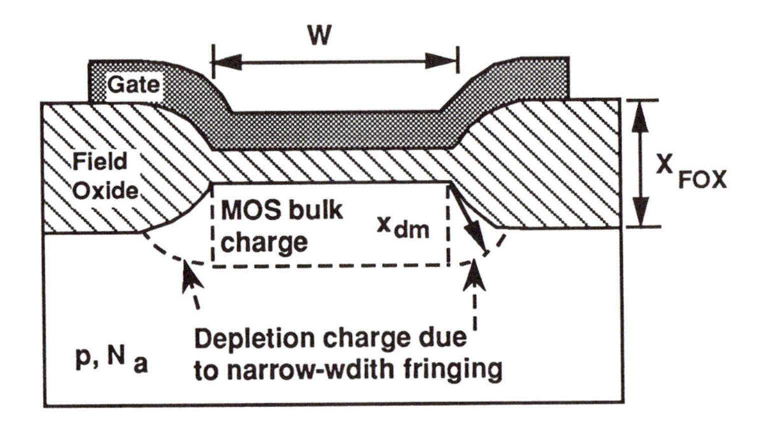

Figure 2.25: Narrow-Width V_T Geometry

SPICE uses a similar model to include narrow-width effects. The threshold voltage expression is modified to

$$(\Delta V_{T0})_{NWE} = \frac{\delta \epsilon_{Si} \pi}{4 C_{ox} W}(2|\phi_F| + V_{SB}) \tag{2.124}$$

where δ (or DELTA) is an empirical parameter which is used to describe the geometry of the depletion edge.

2.14 Hot Electrons

The velocity v of an electron in silicon increases with the electric field $\mathcal{E}$. This is due to the Lorentz force in conjunction with particle scattering events. For low electric fields, we approximate this dependence by the linear relationship

$$v = \mu \mathcal{E}, \tag{2.125}$$

where μ is the mobility. As $\mathcal{E}$ increases, scattering and other transport mechanisms induce nonlinearities in $v(\mathcal{E})$ as shown in Figure 2.26. In particular, the electron velocity saturates at a value of approximately $v_s \simeq 10^7$ [cm/s] in silicon at room temperature.

The concept of **electron temperature** is introduced to classify the velocity regions. If the excitation were due solely to thermal means, then $v \sim T$ with T the temperature; we use this view as an analogy for describing field-aided transport. For low electric fields, v is small which defines **cold electrons**. As the velocity increases and the $v - \mathcal{E}$ curve goes nonlinear, we

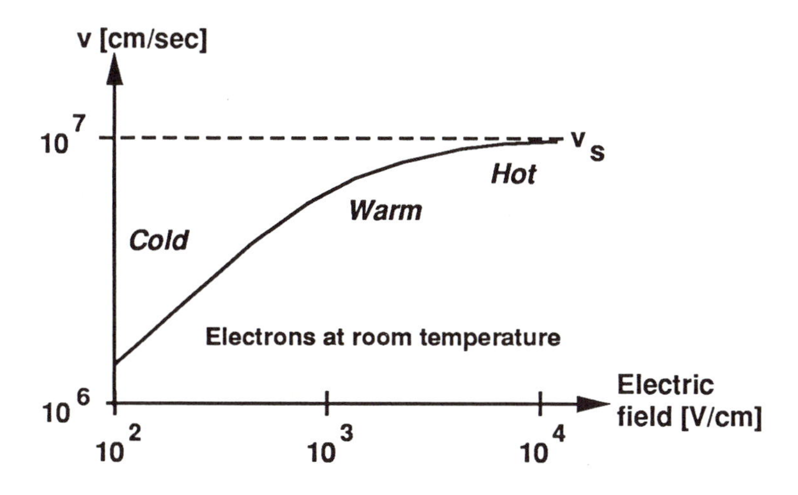

Figure 2.26: Velocity-Field Relation

enter the **warm electron region**. Finally, velocity saturation to a value of $v_s \sim 10^7$ [cm/sec] indicates the **hot electron** region where the transport is complicated by high-field effects.

Hot electron effects have been observed in MOSFETs, particularly in devices with channel lengths smaller than 1 micron [7]. Standard transistors can be degraded by tunnelling effects as illustrated in Figure 2.27. Highly energetic particles can leave the silicon and enter the gate oxide. Trapped electrons increase the oxide charge Q_{ox}, leading to instability of the threshold voltage. Long-term reliability problems may result from this mechanism. In addition, hot electrons may induce leakage gate currents I_g and excessive substrate current I_s.

Reliable design of small MOSFETs usually requires that hot electron effects be minimized. This is accomplished by reducing the magnitude of the electric field that acts on the channel charge carriers. This is particularly important in short-channel devices, since the drain-source voltage V_{DS} must be dropped along the current flow path. Using the simple average field approximation

$$\mathcal{E}_{lat} \approx \frac{V_{DS}}{L} \tag{2.126}$$

to estimate the lateral (horizontal) field strength, we see that, with $V_{DS} = 5$ [V], a channel length of $L = 1$ [μm] gives an average field of $\mathcal{E} \approx 5 \times 10^4$ [V/cm]; this value is sufficient to induce hot electron effects. Using the gradual channel analysis shows that the problem is much worse than this

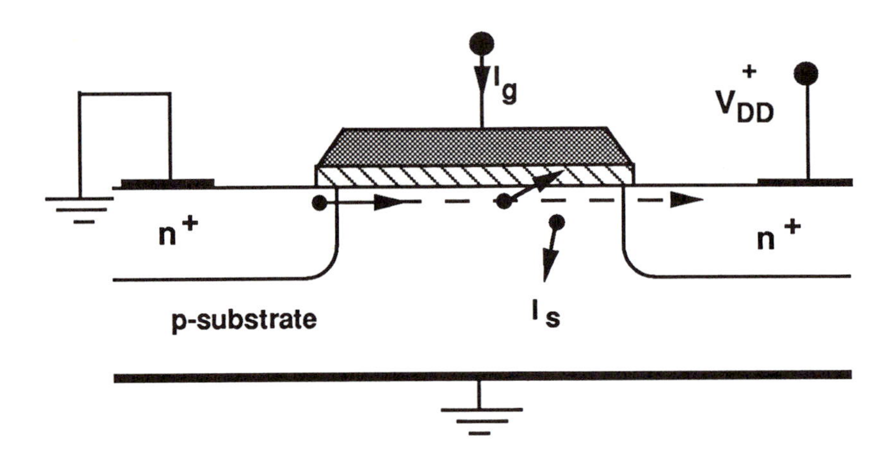

Figure 2.27: Hot Electrons in MOSFETs

on the drain side of the transistor where the electric field is a maximum. One obvious solution is to reduce the operating voltage, but even a power supply standard of $V_{DD} = 3.3$ [V] can induce hot electron effects. Owing to this, device engineers have developed structures which are designed to minimize hot electron effects. Various approaches have been presented in the literature. In the present discussion we will examine using lightly-doped drains (LDD) structures to deal with hot electrons [5].

An LDD structure is shown in shown in Figure 2.28. The drain and source regions have been modified by inserting lightly doped n^- regions between the channel and the low resistance n^+ areas. This reduces the effective built-in electric field on the drain side of the channel where the hot electron tunnelling probability is the largest. To understand this, denote the n^- donor doping by N_d^- and approximate the n^-p boundary as a step junction. The maximum (vertical) electric field occurs at the junction with

$$max(\mathcal{E}_{vert}) = \frac{qN_d^- x_d}{\epsilon_{Si}}, \tag{2.127}$$

where x_d is the depletion width. Setting N_d^- to be 2 or 3 orders of magnitude smaller than the doping in the n^+ regions decreases the tunneling probability at the drain. Both I_g and I_s are reduced accordingly. The price paid for this design is the fact that the drain and source resistances are larger and the processing is more complicated since it requires another masking step. (The increase in drain and source resistance actually reduces the channel electric field even further, but they degrade the transient response.)

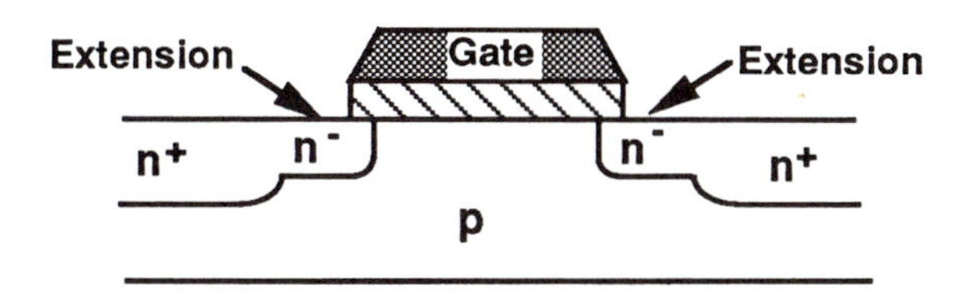

Figure 2.28: LDD MOSFET

2.15 MOSFET Modelling in SPICE

SPICE (Simulation Program with Integrated Circuit Emphasis) and its relatives are useful in chip design and analysis. The accuracy of a computer simulation depends on the device equations. Increased precision generally requires better modelling and longer run times.

2.15.1 SPICE2 MOSFET Model

Three levels of accuracy are allowed for MOSFET models in standard SPICE. The choice of device equations is accomplished by specifying

- LEVEL: Specifies the device equation set
- **Parameters:** Values for each transistors

for use in the .MODEL description line of an input circuit file. LEVEL can be set to 1, 2, or 3 to choose the desired model. For LEVEL=1, the basic square-law equations are used to simulate the device. The LEVEL=2 treatment is a modified bulk-charge model, and LEVEL=3 is an empirical curve-fit.

Table 2.4 lists the device parameters which can be specified, while Table 2.5 and Table 2.6 list the main model parameters and the parasitic elements. Symbols used in this book are provided when there is a direct correlation. It is important to note SPICE uses the meter as the basic length unit, and some parameters must be changed from [cm] to [m] with proper conversion factors.

To describe a MOSFET in SPICE, a device line is included with the basic listing of

- MNAME ND NG NS NB MODNAME < Parameters>

where ND, NG, NS, NB are the node numbers for the drain, gate, source, and bulk, respectively, and MODNAME specifies the model line which provides the device parameters. Optional (but usually important) <Parameters> in the device listing include device dimensions W, L, AD, AS, PD, PS; these are listed and explained in Table 2.4.

Symbol	Meaning	Units
MNAME	Device Name	
MODNAME	.MODEL reference	
W	Channel width	m
L	Channel length	m
AD	Area of drain	m^2
AS	Area of source	m^2
PD	Perimeter of drain	m
PS	Perimeter of source	m
NRD	Number of squares for drain	
NRS	Number of squares for source	
OFF	Specify IC of OFF	
IC=V1,V2,V3	Specify IC for VDS,VGS,VBS	

Table 2.4: SPICE Device Parameters

The .MODEL description is used for parametric specifications and is required for completing the input file. Base process information is included here with the general format of

- .MODEL MODNAME ...

where the ellipses denotes the listing of desired input parameters. SPICE uses default values for parameters which are not specified. Each transistor in the circuit requires its own device statement, but a .MODEL line can be shared by any number of MOSFETs. The listing in Table 2.7 illustrates typical SPICE information listings.

A complete discussion of SPICE device modelling can be found in Antognetti and Massobrio [2] or commercial user manuals. Some of the more important features are summarized below.

Text Symbol	SPICE NAME	Level	Description and Units
V_{T0}	VTO	1-3	Zero-bias threshold, V
k'	KP	1-3	Transconductance, A/V^2
γ	GAMMA	1-3	Body-bias parameter, $V^{1/2}$
$2\|\phi_F\|$	PHI	1-3	Surface inversion potential, V
λ	LAMBDA	1,2	Channel modulation, 1/V
μ	UO	1-3	Surface mobility, cm^2/V-sec
x_{ox}	TOX	1-3	Oxide thickness, m
N_a, N_d	NSUB	1-3	Substrate doping, cm^{-3}
x_j	XJ	2,3	Junction depth , m
L_o	LD	1-3	Lateral diffusion length, m
Q_{ss}/q	NSS	2,3	Surface state density, cm^{-3}
	NFS	2,3	Fast Surface state density cm^{-3}
	NEFF	2	Total channel charge coefficient
	TPG	2,3	Type of gate material
U_c	UCRIT	2	Mobility critical field, U_c V/cm
U_e	UEXP	2	Mobility coefficient
U_t	UTRA	2	Transverse field coefficient
v_s	VMAX	2,3	Maximum drift velocity, m/sec
δ	DELTA	2,3	Narrow-width parameter
η	ETA	3	DIBL factor, 1/V
θ	THETA	3	Mobility modulation factor
	AF	1-3	Flicker-noise exponent
	KF	1-3	Flicker-noise coefficient

Table 2.5: SPICE .MODEL Parameters

Turn-on Voltage

The threshold voltage V_T represents the voltage needed to induce surface inversion and channel formation. The **on voltage** V_{on} can be viewed as the boundary between weak and strong inversion levels. SPICE models the on-voltage using

$$V_{on} = V_T + \frac{nkT}{q}, \tag{2.128}$$

where

$$n = 1 + \frac{qN_{FS}}{C_{ox}} + \frac{C_d}{C_{ox}}. \tag{2.129}$$

Text Symbol	SPICE NAME	Description and Units
V_{bi}	PB	Junction built-in voltage , V
C_{j0}	CJ	Zero-bias junction capacitance, F/m^2
m	MJ	Junction grading coefficient
C_{jsw}	CJSW	Zero-bias perimeter capacitance, F/m
m_{sw}	MJSW	Sidewall grading coefficient
C_{GBO}	CGBO	Gate-bulk overlap capacitance , F/m
C_{GDO}	CGDO	Gate-drain overlap capacitance, F/m
C_{GSO}	CGSO	Gate-source overlap capacitance, F/m
I_s	IS	Junction leakage current , A
J_s	JS	Junction leakage current density, A/m^2
R_D	RD	Drain resistance , Ω
R_S	RS	Source resistance , Ω
R_s	RSH	Sheet resistance (source/drain), Ω

Table 2.6: Parasitic SPICE Parameters

In this expression, N_{FS} represents the fast surface-state density and

$$C_d = \frac{\gamma}{2\sqrt{2|\phi_F| + V_{SB}}} C_{ox} \tag{2.130}$$

is the depletion capacitance per unit area.

Subthreshold Current

For $V_{GS} < V_T$, subthreshold current flow is described using the simple model

$$I_{sub} = I_{on} e^{q(V_{GS}-V_{on})/nkT} \tag{2.131}$$

where V_{on} and n are defined above. This exponential dependence is only a low-order approximation, and should be used with caution.

Device Capacitances

SPICE generates capacitances using separate equation sets for cutoff, saturation, and non-saturation. Both depletion and gate-channel contributions are included in the calculations with zero-bias values set by CJ, CJSW, CGBO, CGDO, and CGSO. The grading parameters MJ and MJSW can be adjusted to model the doping profile. Area and perimeter information is

```
.MODEL CMOSN NMOS LEVEL=2 LD=0.25U TOX=382E-10
+ NSUB=2.30325E+16 VTO=0.9423 KP=5.571E-5 GAMMA=0.9673
+ PHI=0.6 UO=616.32 UEXP=0.252824 UCRIT=141868 DELTA=1.151392
+ VMAX=100000 XJ=0.25U LAMBDA=2.23025E-2 NFS=2.448399E+12
+ NEFF=1 NSS=1.0E+12 TPG=1.00 RSH=19.4800 CGDO=3.389866E-10
+ CGSO=3.389866E-10 CGBO=6.879548E-10 CJ=3.984E-4
+ MJ=0.46230 CJSW=5.4980E-10 MJSW=0.37140 PB=0.8000
.MODEL CMOSP PMOS LEVEL=2 LD=0.25U TOX=382E-10
+ NSUB=6.21260E+15 VTO=-0.7022 KP=2.457E-5 GAMMA=0.5024
+ PHI=0.6 UO=271.78 UEXP=0.343785 UCRIT=50597.6 DELTA=1.0E-6
+ VMAX=47136 XJ=0.25U LAMBDA=6.458996E-2 NFS=5.173338E+11
+ NEFF=1 NSS=1.0E+12 TPG=-1.00 RSH=73.6200 CGDO=3.389866E-10
+ CGSO=3.389866E-10 CGBO=6.134733E-10 CJ=1.923E-4
+ MJ=0.44720 CJSW=2.2370E-10 MJSW=0.17500 PB=0.7000
```

Table 2.7: SPICE .MODEL Statements for a Sample 2 [μm] Process

input into the device descriptions using AD, AS, PD, and PS in the device line. In general, the division of area among devices with common drain or source regions is arbitrary.

The depletion capacitances for drain and source regions are computed using

$$\begin{aligned} \text{CD} &= CJ * AD(1 - \tfrac{V}{PB})^{-MJ} + CJSW * PD(1 - \tfrac{V}{PB})^{-MJSW} \\ \text{CS} &= CJ * AS(1 - \tfrac{V}{PB})^{-MJ} + CJSW * PS(1 - \tfrac{V}{PB})^{-MJSW} \end{aligned} \tag{2.132}$$

Basic implementations of SPICE only allow one built-in potential PB to be specified; in this case, the difference between the bottom and sidewall doping values is ignored.

In a realistic parameter list, the values of the overhang capacitances CGDO, CGSO, CGBO include fringing field effects. If one attempts to compute these values using, for example, CGSO=COX*LD with COX the SPICE-equivalent value of $C_{ox} = (\epsilon_{ox}/x_{ox})$, the two values should be inconsistent since the simple analytic approach ignores the fringing fields.

2.15.2 BSIM

The **Berkeley short-channel IGFET model** (BSIM) provides a more accurate description for use in SPICE [15] simulations. It has the nice properties that it is analytically simple and is based on a small number of parameters which can be extracted experimentally from a sample die. The BSIM parameter set consists of the seven quantities summarized in Table

2.8. In this section we will briefly summarize the main features of MOSFET

Symbol	Meaning	Units
V_{FB}	Flatband Voltage	V
ϕ_S	Surface inversion potential	V
K_1	Body-bias coefficient	$V^{1/2}$
K_2	Charge-sharing coefficient	$V^{1/2}$
η	DIBL coefficient	
U_0	Vertical mobility degradation	V^{-1}
U_1	Velocity saturation coefficient	m/V

Table 2.8: BSIM Device Parameters

performance included in this model.

BSIM calculates the device threshold voltage using

$$V_T = V_{FB} + \phi_S + K_1\sqrt{\phi_S + V_{SB}} - K_2(\phi_S + V_{SB}) - \eta V_{DS}, \qquad (2.133)$$

which includes body bias ($K_1 = \gamma$), short-channel effects (K_2), and drain-induced barrier lowering[6] (η). The three MOSFET operational modes are defined relative to V_T as in a standard model.

Cutoff

When $V_{GS} \leq V_T$, BSIM uses

$$I_D = I_{D,S} + I_{D,W}, \qquad (2.134)$$

where $I_{D,S}$ is the drain current in strong inversion, while $I_{D,W}$ accounts for leakage currents in weak inversion. In this expression,

$$I_{D,W} = \frac{I_{exp} I_{limit}}{I_{exp} + I_{limit}}, \qquad (2.135)$$

with

$$I_{exp} = \beta\left(\frac{kT}{q}\right)^2 e^{1.8} e^{q(V_{GS}-V_T)/nkT}[1 - e^{-qV_{DS}/kT}] \qquad (2.136)$$

as the subthreshold current, n as the slope coefficient,

$$I_{limit} = \frac{\beta}{2}\left(\frac{3kT}{q}\right)^3, \qquad (2.137)$$

[6] Also known as DIBL

and

$$\beta = \mu C_{ox}\left(\frac{W}{L}\right). \tag{2.138}$$

This provides a smooth transition between conducting and non-conducting states.

Saturation Voltage

The saturation voltage is computed from

$$V_{DS,sat} = \frac{V_{GS} - V_T}{a\sqrt{K}}, \tag{2.139}$$

where a accounts for body bias and is computed using

$$a = 1 + \frac{gK_1}{2\sqrt{\phi_s + V_{SB}}} \tag{2.140}$$

with

$$g = 1 - \frac{1}{1.744 + 0.8364(\phi_S + V_{SB})}. \tag{2.141}$$

The parameter K is defined by

$$K = \frac{1}{2}(1 + v_c + \sqrt{1 + 2v_c}) \tag{2.142}$$

and is used to describe velocity saturation in the equation

$$v_c = \frac{U_1}{L}\left(\frac{V_{GS} - V_T}{a}\right). \tag{2.143}$$

The value of $V_{DS,sat}$ thus includes small device effects.

Non-saturation

Triode or non-saturated operation occurs with $V_{GS} \geq V_T$ and $V_{DS} \leq V_{DS,sat}$. In this case,

$$I_D = \beta_1[2(V_{GS} - V_T)^2 - aV_{DS}^2], \tag{2.144}$$

where β_1 is given by

$$\beta_1 = \frac{\beta}{[1 + U_0(V_{GS} - V_T)][1 + (U_1/L)V_{DS}]} \tag{2.145}$$

which is similar to the modeling discussed earlier.

Saturation

Saturated current flow occurs when $V_{GS} \geq V_T$ and $V_{DS} \geq V_{DS,sat}$. The BSIM expression is

$$I_D = \frac{\beta_2}{2aK}(V_{GS} - V_T)^2, \tag{2.146}$$

where now

$$\beta_2 = \frac{\beta}{[1 + U_0(V_{GS} - V_T)]} \tag{2.147}$$

describes the device transconductance.

The BSIM model has the advantage that it accounts for several small-device effects and accepts experimentally derived input values. Since it is empirically-based, BSIM allows SPICE to overcome the limitations of the simpler device models. As such, it can provide very accurate circuit simulations.

2.16 References

MOSFETs are discussed in a large number of books and even larger number of journal articles. All of the books listed below contain excellent discussions of MOSFET characteristics. A few selected journal articles have also been included for more in-depth coverage.

[1] L.A. Akers and J.J. Sanchez, "Threshold Voltage Models of Short, Narrow and Small Geometry MOSFET's: A Review", Solid-State Electronics, vol. 25, pp. 621-641, July, 1982.

[2] P. Antognetti and G. Massobrio (eds.), **Semiconductor Device Modelling with SPICE**, McGraw-Hill, New York, 1988.

[3] J.R. Brews, W. Fichtner, E.H. Nicollian and S.M. Sze, "Generalized guide for MOSFET Miniaturization", IEEE Electron Device Letters, vol. EDL-1, pp. 2-4, 1980.

[4] J.Y. Chen, **CMOS Devices and Technology for VLSI**, Prentice-Hall, Englewood Cliffs, NJ, 1990.

[5] Y-Z. Chen and T-W. Tang, "Computer Simulation of Hot-Carrier Effects in Asymmetric LDD and LDS MOSFET Devices", IEEE Trans. Electron Devices, vol. 36, pp. 2492-2498, November, 1989.

[6] J.Y. Chi and R.P. Holstrom, "Constant voltage scaling of FET's for high frequency and high power applications", Solid-State Electronics, vol. 26, pp. 667-670, July, 1983.

[7] P.E. Cottrell, R.R. Troutman, and T.H. Ning, "Hot-Electron Emission in N-Channel IGFET's", IEEE Trans. Electron Devices, vol. ED-26, pp. 520-532, April, 1979.

[8] M.J. Deen and Z.P. Zuo, "Edge Effects in Narrow-Width MOSFET's", IEEE Trans. Electron Devices, vol. 38, pp. 1815-1819, August, 1991/

[9] R.H. Dennard, et. al , "Design of ion-implanted MOSFETs with very small physical dimensions", IEEE J. Solid-State Circuits, vol. SC-9, pp. 256-268, October, 1974.

[10] D. A. Divekar, **FET Modeling for Circuit Simulation**, Kluwer Academic Publishers, Boston, 1988.

[11] D.K. Ferry, L. A. Akers, and E.W. Greeneich, **Ultra Large Scale Integrated Microelectronics**, Prentice-Hall, Englewood Cliffs, NJ, 1988.

[12] G. Krieger, R. Sikora, P.P. Ceuvas, and M.N. Misheloff, " Moderately Doped NMOS (M-LDD)-Hot Electron and Current Drive Optimization", IEEE Trans. Electron Devices, vol. 38, pp. 121-127, January, 1991.

[13] R.S. Muller and T.I Kamins, **Device Electronics for Integrated Circuits**, 2nd ed., John Wiley & Sons, New York, 1986.

[14] E.H. Nicollian and J.R. Brews, **MOS Physics and Technology**, Wiley-Interscience, New York, 1982.

[15] B.J. Sheu, D.L. Scharfetter, P-K. Ko, M-C Jeng, "BSIM: Berkeley Short-Channel IGFET Model for MOS Transistors", *IEEE J. Solid-State Circuits*, vol. SC-22, No. 4, pp. 558-565, August, 1987.

[16] S. Sze, **Physics of Semiconductor Devices**, 2nd ed., John Wiley & Sons, New York, 1981.

[17] K-Y. Toh, P-K. Ko, and R.G. Meyer, " An Engineering Model for Short-Channel MOS Devices", *IEEE J. Solid-State Circuits*, vol. 23, No. 4, pp. 950-957, August, 1988.

[18] Y. P. Tsividis, **Operation and Modeling of The MOS Transistor**, McGraw-Hill, New York, 1987.

[19] J. P. Uyemura, **Fundamentals of MOS Digital Integrated Circuits**, Addison-Wesley, Reading, MA, 1988.

[20] E.S. Yang, **Microelectronic Devices**, McGraw-Hill, New York, 1988.

Chapter 3

The CMOS Inverter

Complementing a logical variable A to give $\overline{A}$ is accomplished using a basic inverter circuit. A standard CMOS inverter is quite simple and is built using two opposite-polarity MOSFETs in a complementary manner. The circuit gives a large output voltage swing and only dissipates significant power when the input is switched; these are two important properties of CMOS logic circuits. This chapter provides a detailed examination of a CMOS inverter and sets the foundations for most higher-level CMOS designs.

3.1 Circuit Operation

Figure 3.1 shows a CMOS inverter circuit. The input voltage V_{in} is connected to the gate of both an nMOS and a pMOS transistor. The output voltage V_{out} is taken from the common drain terminals. Transistor placement is chosen in a manner that ensures only one of the MOSFETs conducts when the input is at a stable low or high voltages. Although somewhat superfluous, we will refer to this type of transistor arrangement as **fully-complementary CMOS structuring** to distinguish it from other approaches to CMOS circuits.

The inverter circuit operation can be understood by examining the relationship between V_{in} and the gate-source voltages of the MOSFETs. We see that

$$\begin{aligned} V_{GSn} &= V_{in} \\ V_{SGp} &= V_{DD} - V_{in}, \end{aligned} \tag{3.1}$$

where V_{in} is assumed to be in the voltage range $[0, V_{DD}]$.

A high input voltage of $V_{in} = V_{DD}$ gives

$$\begin{aligned} V_{GSn} &= V_{DD} \\ V_{SGp} &= 0, \end{aligned} \tag{3.2}$$

so that the p-channel MOSFET Mp is in cutoff while the n-channel MOSFET Mn is conducting in the non-saturated mode. Mn thus provides a current path to ground giving

$$min[V_{out}] = V_{OL} \simeq 0. \tag{3.3}$$

Conversely, a low input voltage of $V_{in} = 0$ results in

$$\begin{aligned} V_{GSn} &= 0 \\ V_{SGp} &= V_{DD} \end{aligned} \tag{3.4}$$

which shows that Mn is in cutoff, while Mp conducts in the non-saturated mode. The pMOS transistor Mp then provides a path to the power supply so that

$$max[V_{out}] = V_{OH} \simeq V_{DD}. \tag{3.5}$$

Because of the placement and operation of each MOSFET, Mn is often

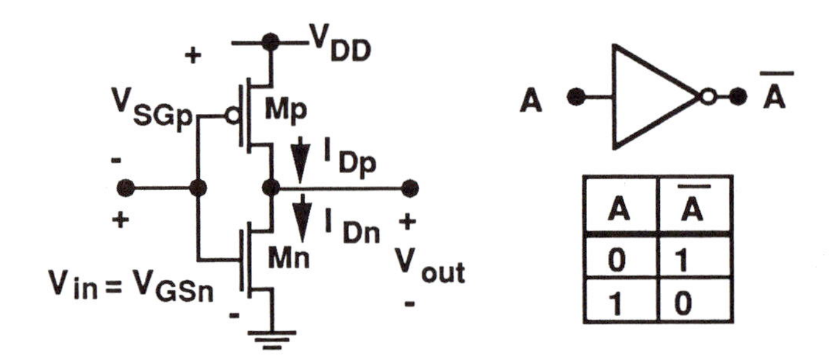

Figure 3.1: CMOS Inverter

called a **pull-down** transistor, while Mp is termed a **pull-up** device.

The DC input-output characteristics are portrayed graphically using the **Voltage-Transfer Curve (VTC)** shown in Figure 3.2. This is simply a plot of V_{out} as a function of V_{in}. The inversion operation is seen directly from the curve: when V_{in} is small, V_{out} is large, and vice-versa[1]. Qualitatively, the sharpness of the transition is a measure how well the circuit is

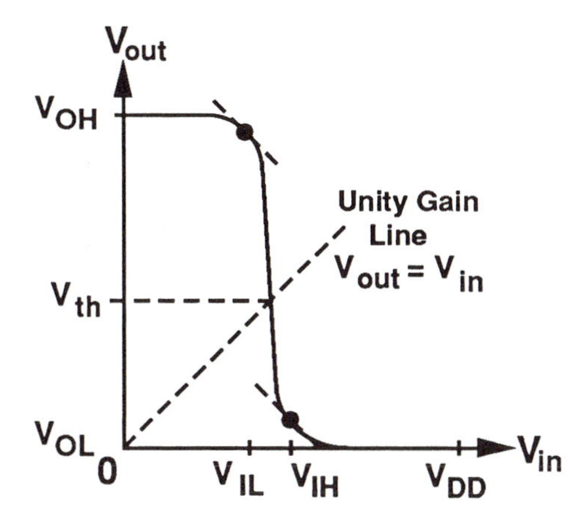

Figure 3.2: Inverter Voltage Transfer Curve

able to perform digital operations. All of the important DC circuit characteristics can be extracted from the VTC. In a digital logic circuit, logic 0 and logic 1 states are defined by a *range of voltages* for each logic level. Moreover, the limits used to define logic 0 and logic 1 values are different for the input and output terminals. We may use the VTC to obtain a set of **critical voltages** to work with. The important values are denoted by V_{OH}, V_{OL}, V_{IL}, V_{IH}, and V_{th}, and are listed in Table 3.1.

The input and output voltage ranges used to define logic 0 and logic 1 states are shown in Figure 3.3. For the input voltage V_{in}, we make the association

$$\begin{aligned} \text{Logic } 0: &\quad V_{in} \in [0, V_{IL}], \\ \text{Logic } 1: &\quad V_{in} \in [V_{IH}, V_{DD}]. \end{aligned} \tag{3.6}$$

At the output, logical values are defined by the voltage ranges

$$\begin{aligned} \text{Logic } 0: &\quad V_{out} \in [0, V_{OL}], \\ \text{Logic } 1: &\quad V_{out} \in [V_{OH}, V_{DD}]; \end{aligned} \tag{3.7}$$

in a CMOS circuit, we will find values of $V_{OL} = 0$ and $V_{OH} = V_{DD}$, as discussed below.

Several other voltages are used to characterize a digital circuit. Of particular interest are the following values.

[1]The DC VTC assumes that transient effects have decayed away.

Symbol	Name	Meaning
V_{OH}	Output-High Voltage	Maximum V_{out}
V_{OL}	Output-Low Voltage	Minimum V_{out}
V_{IL}	Input-Low Voltage	Maximum Logic 0 Input
V_{IH}	Input-High Voltage	Minimum Logic 1 Input
V_{th}	Gate Threshold Voltage	Switching Voltage

Table 3.1: Critical VTC Voltages

(a) The **Logic Swing**: This is defined by

$$V_\ell = V_{OH} - V_{OL} \tag{3.8}$$

and gives the maximum output voltage variation.

(b) The **Transition Width** characterizes the input variations such that

$$\text{TW} = V_{IH} - V_{IL}. \tag{3.9}$$

This provides a measure of the input sensitivity.

(c) **Voltage Noise Margins**. These are defined for high and low logic states by

$$\begin{aligned} \text{VNM}_\text{L} &= V_{IL} - V_{OL}, \\ \text{VNM}_\text{H} &= V_{OH} - V_{IH}, \end{aligned} \tag{3.10}$$

and represent level of voltage separation when identical stages are cascaded. The voltage noise margins must be greater than zero for a functional digital circuit.

3.1.1 DC Inverter Calculations

The critical voltages are established by the MOSFET parameters summarized in Table 3.2. Once these values are chosen, the VTC characteristics of the circuit are set. In general, circuit designers may only vary the device aspect ratios $(W/L)_n$ and $(W/L)_p$, since the electrical parameters are a result of the fabrication and cannot be changed. The critical voltages may be computed by equating drain currents $I_{Dn} = I_{Dp}$ along with any other necessary conditions. The results are summarized below where we denote

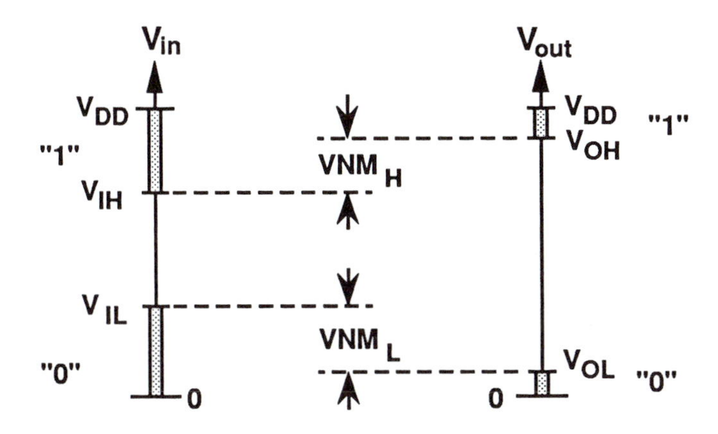

Figure 3.3: Logic-Voltage Relations

MOSFET	Parameters
Mn	k'_n, V_{Tn}, $(W/L)_n$, λ_n, γ_n, $2\|\phi_{Fp}\|$
Mp	k'_p, V_{Tp}, $(W/L)_p$, λ_p, γ_p, $2\phi_{Fn}$

Table 3.2: CMOS Inverter Device Parameters

device transconductance parameters of nMOS and pMOS transistors by

$$\beta_n = k'_n \left(\frac{W}{L}\right)_n, \qquad \beta_p = k'_p \left(\frac{W}{L}\right)_p, \tag{3.11}$$

respectively. Square-law current flow equations from Section 2.2.1 are used to obtain closed form approximations in our calculations. Channel-length modulation effects are ignored for simplicity, but are usually included in computer simulations.

(a) $\mathbf{V_{OH}}$: The output-high voltage is the largest value of V_{out}, and occurs when $V_{in} < V_{Tn}$. In this case, Mp is biased into the active region while Mn is in cutoff. Ideally, the simplified MOSFET equations give $I_{Dp} = 0$ which implies that the source-drain voltage V_{SDp} is 0 [V]; thus,

$$V_{out} = V_{OH} = V_{DD}, \tag{3.12}$$

with V_{DD} the power supply voltage. In realistic circuits, leakage currents (which are ignored in the square-law model) reduce the value slightly.

(b) $\mathbf{V_{OL}}$: The output-low voltage represents the smallest value of V_{out} from the circuit. Setting the input voltage to a value $V_{in} > (V_{DD} - |V_{Tp}|)$ places Mp in cutoff, and gives $V_{out} = V_{OL}$. Since Mn is biased active but has $I_{Dn} = 0$, the drain-source voltage across the nMOSFET is 0 [V]. At this point, the inverter output is given by

$$V_{out} = V_{OL} = 0. \tag{3.13}$$

Leakage currents increase the actual value of V_{OL} slightly.

As mentioned above, an important property of CMOS is that the output logic swing is given by

$$V_\ell \simeq V_{DD}. \tag{3.14}$$

This **full-rail** output automatically helps achieve large noise margin values.

(c) $\mathbf{V_{IL}}$: The input-low voltage V_{IL} represents the largest value of V_{in} which can be interpreted as a logic 0 input. If V_{in} is increased above V_{IL}, the circuit moves into the transition region as shown in Fig. 3.2. Using stability arguments, we define V_{IL} as the point where the slope of the VTC is -1, i.e.,

$$\frac{dV_{out}}{dV_{in}} = -1. \tag{3.15}$$

V_{IL} can be computed by combining this condition with $I_{Dn} = I_{Dp}$, where we note that Mn is saturated and Mp is non-saturated. The analysis yields two simultaneous equations in the form

$$\begin{aligned} \beta_n(V_{IL} - V_{Tn})^2 &= \beta_p[2(V_{DD} - V_{IL} - |V_{Tp}|) - (V_{DD} - V_{out})] \\ &\qquad \times (V_{DD} - V_{out}) \\ V_{IL}(1 + \beta_n/\beta_p) &= 2V_{out} + (\beta_n/\beta_p)V_{Tn} - V_{DD} - |V_{Tp}|, \end{aligned} \tag{3.16}$$

which can be solved for V_{IL} by eliminating V_{out}.

(d) $\mathbf{V_{IH}}$: The input-high voltage V_{IH} is the smallest value of V_{in} that can be interpreted as a logic 1 level. It is calculated from the current flow equations and the same slope requirement, except that at this point, Mn is non-saturated while Mp is saturated. The resulting simultaneous equations are

$$\beta_p(V_{DD} - V_{IH} - |V_{Tp}|)^2 = \beta_n[2(V_{IH} - V_{Tn})V_{out} - V_{out}^2] \tag{3.17}$$

$$V_{IH}(1 + \beta_p/\beta_n) = 2V_{out} + (\beta_p/\beta_n)(V_{DD} - |V_{Tp}|) + V_{Tn}, \tag{3.18}$$

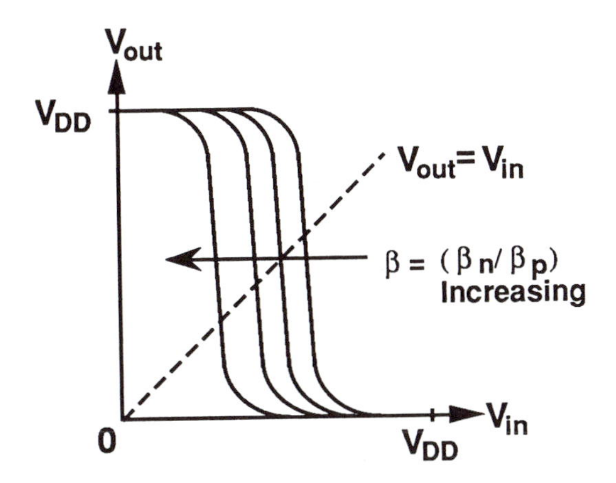

Figure 3.4: VTC Shift with Device Parameters

with V_{IH} calculated by eliminating V_{out}.

(e) $\mathbf{V_{th}}$: The gate threshold voltage is defined by the point where $V_{in} = V_{th} = V_{out}$, and represents the midpoint of the switching characteristics. In terms of the plot in Fig. 3.2, V_{th} occurs at the intersection of the VTC and the unity gain line. Noting that both Mn and Mp are saturated at this point, equating drain currents $I_{Dn} = I_{Dp}$ and solving gives

$$V_{th} = \frac{\sqrt{\beta_n/\beta_p}V_{Tn} + (V_{DD} - |V_{Tp}|)}{1 + \sqrt{\beta_n/\beta_p}}, \tag{3.19}$$

which is a simple closed-form expression.

The above discussion shows that the full-rail logic swing $V_\ell \simeq V_{DD}$ is due to the complementary placement of the nMOS and pMOS transistors. This property can be preserved in more complicated logic gates by ensuring that the opposite polarity devices are always used in complementary pairs.

Critical DC switching voltages are set by the designer through the devices parameters β_n and β_p. Varying the ratio $\beta = (\beta_n/\beta_p)$ affects the gate threshold voltage V_{th} as described by

$$\frac{dV_{th}}{d\beta} = -\frac{(V_{th} - V_{Tn})}{(\beta + \sqrt{\beta})}. \tag{3.20}$$

In particular, increasing β decreases V_{th}. Figure 3.4 illustrates the dependence of the VTC on the device transconductance ratio (β_n/β_p). Note in particular that the gate threshold voltage can be set by the circuit design using the aspect ratios $(W/L)_n$ and $(W/L)_p$. Transient switching characteristics are also affected by the choice of β values as discussed in the next section.

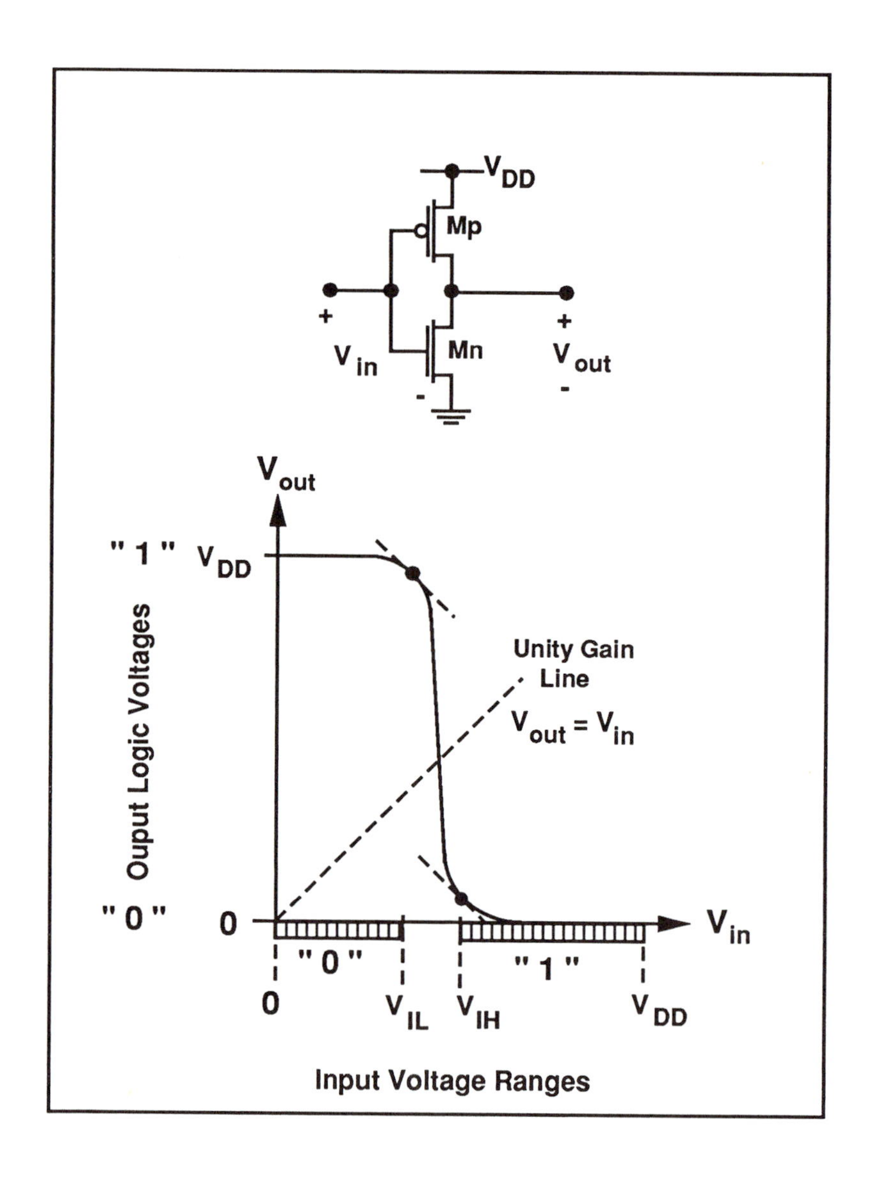

EXAMPLE 3.1: CMOS Inverter Characteristics

To illustrate the calculations, let us assume a circuit designed with $(W/L)_n = 4 = (W/L)_p$ which is fabricated in a process where

$$k'_n = 35\ [\mu A/V^2], \qquad V_{TOn} = 0.75\ [V],$$
$$k'_p = 20\ [\mu A/V^2], \qquad V_{TOp} = -0.8\ [V].$$

We assume a power supply of $V_{DD} = 5$ [V].

First note that the output voltages are given by

$$\begin{aligned} V_{OL} &= 0\ [V], \\ V_{OH} &= 5\ [V], \end{aligned}$$

due to the complementary structuring.

To compute V_{IL} we must solve the two equations

$$\begin{aligned} 35(V_{IL} - .75)^2 &= 20[2(5 - V_{IL} - .8) - (5 - V_{out})](5 - V_{out}) \\ V_{IL}(1 + 1.75) &= 2V_{out} + (1.75)(.75) - 5 - .8 \quad . \end{aligned}$$

The algebra gives

$$V_{IL} \simeq 1.73\ [V]$$

at an output voltage of $V_{out} = 4.63$ [V]. Similarly, V_{IH} is obtained by solving

$$\begin{aligned} 20(5 - V_{IL} - .8)^2 &= 35[2(V_{IH} - .75)V_{out} - V_{out}^2] \\ V_{IL}[1 + (1/1.75)] &= 2V_{out} + (1/1.75)(5 - .8) + .75 \quad , \end{aligned}$$

yielding

$$V_{IH} = 2.59\ [V]$$

with $V_{out} = 0.46$ [V] at this point.

Finally, the gate threshold voltage is determined by

$$\begin{aligned} V_{th} &= \frac{0.75 + \sqrt{1/1.75}\ (5 - .8)}{1 + \sqrt{1/1.75}} \\ &\simeq 2.24\ [V], \end{aligned}$$

which completes the calculation of the critical voltages for the circuit.

3.1.2 Symmetrical Inverter

A symmetrical CMOS inverter is defined to have

$$V_{th} = \frac{1}{2} V_{DD}. \tag{3.21}$$

Assuming that the process is polarity-symmetric with $V_{Tn} = |V_{Tp}| = V_T$, this condition can be achieved by designing the circuit with

$$\beta_n = \beta_p, \tag{3.22}$$

or, equivalently,

$$k'_n (W/L)_n = k'_p (W/L)_p. \tag{3.23}$$

Symmetrical switching characteristics can be achieved if the aspect ratios are adjusted to satisfy

$$\frac{(W/L)_p}{(W/L)_n} = \frac{k'_n}{k'_p}. \tag{3.24}$$

Since $k' = \mu C_{ox}$ and C_{ox} is generally the same for both n- and p- channel devices, this equation reduces to the requirement that

$$\left(\frac{W}{L}\right)_p = \frac{\mu_n}{\mu_p} \left(\frac{W}{L}\right)_n. \tag{3.25}$$

In general $\mu_n > \mu_p$, so that the pMOSFET must be larger than the nMOSFET. Specific mobility values vary with substrate doping and oxide quality. In a CMOS process, $(k'_n/k'_p) = (\mu_n/\mu_p) \sim 2.5$ is typical, and gives

$$\left(\frac{W}{L}\right)_p \simeq (2.5) \left(\frac{W}{L}\right)_n \tag{3.26}$$

for the design.

A symmetrical inverter has critical input voltages given by

$$\begin{aligned} V_{IL} &= \frac{3}{8} V_{DD} + \frac{1}{4} V_T \\ V_{IH} &= \frac{5}{8} V_{DD} - \frac{1}{4} V_T \end{aligned} \tag{3.27}$$

such that

$$V_{IL} + V_{IH} = V_{DD} \tag{3.28}$$

is valid. The noise margins are equal with

$$\mathrm{NM_H} = \mathrm{NM_L} = V_{IL}, \tag{3.29}$$

as verified by direct calculation.

EXAMPLE 3.2: Symmetrical CMOS Inverter

Consider a CMOS inverter circuit where

$$k_n' = 55\ [\mu\text{A/V}^2], \qquad k_p' = 25\ [\mu\text{A/V}^2],$$
$$(W/L)_n = 2,\ (W/L)_p = 4.4,\ V_T = 0.6\ [\text{V}]$$

The device transconductance values are

$$\beta_n = 110\ [\mu\text{A/V}^2] = \beta_p.$$

The results for $V_{DD} = 3.3$[V] and $V_{DD} = 5$ [V] are summarized in the table below.

Quantity	$V_{DD} = 3.3$ V	$V_{DD} = 5$ V
V_{OL}	0	0
V_{OH}	3.3	5
V_{IL}	1.39	2.03
V_{IH}	1.91	2.98
V_{th}	1.65	2.5

3.2 Inverter Switching Characteristics

Transient switching times are used to calculate data throughput rates and are also important in system timing. Table 3.3 lists the important time intervals which will be examined in this section.

Switching times are a result from two circuit properties: transistor current flow levels and parasitic capacitances. Both are set by the chip design parameters, and are sensitive to the transistor aspect ratios, layout geometry, and logic routing. To model the basic problem, we introduce the **output capacitance** C_{out} as shown in Figure 3.5.

C_{out} is assumed to be a linear, time-invariant (LTI) quantity for the hand calculations. As will be seen later in this section, the actual output node capacitance is a non-linear function of the circuit voltages.

Symbol	Name
t_{HL}	High-to-Low Time
t_{LH}	Low-to-High Time
t_{PHL}	High-to-Low Propagation Time
t_{PLH}	Low-to-High Propagation Time
t_P	Gate Propagation Delay Time

Table 3.3: Gate Switching Times

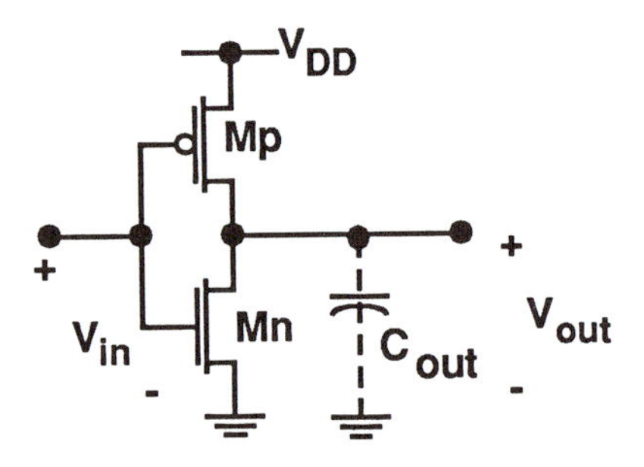

Figure 3.5: Output Capacitance

3.2.1 Switching Intervals

Transient characteristics of digital circuits are set by the times required to charge and discharge capacitors. CMOS inverters use transistors to provide current flow paths between the power supply (Mp) and ground (Mn). All switching times are thus set by the current levels and the value of C_{out}. The important switching times are shown in Figure 3.6 when the input voltage is assumed to be a step transition.

RC Model

A simple RC network model can be used to obtain first-order estimates of the switching time. In Figure 3.7 the MOSFETs are replaced by resistor-switch subnetworks; a conducting transistor is modelled by a closed switch. Denoting the drain-source equivalent resistances by R_n and R_p, we see that logic transfer is based on the charging and discharging of C_{out}.

Consider first the charging circuit. This corresponds to having a low input voltage $V_{in} \simeq 0$, so that Mp is on and Mn is off. Assuming the

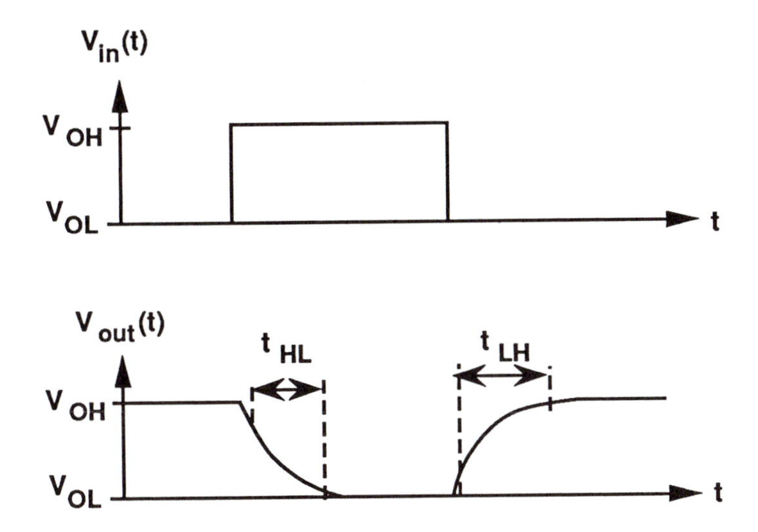

Figure 3.6: Switching Time Intervals

worst-case situation where $V_{out}(t = 0) = 0$, the voltage buildup is given by

$$V_{out}(t) = V_{DD}[1 - e^{-(t/\tau_p)}], \tag{3.30}$$

where $\tau_p = R_p C_{out}$ is the time constant. Since the MOSFET is at best a nonlinear resistor, R_p can only be approximated. The best-case value of the resistance is where Mp is assumed to be saturated. With a drain-srouce voltage of V_{DD}, the pMOS resistance is approximated by

$$\begin{aligned} R_p &\simeq \frac{V_{DD}}{I_{Dp}} \\ &\simeq \frac{2V_{DD}}{\beta_p(V_{DD} - |V_{T0p}|)^2} \end{aligned} \tag{3.31}$$

for an order-of-magnitude estimate.

The discharge event may be computed in a similar manner. With a high input $V_{in} > (V_{DD} - |V_{T0p}|)$, Mn is conducting while Mp is off. The output voltage is approximated by

$$V_{out}(t) = V_{DD} e^{-(t/\tau_n)}, \tag{3.32}$$

where we have assumed an initial condition of $V_{out} = V_{DD}$. The discharge time constant through Mn is given by $\tau_n = R_n C_{out}$ such that

$$R_n \simeq \frac{2V_{DD}}{\beta_n(V_{DD} - V_{T0n})^2} \tag{3.33}$$

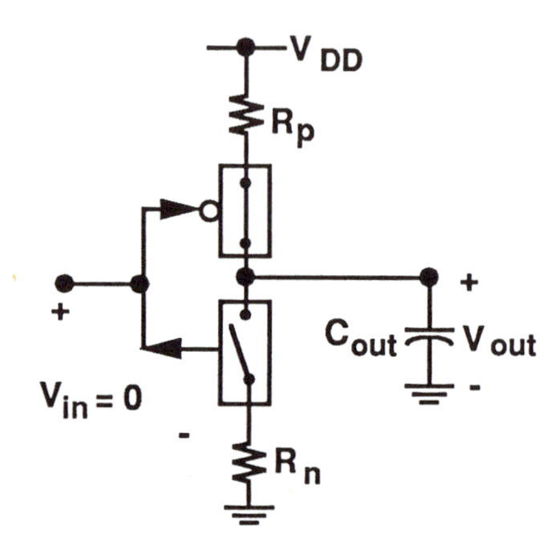

Figure 3.7: Transient Switch Model

is the best-case value of the nMOS drain-source resistance.

Exponential models provide reasonable first-order approximations for estimating the delay. Simplified networks based on RC time constants are particularly useful for evaluating complex high-performance designs and also provide valuable insight into the operation. Logic simulation tools are often based on switched networks of this type. Individual circuits, however, can be more accurately characterized by including device conduction properties. Although this requires substantially more work, gate-level optimization is important for custom circuits, ASIC cell design, and gate-array transistor networks.

High-to-Low Time

The output high-to-low time is calculated using the subcircuit in Figure 3.8. It represents the time interval needed for the output capacitor to discharge through the n-channel MOSFET Mn when Mp is in cutoff [2]. t_{HL} is also known as the **fall time** for the circuit since it gives the time needed for the output to decay from a well-defined logic 1 state to a well-defined logic 0 state. The discharge is described by the capacitor equation

$$I_{Dn} = C_{out}\frac{dV_{out}}{dt}; \tag{3.34}$$

[2] All of the transient time calculations are based on the assumption of a step function input voltage.

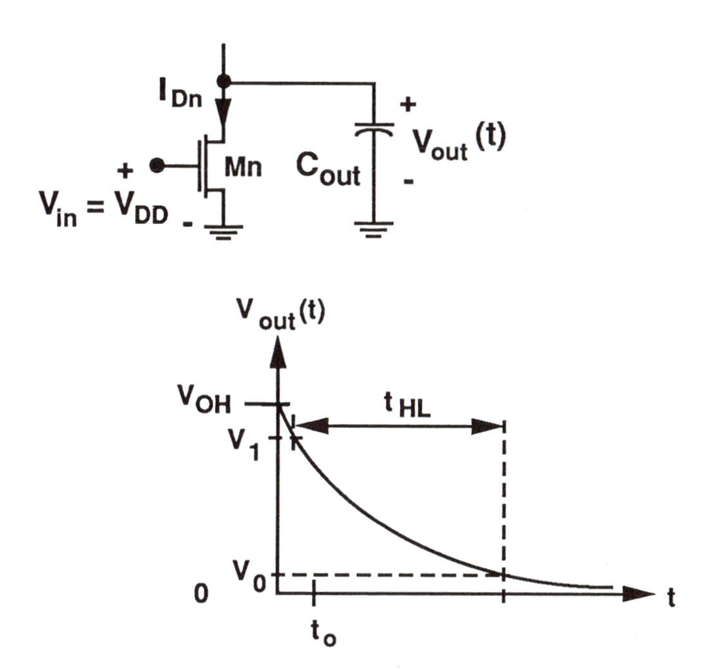

Figure 3.8: Subcircuit for calculating t_{HL}

we will assume an initial condition of $V_{out}(t=0) = V_{DD}$. At the beginning of the discharge, Mn is saturated, giving a linear voltage decay as described by

$$V_{out}(t) = V_{DD} - \frac{\beta_n}{2C_{out}}(V_{DD} - V_{Tn})^2 t. \tag{3.35}$$

When the output voltage drops to $V_{out} = (V_{DD} - V_{Tn})$, the MOSFET enters the non-saturated conduction region. This occurs at a time

$$t_o = \frac{2C_{out}V_{Tn}}{\beta_n(V_{DD} - V_{Tn})^2} \tag{3.36}$$

so that, for $t \geq t_o$,

$$V_{out}(t) = (V_{DD} - V_{Tn})\left[\frac{2e^{-(t-t_o)/\tau_n}}{1 + e^{-(t-t_o)/\tau_n}}\right]. \tag{3.37}$$

In this equation,

$$\tau_n = \frac{C_{out}}{\beta_n(V_{DD} - V_{Tn})} \tag{3.38}$$

is the **discharge time constant** for the circuit.

The value of t_{HL} is usually defined between **90 %** and **10%** voltage points, i.e., from $V_1 = 0.9V_{DD}$ to $V_0 = 0.1V_{DD}$ in the CMOS circuit. The analysis gives

$$t_{HL} = \tau_n\left[\frac{2(V_{Tn} - V_0)}{(V_{DD} - V_{Tn})} + \ln\left(\frac{2(V_{DD} - V_{Tn})}{V_0} - 1\right)\right] \tag{3.39}$$

as the result. The first term represents the time when Mn is saturated, while the second term is due to non-saturated conduction. It is important to note that $t_{HL} \propto \tau_n$, as the time constant is often used as a first estimate of the fall time.

As a final point, note that the definition of the time constant τ_n implies that we may approximate the MOSFET drain-to-source resistance by

$$R_n \simeq \frac{1}{\beta_n(V_{DD} - V_{Tn})} \; [\Omega] \tag{3.40}$$

since this gives the form

$$\tau_n = R_n C_{out}. \tag{3.41}$$

This provides a simple rule-of-thumb for many circuit performance estimates and is more realistic than the best-case value discussed in the simple RC equivalent circuit model.

Low-to-High Time

The low-to-high time, also known as the **rise time**, is found in the same manner. During this time interval, Mn is in cutoff while Mp is conducting from the power supply. As shown in Figure 3.9, t_{LH} is the time required to charge C_{out} through Mp. Charging is described by

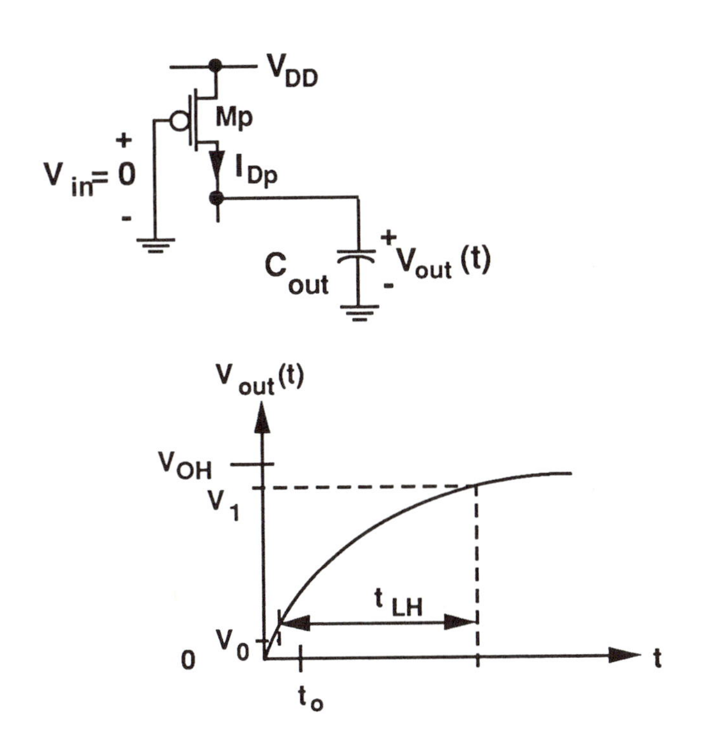

Figure 3.9: Subcircuit for calculating t_{LH}

$$I_{Dp} = C_{out} \frac{dV_{out}}{dt} \tag{3.42}$$

with the initial condition $V_{out}(t = 0) = 0$. Mp is initially saturated, which gives linear charging as described by

$$V_{out}(t) = \frac{\beta_p}{2C_{out}}(V_{DD} - |V_{Tp}|)^2 t. \tag{3.43}$$

This is valid until a time

$$t_o = \frac{2C_{out}|V_{Tp}|}{\beta_p(V_{DD} - |V_{Tp}|)^2} \tag{3.44}$$

when Mp enters the non-saturated region of conduction. For times $t \geq t_o$,

$$V_{out}(t) = V_{DD} - (V_{DD} - |V_{Tp}|)\left[\frac{2e^{-(t-t_o)/\tau_p}}{1 + e^{-(t-t_o)/\tau_p}}\right], \tag{3.45}$$

where the **charging time constant** is

$$\tau_p = \frac{C_{out}}{\beta_p(V_{DD} - |V_{Tp}|)}. \tag{3.46}$$

Defining t_{LH} as the time to charge C_{out} from V_0 (the 10% point) to V_1 (the 90 % point) gives

$$t_{LH} = \tau_p\left[\frac{2(|V_{Tp}| - V_0)}{(V_{DD} - |V_{Tp}|)} + \ln\left(\frac{2(V_{DD} - |V_{Tp}|)}{0.1V_{DD}} - 1\right)\right] \tag{3.47}$$

for the circuit rise time. Note that t_{LH} has the same form as the fall time t_{HL}, except that pMOSFET parameters appear instead of the nMOS quantities. This is expected from the complementary symmetry of the circuit.

The pMOS resistance may be approximated by

$$R_p \simeq \frac{1}{\beta_p(V_{DD} - |V_{Tp}|)} \; [\Omega] \tag{3.48}$$

such that $\tau_p = R_pC_{out}$ gives the charging time constant. Note that both R_n and R_p are inversely proportional to (W/L); increasing the aspect ratio decreases the equivalent resistance.

Maximum Switching Frequency

The sum of the transient times $(t_{HL} + t_{LH})$ represents the minimum time needed for a gate to undergo a complete switching cycle. We may use this to define the **maximum switching frequency** by

$$f_{max} = \frac{1}{(t_{HL} + t_{LH})}. \tag{3.49}$$

This represents the maximum rate of data transfer for the gate. In system design, the working value of f_{max} is set by the slowest gate in the network.

Propagation Delay

Logic delay through a gate is conveniently described by the **propagation delay time** t_P. Physically we interpret t_P as the average time needed for the output to respond to a change in the input logic state. By definition,

$$t_P = \frac{1}{2}(t_{PHL} + t_{PLH}), \tag{3.50}$$

where t_{PHL} and t_{PLH} are the propagation delays for a high-to-low and a low-to-high transition, respectively. The high-to-low propagation delay represents the time needed for the output to fall from V_{OH} to V_{th}; to simplify the calculations, we usually approximate $V_{th} \simeq (V_{DD}/2)$. Similarly, t_{PLH} gives the time that it takes for the output node to rise from V_{OL} up to V_{th}. Using the same analyses as above with different voltage limits gives

$$\begin{aligned} t_{PHL} &= \tau_n \left[\frac{2V_{Tn}}{(V_{DD} - V_{Tn})} + \ln\left(\frac{4(V_{DD} - V_{Tn})}{V_{DD}} - 1 \right) \right], \qquad (3.51) \\ t_{PLH} &= \tau_p \left[\frac{2|V_{Tp}|}{(V_{DD} - |V_{Tp}|)} + \ln\left(\frac{4(V_{DD} - |V_{Tp}|)}{V_{DD}} - 1 \right) \right]. \end{aligned}$$

In a symmetrical inverter with $\beta_n = \beta_p$ and $V_{Tn} = |V_{Tp}|$, $t_{PHL} = t_{PLH} = t_P$.

3.3 Output Capacitance

A quick examination of the calculations above shows that all of the transient times are proportional to the output capacitance C_{out}. A major flaw in the approach is the assumption that C_{out} is a linear, time-invariant (LTI) element. Instead, the gate-channel and depletion contributions make C_{out} a nonlinear function of the voltages.

All is not lost, however. We may still use the formulas by defining C_{out} to be an average value over the voltage range. So long as we exercise caution in interpreting the final results, the analytic approach can be used for initial design and performance estimates. Increased accuracy can be obtained by a computer simulation when it is needed. In this section we will examine the contributions to C_{out} and illustrate the averaging process.

Figure 3.10 shows the main contributions to C_{out}. Only those capacitors that are directly driven by the output node and also experience a change in voltage during a switching event have been included. The effective value of C_{out} can be obtained by examining the load presented to the output node during a switching event. Consider, for example, the case where V_{in} is initially high, and then falls to a value of $V_{in} = 0$ at time $t = 0$. All of the capacitors shown in the drawing change voltage as V_{out} charges from 0 volts to V_{DD}. Thus, we estimate the output capacitance by

$$C_{out} = C_{GDn} + C_{GDp} + C_{DBn} + C_{DBp} + C_{line} + C_{in}, \qquad (3.52)$$

which is equivalent to having all of the contributions in parallel.[3] The gate-

[3] Note that C_{DBp} actually discharges in this case. This implies that the formula for C_{out} gives pessimistic values.

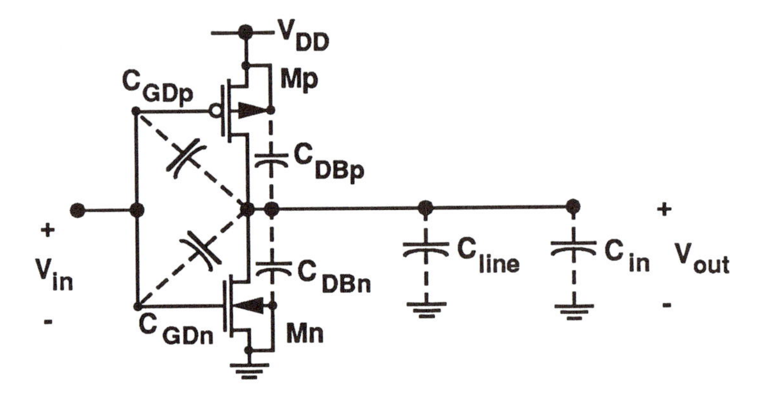

Figure 3.10: Capacitance in the CMOS Inverter

drain contributions C_{GD} are due to the gate coupling capacitances[4], while the drain-bulk terms C_{DB} are nonlinear depletion capacitances. C_{line} is the interconnect line contribution, and C_{in} represents the input capacitance into the next stage(s).

To obtain an average value for C_{out}, we will approximate the nonlinear terms using simple formulas. First, the gate-drain contributions are taken to be

$$\begin{aligned} C_{GDn} &= \frac{1}{2}C_{ox}(WL)_n + C_{on}, \\ C_{GDp} &= \frac{1}{2}C_{ox}(WL)_p + C_{op}, \end{aligned} \tag{3.53}$$

as discussed in Section 2.4.4 of the previous chapter; these represent worst-case values.

Consider next the depletion capacitances C_{DBn} and C_{DBp}. Since the output node voltage reverse biases both drain-bulk junctions, these terms vary during a switching event. An average value is obtained by defining

$$C_{av} = \frac{1}{V_\ell} \int_{V_{OL}}^{V_{OH}} \frac{C_{j0}A}{[1 + (V/V_{bi})]^m} dV \tag{3.54}$$

where $V_\ell = (V_{OH} - V_{OL})$ is the logic swing and A is the junction area. When we apply this formula to a MOSFET junction, both bottom and sidewall contributions must be included.

[4] The additional capacitance introduced by the Miller effect is neglected due to the small value of the gain.

To calculate the average capacitance on the bottom we assume that $m = 1/2$ corresponding to a step-like doping profile. In this case,

$$C_{av,bot} = K_{1/2}(V_\ell)C_{j0}A_{bot}, \tag{3.55}$$

with

$$K_{1/2}(V_\ell) = \frac{2V_{bi}}{V_{OH}}\left[\left(1 + \frac{V_{DD}}{V_{bi}}\right)^{1/2} - 1\right]; \tag{3.56}$$

note that $K_{1/2} < 1$ since the depletion capacitance is maximum at zero-bias. Using this formula, we then average the drain-bulk contributions by means of

$$\begin{aligned} C_{DBn,bot} &= K(V_\ell)C_{j0n}A_{Dn}, \\ C_{DBp,bot} &= K(V_\ell)C_{j0p}A_{Dp}, \end{aligned} \tag{3.57}$$

where A_{Dn} and A_{Dp} are the bottom areas for the n- and p-channel drain regions, respectively.

Sidewall doping profiles can often be modelled by $m = 1/3$. Integrating gives

$$C_{av,sw} = K_{1/3}(V_\ell)C_{jsw}\ell, \tag{3.58}$$

where C_{jsw} is the zero-bias perimeter capacitance per unit length, ℓ is the perimeter length, and

$$K_{1/3}(V_\ell) = \frac{3V_{bi}}{2V_{OH}}\left[\left(1 + \frac{V_{DD}}{V_{bi}}\right)^{2/3} - 1\right] < 1 \tag{3.59}$$

is the averaging factor. The sidewall contributions are

$$\begin{aligned} C_{DBn,sw} &= K_{1/3}(V_\ell)C_{jswn}\ell_{Dn}, \\ C_{DBp,sw} &= K_{1/3}(V_\ell)C_{jswp}\ell_{Dp}, \end{aligned} \tag{3.60}$$

with ℓ_{Dn} and ℓ_{Dp} the appropriate drain perimeter lengths. Summing the bottom and sidewall contributions results in the final form

$$\begin{aligned} C_{DBn} &= C_{DBn,bot} + C_{DBn,sw}, \\ C_{DBp} &= C_{DBp,bot} + C_{DBp,sw}. \end{aligned} \tag{3.61}$$

It is important to note the dependence on layout geometry.

Line capacitance C_{line} is due to the interconnect wiring. For a simple straight-line geometry, an interconnect with length ℓ gives

$$C_{line} = C_{int}\ell \tag{3.62}$$

where

$$C_{int} = \frac{\epsilon_{int}}{x_{int}} w \text{ [F/cm]} \tag{3.63}$$

is the interconnect capacitance per unit length; w is the width of the line, while the insulator has a thickness x_{int} and permittivity ϵ_{int}. For example, if the interconnect is over a field-oxide region with an oxide thickness of X_{FOX},

$$C_{int} = C_{FOX} w = \frac{\epsilon_{ox} w}{X_{FOX}} \tag{3.64}$$

is the appropriate quantity[5].

Finally, the input capacitance C_{in} represents the capacitance of the MOSFETs in the next stage. This is directly related to the **fan-out** FO circuit. For fully-complementary structuring in which every input is connected to the gate of both an n-channel and a p-channel transistor, we approximate

$$C_{in} = \sum_{\alpha=1}^{FO} (C_{Gn} + C_{Gp})_{\alpha}, \tag{3.65}$$

where

$$\begin{aligned} C_{Gn} &= C_{ox} L_n W_n + 2C_{on} \\ C_{Gp} &= C_{ox} L_p W_p + 2C_{op} \end{aligned} \tag{3.66}$$

include both the gate (C_g) and overlap contributions for each device.

EXAMPLE 3.3: Capacitance Calculation

Consider the inverter layout in Figure 3.11; all dimensions are in units of microns [μm]. We will use the simple parameter set listed below.

nMOS Parameters

$$\begin{aligned} &x_{ox} = 350 \text{ [Å]}, \quad V_{T0n} = +0.70 \text{ [V]}, \quad V_{bi} = 0.85 \text{ [V]}, \\ &k'_n = 35 \text{ [}\mu\text{A/V}^2\text{]}, \quad C_{j0n} = 0.40 \text{ [fF/}\mu\text{m}^2\text{]}, \quad C_{jsw} = 0.60 \text{ [fF/}\mu\text{m]}. \end{aligned}$$

pMOS Parameters

$$\begin{aligned} &x_{ox} = 350 \text{ [Å]}, \quad V_{T0p} = -0.70 \text{ [V]}, \quad V_{bi} = 0.95 \text{ [V]}, \\ &k'_p = 20 \text{ [}\mu\text{A/V}^2\text{]}, \quad C_{j0p} = 0.22 \text{ [fF/}\mu\text{m}^2\text{]}, \quad C_{jsw} = 0.28 \text{ [fF/}\mu\text{m]}. \end{aligned}$$

---

[5] Details of interconnect properties are discussed more thoroughly in Chapter 6.

Interconnect Capacitance

$$C_{Poly-Sub} = 0.06 \text{ [fF}/\mu\text{m}^2],$$
$$C_{Metal-Sub} = 0.04 \text{ [fF}/\mu\text{m]}.$$

The power supply is taken to be V_{DD}=5[V], and the inverter is cascaded into the (poly-level) input of an identical stage. Both devices have a gate overlap of $L_o = 0.25$ [μm] which is not shown explicitly on the layout. The bottom junction has a grading coefficient of $m = 0.5$, while the sidewalls are described by $m = 0.33$.

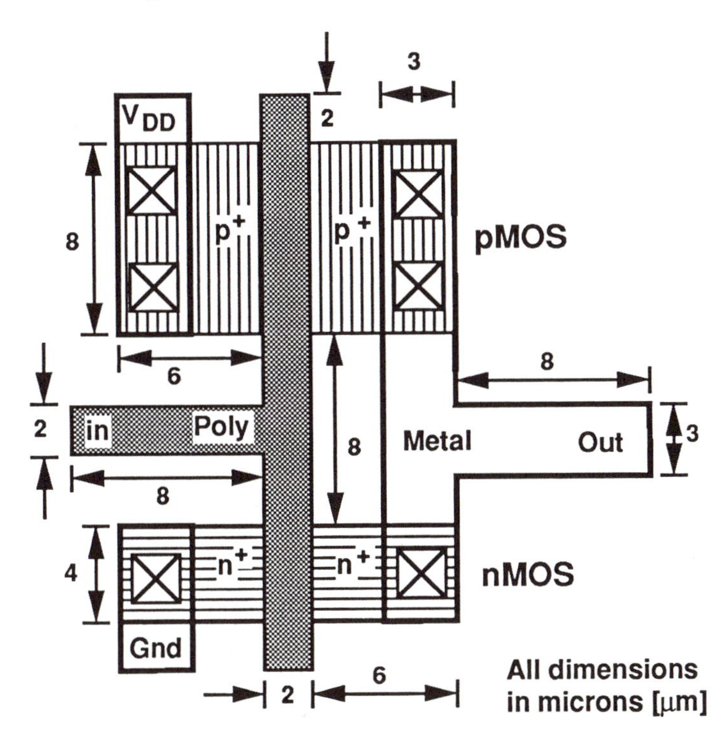

Figure 3.11: Inverter Layout for Example

Gate Capacitance: First we compute

$$\begin{aligned} C_{ox} &= \frac{\epsilon_{ox}}{x_{ox}} \\ &= \frac{(3.9)(8.854 \times 10^{-14})}{3.5 \times 10^{-6}} \\ &\simeq 0.987 \text{ [fF}/\mu\text{m}^2] \end{aligned}$$

where the final answer has been converted from [F/cm^2] to [fF/μm^2] for convenience in hand calculations.

Overlap Capacitance: The overlap capacitance is estimated from

$$\begin{aligned} C_o &= C_{ox} L_o \simeq (0.987)(.25) \\ &\simeq 0.247 \text{ [fF}/\mu\text{m]}. \end{aligned}$$

From the layout geometry, $W_n = 4$ [μm] and $W_p = 8$ [μm], so

$$\begin{aligned} C_{on} &= 0.988 \text{ [fF]} \\ C_{op} &= 1.976 \text{ [fF]}. \end{aligned}$$

Gate-Drain Capacitance: The actual channel lengths are the **drawn** values on the layout minus the overlap total $2L_o$. Computing $L_n = (2 - .5) = 1.5$ [μm] and $L_p = (2 - .5) = 1.5$ [μm] we have

$$\begin{aligned} C_{GDn} &\simeq (0.987)(4)(1.5) + 2(0.988) \\ &\simeq 7.898 \text{ [fF]} \\ C_{GDp} &\simeq (0.987)(8)(1.5) + 2(1.976) \\ &\simeq 15.796 \text{ [fF]} \end{aligned}$$

for the gate-drain contributions.

Depletion Capacitances: We will start with the nMOS transistor drain region. Consider first the bottom contribution. Noting that $V_{DD} = 5$ [V] for the circuit gives an averaging factor of

$$\begin{aligned} K_{1/2} &= \frac{2(.85)}{5}\left[\left(1 + \frac{5}{.85}\right)^{1/2} - 1\right] \\ &\simeq 0.552. \end{aligned}$$

The area of the drain n^+ bottom region (including the overlap underneath the gate) is $(4 \times 6.25) = 25$[μm^2] so that the average value is

$$C_{DBn,bot} \simeq (.552)(.40)(25) \simeq 5.52 \text{ [fF]}.$$

The sidewall calculation proceeds along identical lines. First we compute

$$\begin{aligned} K_{1/3} &= \frac{3(.85)}{2(5)}\left[\left(1 + \frac{5}{.85}\right)^{2/3} - 1\right] \\ &\simeq 0.668. \end{aligned}$$

Then, since the sidewall perimeter length (including the region underneath the gate) is $\ell = 20.5$ [μm] we have

$$C_{DBn,sw} \simeq (.668)(.60)(20.5) \simeq 8.216 \text{ [fF]}$$

completing the depletion capacitance calculation.

Calculations for the pMOS drain region are analogous and result in

$$\begin{aligned} C_{DBp,bot} &\simeq 6.281 \text{ [fF]} \\ C_{DBp,sw} &\simeq 5.450 \text{ [fF]} \end{aligned}$$

as is easily verified.

Interconnect: The interconnect can be split into METAL and POLY contributions

$$C_{int} = C_M + C_P$$

where the metal is at the output. We assume that the inverter is cascaded into an identical circuit, so that the poly contribution C_P can be computed using the input poly geometry shown in the layout. For simplicity we ignore the area where metal overlaps n^+ or p^+. In addition, we assume that any metal-to-poly capacitance C_{M-P} which may occur when cascading the stages is negligible.

The area of the output metal is 84 [μm^2], and the area of the input poly line is 32 [μm^2]. This gives

$$\begin{aligned} C_{int} &= (.04)(84) + (.06)(32) \\ &\simeq 5.28 \text{ [fF]} \end{aligned}$$

Input Gate Capacitance: The inverter is connected to the input of an identical stage. Since L includes the overlap distance L_o, the gate capacitance is $C_G = C_{ox}LW$ for both MOSFETs and

$$C_G = (.987)[(2)(8) + (2)(4)] \simeq 23.688 \text{ [fF]}$$

gives the needed value.

Total Capacitance: Summing the contributions above gives

$$C_{out} \simeq 76.689 \text{ [fF]}$$

as our estimate for the circuit.

Switching Times: We may use C_{out} to estimate the important transient time intervals in Section 3.2. The time constants of interest are

$$\tau_n = \frac{(76.89 \times 10^{-15})}{(35 \times 10^{-6})(4/1.5)(5-0.7)} \simeq 0.191 \text{ [ns]}$$
$$\tau_p = \frac{(76.89 \times 10^{-15})}{(20 \times 10^{-6})(8/1.5)(5-0.7)} \simeq 0.167 \text{ [ns]}$$

which gives

$$t_{HL} \simeq 0.57 \text{ [ns]},$$
$$t_{LH} \simeq 0.51 \text{ [ns]},$$

for the circuit. The maximum switching frequency for the inverter is

$$f_{max} \simeq \frac{10^9}{(0.57+0.51)} \simeq 926 \text{ [MHz]}.$$

which represents the theoretical upper limit of the single inverter.

It is significant to note that the gate contribution C_G, which represents the input capacitance into the next stage, accounts for about 31 % of the total output capacitance. Increasing the FO to a value greater than unity slows down the circuit accordingly.

3.4 Secondary Parasitic Effects

Parasitic capacitances limit the overall performance of the inverter circuit and must be included in the analysis. However, secondary parasitic effects of leakage and resistance contributions are usually negligible in static logic circuits and are often ignored. We will briefly introduce both problems here for completeness in the discussion[6].

3.4.1 Leakage Currents

Leakage currents across reverse-biased drain-bulk and/or source-bulk pn-junctions occur in both the pMOS and nMOS transistors. The primary current flow paths are illustrated in Figure 3.12. The drain of the p-channel

[6] A detailed examination of leakage currents and its importance to the operation of the dynamic logic circuits is covered in Chapter 7.

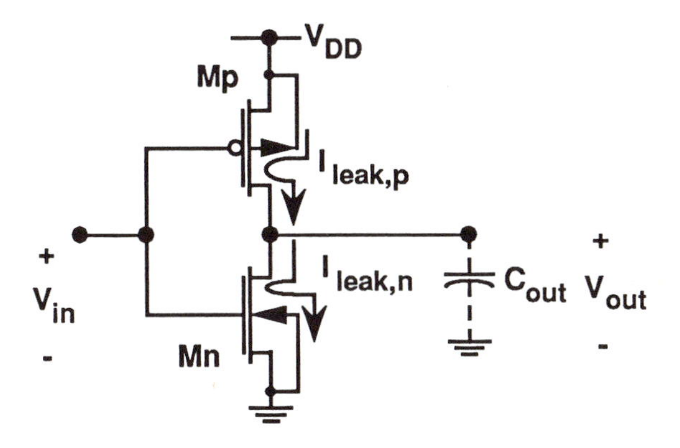

Figure 3.12: Junction Leakage Currents

MOSFET Mp provides a leakage path from the power supply V_{DD} to the output node, while the n-channel transistor induces parasitic conduction from C_{out} to ground. Generally, the magnitudes of both leakage currents are below the nanoampere range, so that they have a negligible effect on the output characteristics. This insensitivity is due to the full-complementary structuring which provides a strong conducting path from the output node to either the power supply or to ground.

3.4.2 Parasitic Resistances

Although the n^+ and p^+ drain/source regions are very heavily doped, they still exhibit non-zero sheet resistances with typical values in the range of $R_s \simeq 20 - 100$ [$\Omega/\square$]. Contact resistance (for example, metal- to-n^+) also contributes to the parasitic resistance in the circuit.

Figure 3.13 provides a simple lumped-element placement for the basic parasitic resistances R_{DN}, R_{SN}, R_{DP}, R_{SP} of the inverter. Examining this circuit shows that he values of the critical VTC voltages will change slightly if the nMOS and pMOS values are different.

Applying the inverter RC switching model indicates that all switching times will be increased due to the additional resistances. Referring to the analysis in Section 3.2.1 we see that the time constants are increased to

$$\begin{aligned} \tau_p &\simeq (R_p + R_{DP} + R_{SP})C_{out} \\ \tau_n &\simeq (R_n + R_{DN} + R_{SN})C_{out}. \end{aligned} \quad (3.67)$$

Since the drain-source device resistance generally dominates the equations, the actual changes in the transient times are often neglected.

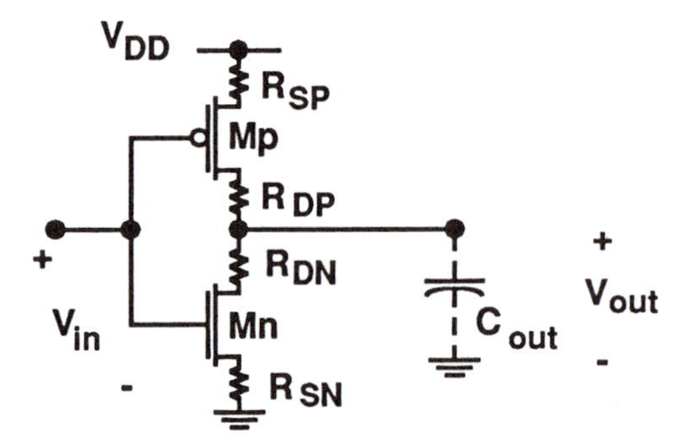

Figure 3.13: Parasitic Resistances

3.5 Comparison with SPICE

Hand-calculations are generally based on the simplest equations possible. Although quite crude, they are useful for estimating the performance of a circuit and providing initial design equations to the engineer. Computer-based circuit simulators such as SPICE can provide much more accurate results due to their ability to handle more complex models. It is interesting to examine the differences between the analytic treatment developed in this chapter and the equivalent simulation using SPICE LEVEL=2 transistor models.

3.5.1 Capacitances

The value of C_{out} used in the analytic treatment is crucial to calculating the transient response. Simulation programs are able to provide more accurate values for the device contributions. Nonlinearities can be included in an incremental manner, resulting in a more precise analysis.

Depletion Capacitances

SPICE analyzes circuits by solving the network state vectors for every increment in voltage (or current). Nonlinear depletion capacitances are computed in each iteration, resulting in a higher accuracy than that obtained by the averaging technique. Usually, the SPICE capacitance values will be lower than those obtained using the hand calculation presented here.

Gate and Overlap Capacitances

Analytic models are usually based on an ideal parallel plate geometry and assume an overlap capacitance of $C_o = C_{ox} L_o$ [F/cm]. This ignores the presence of fringing electric fields, and can lead to significant errors in the calculations.

SPICE values for CGDO and CGSO normally include the contributions from fringing fields[7]. In addition, the gate-bulk (overhang) capacitance CGBO is usually due to a complicated non-parallel plate geometry; even simple estimates are difficult to obtain. Empirical determination of simulation parameters is still the most accurate.

3.5.2 MOSFET Models

The square law used in the analytic calculations corresponds to the SPICE LEVEL=1 case. This automatically limits the accuracy of the hand calculations. Simulations based on LEVEL=2 or LEVEL=3 models provide better results at the expense of computer time.

3.5.3 Input Waveforms

We have modelled $V_{in}(t)$ using step functions to simplify the mathematics in the transient response calculations. An equivalent SPICE simulation uses an input pulse with extremely short (for example, ~ 1 [ps]) rise and fall times. While this provides us with first-order estimates for all important transient response times, it ignores the fact that a realistic input waveform is more complicated. A SPICE analysis can overcome this by using a linear ramp or even EXP time functions.

3.6 Inverter Design

Static CMOS gate design is relatively straightforward. Since rail-to-rail output voltage levels are automatic, the design is directed towards either shaping the voltage transfer curve or setting transient characteristics. Inverter design itself provides the basis for more complex gate structures.

Consider first the DC transfer characteristic. The design variables are β_n and β_p such that the ratio establishes the gate threshold voltage V_{th}

[7] This implies that estimating $C_o = C_{ox} L_o$ in a SPICE data listing will not be consistent with the given value of TOX and LD.

according to

$$V_{th} = \frac{\sqrt{\beta_n/\beta_p}V_{Tn} + (V_{DD} - |V_{Tp}|)}{1 + \sqrt{\beta_n/\beta_p}}. \tag{3.68}$$

Aspect ratio values $(W/L)_n$ and $(W/L)_p$ are then chosen according to the transient specifications or real estate limitations. In general, increasing the aspect ratios decreases the switching time; the actual dependence is complicated by the fact that C_{out} is a sensitive function of drain geometry. The time constant expressions give

$$\left(\frac{W}{L}\right)_n = \frac{C_{out}}{k'_n \tau_n (V_{DD} - V_{Tn})} \tag{3.69}$$

$$\left(\frac{W}{L}\right)_p = \frac{C_{out}}{k'_p \tau_p (V_{DD} - |V_{Tp}|)} \tag{3.70}$$

which can be used to find aspect ratios for desired rise and fall times.

Inverter layout is straightforward since only two transistors are involved. Figure 3.14 shows two basic layouts with connections to the power supply and ground.

The difficulty with the design process outlined above is that C_{out} is set by the layout, which is in turn a function of the design. Capacitance estimates are needed to complete the transient design, but these must be rechecked after the final layout is completed. A simple approach is to just complete a layout and then compute the transient response. If a faster circuit is needed, then the layout can be redrawn with different device geometries. In a structured design environment, several predesigned cells are usually available, each having different characteristics.

3.7 The Power-Delay Product

Standard CMOS circuits only use power supply current during a switching event. This low power characteristic is due to the complementary behavior of the nMOS and pMOS transistors, and is one reason that the popularity of CMOS has increased so rapidly.

Figure 3.15 illustrates the main concepts. When V_{in} is in a stable logic 0 or logic 1 state, either the nMOS or the pMOS transistor is in cutoff. In this case, there is no direct current flow path through the transistors between the power supply and ground. However, leakage currents I_{leak} which flow across the reverse-biased the drain-bulk regions of the transistors do exist in the chip. The DC power dissipation is then given by

$$P_{DC} = V_{DD} I_{leak}, \tag{3.71}$$

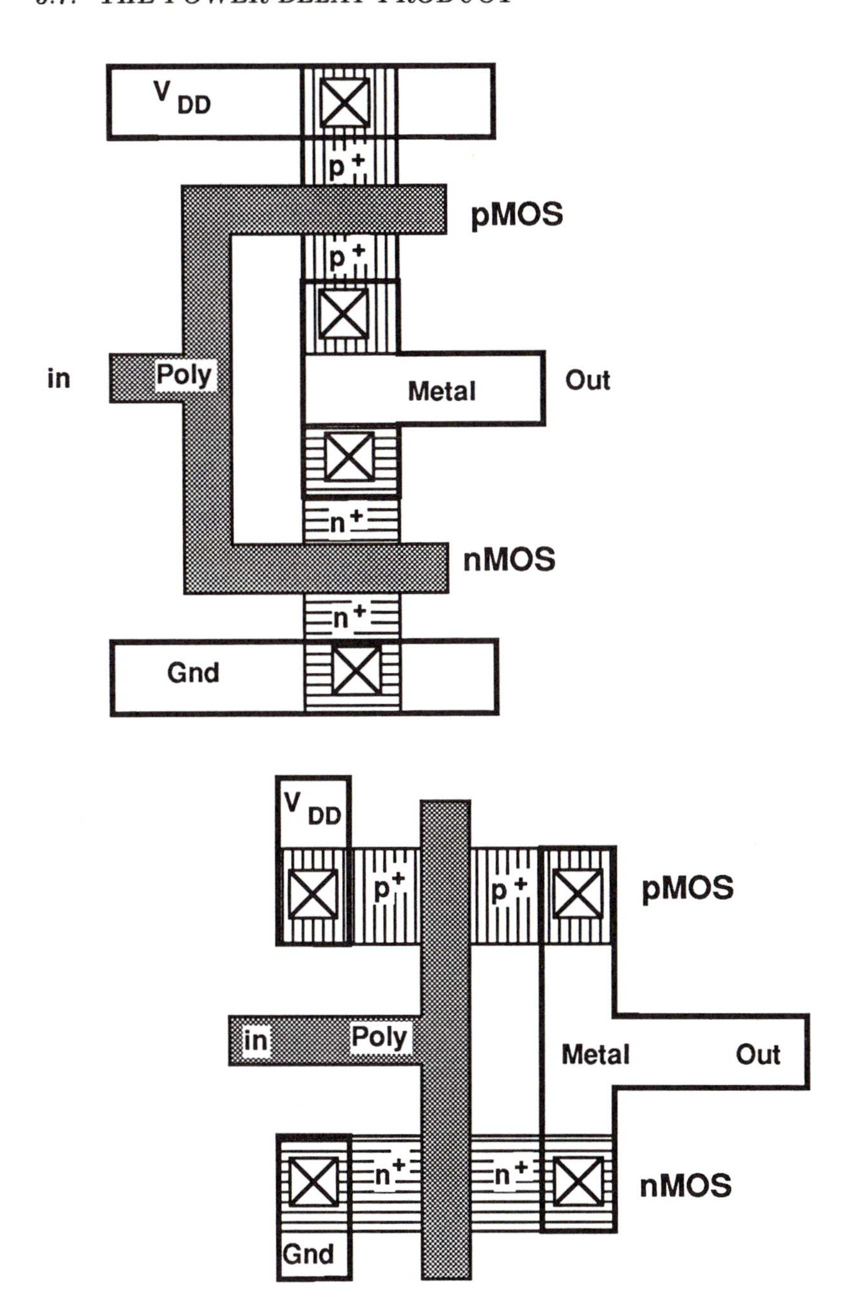

Figure 3.14: Inverter Layout Examples

with I_{leak} typically less than about 0.1 [pA]. This is quite small, particularly when compared with bipolar or nMOS-only circuits.

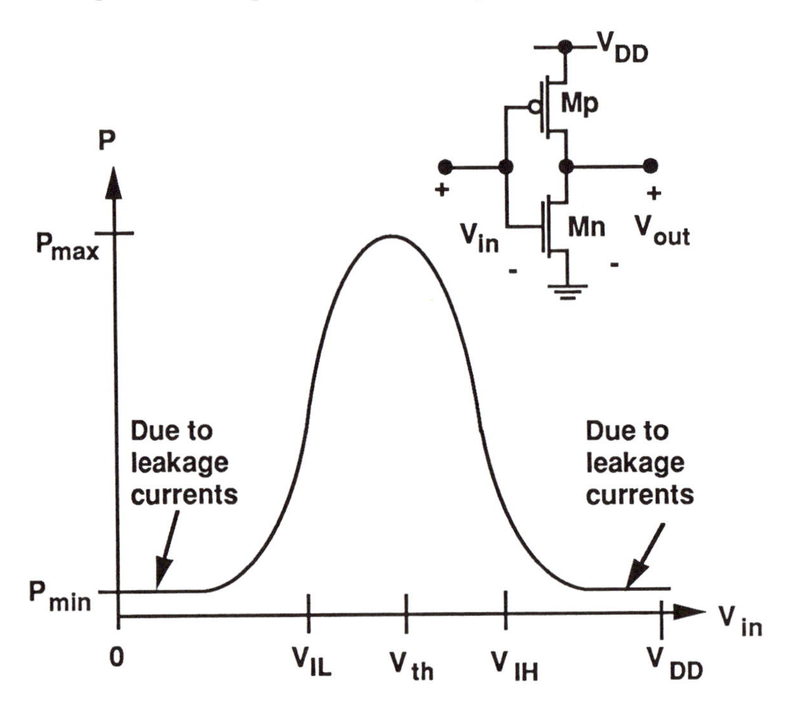

Figure 3.15: Power Dissipation in a CMOS Inverter

The dynamic power dissipation is more complicated. As seen in the drawing, power supply current is used when the inverter input voltage is switched. The maximum value I_{max} is set by the device geometries and power supply, and occurs when both transistors are conducting in the saturated mode. The dynamic power dissipation increases with increasing frequency f.

The **power-delay product** (PDP) is often introduced to compare the performance of competing digital technologies, e.g., CMOS vs. nMOS. It is computed from

$$PDP = P_{av} t_P \tag{3.72}$$

where P_{av} is the average power dissipation over a switching cycle, and t_P is the propagation delay time. The PDP has units of [Watt- sec]=[Joules], and is often interpreted as the average "energy per switch". Small PDP values are desirable, since this implies fast switching and small power dissipation. CMOS circuits are characterized by propagation delay times on the order of a nanosecond. Typically, CMOS exhibits PDP values on the order of picojoules [pJ].

To estimate the PDP for the CMOS inverter, we first note that the DC contribution

$$(PDP)_{DC} = (V_{DD} I_{leak}) t_P \tag{3.73}$$

is negligible due to the small value of the leakage currents.

To find the dynamic contribution to the PDP, we first compute the average power dissipation over a cycle with period T by writing

$$P_{av} = \frac{1}{T} \int_0^T V_{DD} I(\tau) d\tau, \tag{3.74}$$

where the current is given by $I = (dQ/dt)$, and $Q = C_{out} V_{DD}$ is the charge stored on the capacitor. Integrating,

$$\begin{aligned} P_{av} &= \frac{V_{DD} Q}{T} \\ &= \frac{1}{T} C_{out} V_{DD}^2. \end{aligned} \tag{3.75}$$

To account for a switching event during T, we multiply this value by a factor of $(1/2)$. Then, the dynamic contribution to the PDP is

$$(PDP)_{dynamic} = \frac{1}{2} C_{out} V_{DD}^2 \frac{t_P}{T}; \tag{3.76}$$

note that the factor $\frac{1}{2} C_{out} V_{DD}^2$ is the energy stored in the capacitor with a voltage V_{DD}. Finally, noting that

$$\begin{aligned} f &= \frac{1}{T} \\ f_{max} &\simeq \frac{1}{t_P}, \end{aligned} \tag{3.77}$$

we have

$$(PDP)_{dynamic} \simeq \left(\frac{1}{2} C_{out} V_{DD}^2 \right) \frac{f}{f_{max}} \tag{3.78}$$

as a simple estimate. The total PDP is the sum of the DC and transient terms. As the switching frequency increase, so does the PDP. This shows that the often-quoted "low-power property" of CMOS really only holds at low frequencies or when circuit are in stable states.

EXAMPLE 3.4: CMOS *PDP* Estimate

Consider an inverter with $C_{out} = 300$ [fF] which is operated with a 5 [V] supply. For $f = 0.05 f_{max}$, we compute

$$\begin{aligned} PDP &\simeq \frac{1}{2}(300 \times 10^{-15})(5^2)(0.05) \\ &\simeq 0.19 \text{ [pJ]}. \end{aligned}$$

This is a typical order of magnitude.

3.8 Temperature Dependence

The transfer characteristics vary with temperature due to the change in device conduction properties. The effects of increasing T is illustrated by the plot if Figure 3.16. Since the local operating temperatures depend on

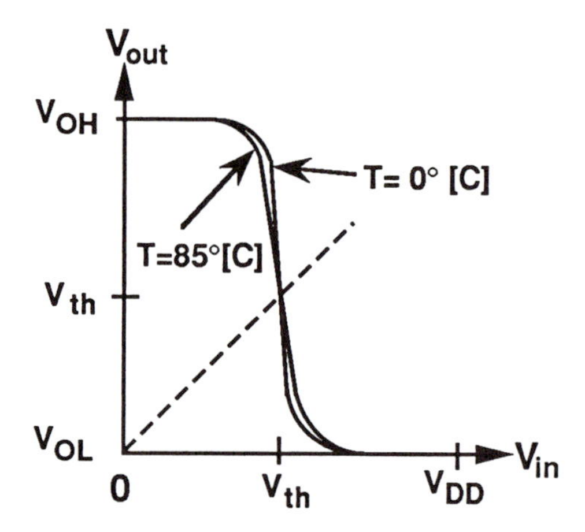

Figure 3.16: Temperature Variation

circuit density, current levels, and switching speeds, the "ambient" should be determined and the circuits simulated accordingly.

3.9 References

The books below examine the CMOS inverter from different perspectives.

[1] M. Annaratone, **Digital CMOS Circuit Design**, Kluwer Academic Publishers, Boston, 1986.

[2] J. Y. Chen, **CMOS Devices and Technology for VLSI**, Prentice-Hall, Englewood Cliffs, NJ, 1990.

[3] E. D. Fabricius, **Introduction to VLSI Design**, McGraw-Hill, New York, 1990.

[4] H. Haznedar, **Digital Microelectonics**, Benjamin Cummings, Redwood City, CA, 1991.

[5] D.A. Hodges and H.J. Jackson, **Analysis and Design of Digital Integrated Circuits**, 2nd. ed., McGraw-Hill, New York, 1988.

[6] A. Mukherjee, **Introduction to nMOS and CMOS VLSI Systems Design**, Prentice-Hall, Englewood Cliffs, NJ, 1986.

[7] M. Shoji, **CMOS Digital Circuit Technology**, Prentice-Hall, Englewood Cliffs, NJ, 1988.

[8] J.P. Uyemura, **Fundamentals of MOS Digital Integrated Circuits**, Addison-Wesley, Reading, MA, 1988.

[9] N. Weste and K. Eshraghian, **CMOS VLSI Design**, Addison-Wesley, Reading, MA, 1985.

Chapter 4

Static Logic Circuits

Logic functions are implemented in CMOS circuits by using MOSFETs to switch current flow paths. The static circuits in this chapter are commonly used for random logic functions, and also provide the basis for more advanced dynamic circuits. Static logic gates are the most widely used CMOS circuit because they are straightforward to design, have controllable characteristics, and exhibit high noise immunity.

The CMOS inverter provides the basic structure for static logic. Using complementary nMOS and pMOS transistors gives a full-rail output voltage swing and purely dynamic power dissipation. These desirable characterisitcs are maintained in more complex logic gates by simply adding transistors to an inverter circuit.

4.1 General Structure

Static CMOS logic gates are constructed using completely symmetric nMOS and pMOS transistor arrays. Figure 4.1 provides a block-diagram view for an arbitrary gate. Each input A, B, C is connected to both an nMOS transistor section ($\overline{F}$) and a pMOS transistor section (F). The functional segments $\overline{F}$ and F may be viewed as mutually exclusive switches. A condition of $\overline{F} = 0$, $F = 1$ indicates that the upper pMOS block is conducting while the lower nMOS block acts as an open circuit. In this case, the output is connected to V_{DD}, giving and output voltage of $V_{out} = V_{OH}$. Conversely, if the inputs give $\overline{F} = 1$, $F = 0$, then the bottom nMOS block conducts to pull the output node to ground; this establishes an output voltage of $V_{out} = V_{OL} = 0$ [V].

Logic gate design centers around three main tasks: forming the logic,

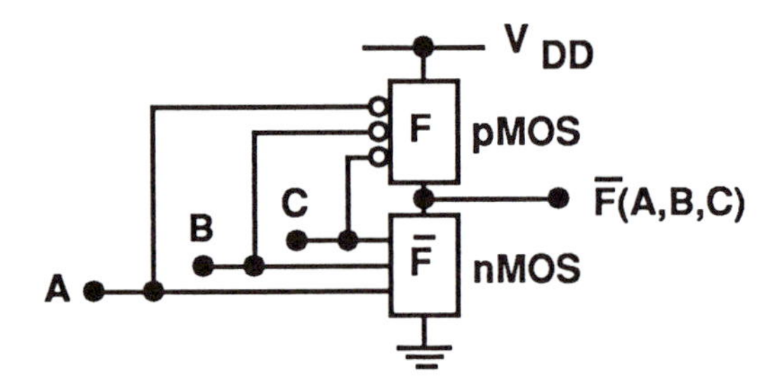

Figure 4.1: General Static CMOS Logic Gate

establishing the DC characteristics, and analyzing the transient response. Logic functions are created by proper placement of series and parallel-connected MOSFETs within the logic blocks Electrical characteristics depend on the device aspect ratios (W/L), the connections of the logic transistors, and the layout geometry.

The next section examines the transient response of series-connected MOSFET chains which are commonly found in logic networks. Although the reader may jump directly to the discussion of various logic gates which begins with Section 4.3, an accurate comparison of logic circuits requires the material in the next section.

4.2 Series-Connected MOSFETs

Transistor chains frequently occur when forming MOS logic circuits. The transient response of such a chain can often be the limiting performance factor so that it is useful to examine simplified modeling of the situation. There are two main problems which need to be studied in order to characterize the logic blocks denoted by $\overline{F}$ and F in Fig. 4.1. First, we will study the problem of discharging C_{out} through a series-connected nMOS chain. Then we will analyze the opposite case where C_{out} is charged through a series-connected chain of p-channel MOSFETs.

4.2.1 Discharging Through an nMOS Chain

Consider the circuit shown in Figure 4.2 where four n-channel MOSFETs are connected in series. Parasitic capacitance exists at each node as indicated in the drawing. We are concerned with the problem where capacitor C_3 is initially charged to a high voltage $V_3(t = 0) = V_{DD}$ and discharges

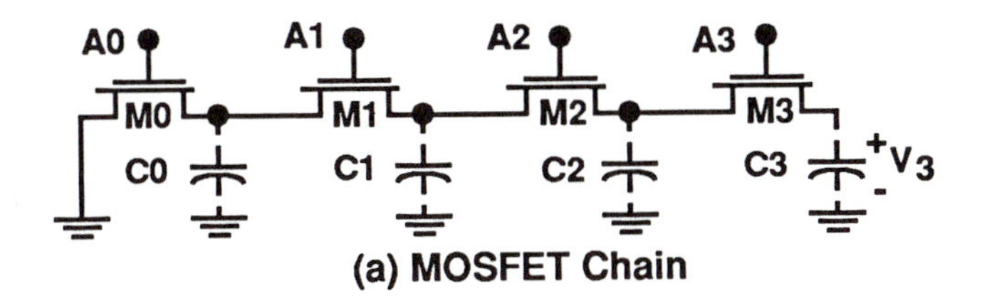

(a) MOSFET Chain

R0 V_0 R1 V_1 R2 V_2 R3

C0 C1 C2 C3 + V3 -

(b) RC Model

Figure 4.2: nMOS Transistor Chain

through the chain. This requires that all inputs be a logic 1 voltages:

$$A_0 = 1 = A_1 = A_2 = A_3. \tag{4.1}$$

To model the effects of the transistors we use the drain-source resistance

$$R_n = \frac{1}{\beta_n (V_{DD} - V_{Tn})} \tag{4.2}$$

to create the equivalent RC ladder network shown in the drawing. Although this ignores some important transistor properties, it provides a first-order simplified approximation to the MOSFET chain problem.

To characterize the RC network we apply KCL to each node. This results in the following equation set:

$$\begin{aligned}
-C_3 \frac{dV_3}{dt} &= \frac{V_3 - V_2}{R_3} \\
-C_2 \frac{dV_2}{dt} &= \frac{V_2 - V_1}{R_2} - \frac{V_3 - V_2}{R_3} \\
-C_1 \frac{dV_1}{dt} &= \frac{V_1 - V_0}{R_1} - \frac{V_2 - V_1}{R_3} \\
-C_0 \frac{dV_0}{dt} &= \frac{V_0}{R_0} - \frac{V_1 - V_0}{R_1}.
\end{aligned} \tag{4.3}$$

Solving this set is an interesting study in network theory; closed-form solutions for a 2-segment ladder are relatively straightforward to find. However, since we are primarily interested in CMOS design, let us forego the details and simply examine the results of the analysis. We will assume the worst-case initial conditions where all of the capacitors are charged to V_{DD} at the beginning of the discharge[1].

Consider the voltage across capacitor C_3. Qualitatively, we might expect that $V_3(t)$ could be approximated by an exponential time dependence of the form

$$V_3(t) \sim V_{DD} e^{-t/\tau_n} \tag{4.4}$$

with τ_n an appropriate time constant. Defining the **delay time** t_D by the point $V_3(t_D) = V_{DD} e^{-1}$ would then give $t_D \simeq \tau_n$. Although the complete analysis of the RC chain is much more complicated than this, a few key approximations may be used to give a simple result for estimating the performance. For N transistors in the chain, the delay time can be viewed as a superposition of time constants in the form

$$t_D \simeq \sum_{i=0}^{N-1} \tau_i \tag{4.5}$$

where

$$\tau_i = (\sum_{j=0}^{i} R_j) C_i. \tag{4.6}$$

Each term τ_i can be viewed as an RC time constant for discharging the ith capacitor C_i through each resistor in the path.

Let us apply this formula to the original 4-ladder RC network. In this case $N = 4$ and we have

$$\begin{aligned} \tau_0 &= R_0 C_0 \\ \tau_1 &= (R_0 + R_1) C_1 \\ \tau_2 &= (R_0 + R_1 + R_2) C_2 \\ \tau_3 &= (R_0 + R_1 + R_2 + R_3) C_3, \end{aligned} \tag{4.7}$$

which gives a total delay time of

$$\begin{aligned} t_D \simeq\ & R_0 C_0 + (R_0 + R_1) C_1 + (R_0 + R_1 + R_2) C_2 \\ & + (R_0 + R_1 + R_2 + R_3) C_3. \end{aligned} \tag{4.8}$$

Although this is only approximate, and it refers to discharge time to a $(1/e)$ voltage (not to 0 [V] or V_0), it is a useful formula for estimating

[1] As shown in the next chapter, only C_3 can actually be charged to V_{DD}. The other capacitors are restricted to a maximum value of $(V_{DD} - V_{Tn})$.

delays through MOSFET chains. We will find many applications of this formula in later treatments, particularly when studying the dynamic ciruits of Chapter 7.

If all of the transistors are identical, then this equation shows that the longest discharge path (through all of the resistors) dominates. In the general case, the time delay depends on the size of the individual transistors. Since the resistance is inversely proportional to the aspect ratio (W/L), large transistors give small values of R. The capacitance is just the opposite, and increases with channel width W. It is possible to adjust the aspect ratios in the chain to provide a minimum delay; this is discussed in Chapter 7 in the context of Domino logic.

4.2.2 Charging Through a pMOS Chain

Consider now a string of pMOSFETs as shown in Figure 4.3. In most CMOS logic gates, this type of circuit connects the power supply to a node, so we will examine the problem of charging the fartherst node (C_3) through the transistor string. To accomplish this task, we will set all of the inputs to logic 0 states with

$$A_0 = 0 = A_1 = A_2 = A_3 \tag{4.9}$$

to insure that all of the MOSFETs are active. We will assume that the pMOS on-resistance can be approximated by

$$R_p = \frac{1}{\beta_p(V_{DD} - |V_{Tp}|)}, \tag{4.10}$$

and that all capacitors are initially uncharged.

Again relying on a qualitative viewpoint, we approximate the voltage V_3 using the simple exponential form

$$V_3(t) \simeq V_{DD}[1 - e^{-t/\tau_p}], \tag{4.11}$$

so that the $(1/e)$ delay time is set by $t_D \simeq \tau_p$. Our problem is reversed from the nMOS discharge problem in that we are now concerned with charging time constants. For the general case with N transistors in the chain we construct

$$t_D \simeq \sum_{i=0}^{N-1}\Big(\sum_{j=0}^{i} R_j\Big)C_i \tag{4.12}$$

as the time delay. With N=4, this gives

$$\begin{aligned} t_D \quad \simeq \quad & R_0C_0 + (R_0 + R_1)C_1 + (R_0 + R_1 + R_2)C_2 \\ & +(R_0 + R_1 + R_2 + R_3)C_3, \end{aligned} \tag{4.13}$$

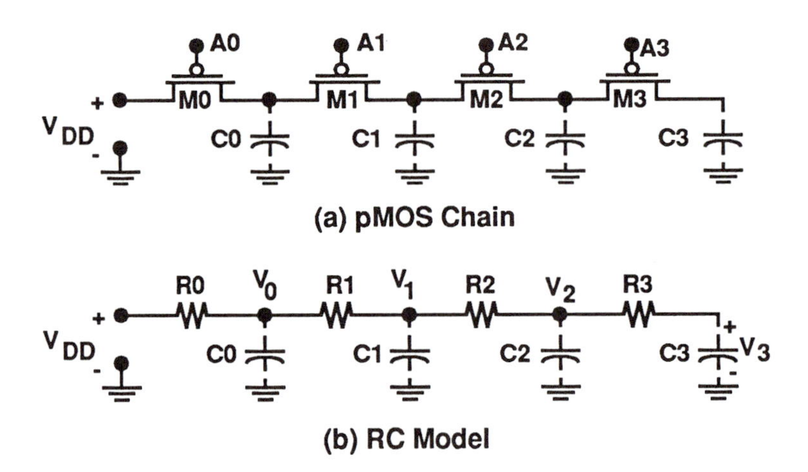

Figure 4.3: pMOS Transistor Chain

which is identical in form to the nMOS charging expression. The important difference is that pMOS transistors are characterized by smaller k' values so that, for identical aspect ratios, $R_p > R_n$. From this we can conclude that long pMOS chains are less desirable than long nMOS chains.

4.2.3 Body-Bias Effects

Series-connected MOSFETs give rise to body-bias effects which alter the conduction characteristics. Consider the two-transistor nMOS chain shown in Figure 4.4. The grounded bulk electrodes are shown to emphasize the point. Body-bias can be seen by noting that

$$V_{SB,1} = V_1 \tag{4.14}$$

where V_1 is the voltage across C_1. This increases the threshold voltage of MOSFET 1 to

$$V_{Tn1} = V_{T0n} + \gamma_n(\sqrt{2|\phi_{Fp}| + V_1} - \sqrt{2|\phi_{Fp}|}\ ); \tag{4.15}$$

it is shown in Section 5.1 of the next chapter that V_1 attains a maximum value of $(V_{DD} - V_{Tn})$. The drain-source resistance is increased from the zero-body bias value since

$$R_{n1} = \frac{1}{\beta_{n1}(V_{DD} - V_{Tn})}, \tag{4.16}$$

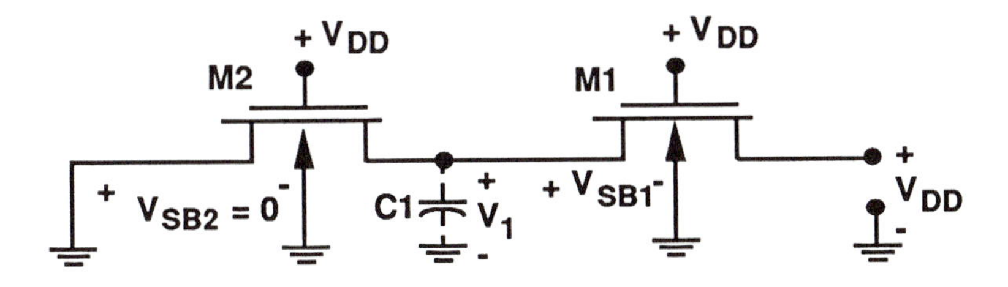

Figure 4.4: Body bias in nMOS Chain

and $V_{Tn} > V_{T0n}$. MOSFET Mn2 does not experience a body bias so that

$$R_{n2} = \frac{1}{\beta_{n2}(V_{DD} - V_{T0n})} \tag{4.17}$$

gives the resistance.

A p-channel MOSFET has an n-type bulk which is connected to the power supply V_{DD}. In a series-connected chain, pMOS transistors which do not have a source connection to the power supply are subject to body bias with

$$V_{Tp} = V_{T0p} - \gamma_p(\sqrt{2\phi_{Fn} + V_{BS,p}} - \sqrt{2\phi_{Fn}}\ ). \tag{4.18}$$

Since $V_{T0p} < 0$, body bias makes V_{Tp} more negative, i.e., $|V_{Tp}|$ increases. The drain-source resistance

$$R_p = \frac{1}{\beta_p(V_{DD} - |V_{Tp}|)} \tag{4.19}$$

increases with increasing $V_{BS,p}$.

Body-bias effects are generally quite important and should be included in circuit calculations if possible. The square-root dependence increases the complexity of the math, often making it impossible to obtain closed form equations. In this case, accuracy is sacrificed in order to simplify the analysis. As demonstrated above, ignoring body bias effects gives smaller equivalent resistance values. Since the RC chains are only rough models to begin with, body bias is often ignored when estimating the transient response of series-connected transistor chains.

4.3 NAND Gate

A 2-input NAND gate (**NAND2**) is illustrated in Figure 4.5. With input variables are A and B, this circuit gives a logical output of

$$F = \overline{A \cdot B}$$

where we have used " · " to denote the Boolean AND operation. The circuit uses full complementary structuring where each input is connected to both an nMOS and a pMOS transistor. Note that only a single polarity transistor is required to perform the logic; for example, if the p-channel MOSFETs are replaced by a simple load resistor, the circuit still functions as a NAND gate. Logic duplication obtained by providing complementary nMOS and pMOS transistors is used to obtain the desirable CMOS characteristics of low power dissipation and a full rail output voltage swing.

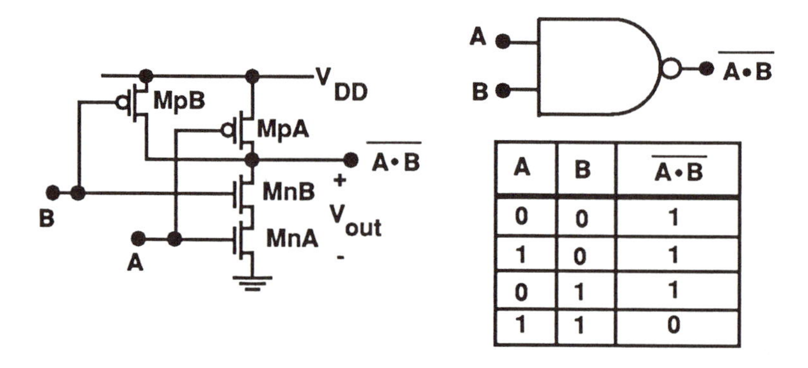

A	B	$\overline{A \cdot B}$
0	0	1
1	0	1
0	1	1
1	1	0

Figure 4.5: NAND2 Gate

Logic formation can be seen by examining the series-connected nMOSFETs MnA and MnB. If both A and B are high, then these transistors conduct current. This in turn provides a strong conduction path to ground, and gives $V_{OL} \simeq 0$ [V]. However, if either A or B is low (or, if both are low) then the path to ground is blocked; in this case, at least one p-channel device is conducting to the power supply, giving $V_{OH} = V_{DD}$. The NAND truth table lists all possible logic combinations.

4.3.1 DC Characteristics

The DC transfer characteristics depend on the input combinations. Figure 4.6 illustrates the VTC for a completely symmetric NAND gate. There are three input combinations which result in the output voltage changing from a high state to a low state. The possibilities are: (a) $A = 1$ while B is switched from 0 to 1; (b) $B = 1$ while A is switched from 0 to 1; and (c) both A and B are simultaneously switched from 0 to 1 states.

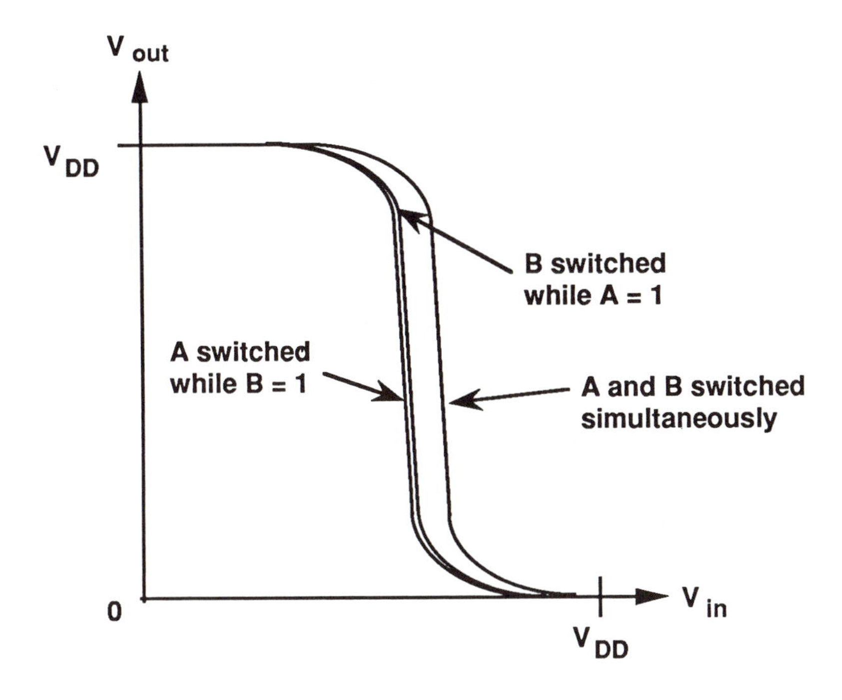

Figure 4.6: VTC for 2-Input NAND Gate

The differences in the curves are due to the electrical structuring of the gate circuit. Since the nMOS transistors are in series, an additional node X enters into the problem as illustrated in Figure. 4.7. Input voltages V_A and V_B are referenced to ground so that the gate-source voltages are

$$\begin{aligned} V_{GS,A} &= V_A \\ V_{GS,B} &= V_B - V_{DS,A}. \end{aligned} \tag{4.20}$$

To establish conduction through the chain, both MnA and MnB must have a voltage $V_{GS} > V_T$. Assuming that the p-bulk is grounded, $V_{SB,A} = 0$, but $V_{SB,B} = V_{DS,A}$ indicates the presence of body-bias on MnB. The transistor threshold voltages are thus given by

$$V_{TA} = V_{T0n}$$

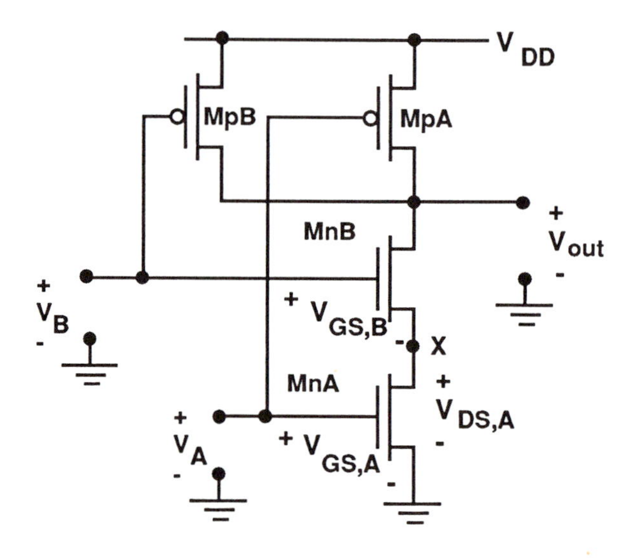

Figure 4.7: NAND nMOS Chain

$$V_{TB} = V_{T0n} + \gamma_n(\sqrt{2|\phi_{Fn}| + V_{DS,A}} - \sqrt{2|\phi_{Fn}|}\ ) \qquad (4.21)$$

which illustrates that MnB is more difficult to turn on than MnA. Combining this observation with the complementary nMOS-pMOS transistor placement accounts for the distinct voltage-transfer characteristics for the three input switching combinations[2].

Now consider the NAND gate threshold voltage where $V_{in} = V_{th} = V_{out}$ for the case of simultaneous switching where both inputs are tied together. The circuit is shown in Figure 4.8. It is assumed that β_n describes both MnA and MnB, while β_p describes to both MpA and MpB. In addition, we will ignore the body bias on MnB for simplicity. To compute V_{th} we first find the currents, apply KCL, and then solve for the voltage.

Consider first the nMOS series chain. With $V_A = V_B = V_{th} = V_{out}$ it is easily established that MnB is saturated while MnA is non-saturated. The drain current through the branch is

$$\begin{aligned} I_D &= \frac{\beta_n}{2}(V_{th} - V_{T0n} - V_{DS,A})^2 \\ &= \frac{\beta_n}{2}[2(V_{th} - V_{T0n})V_{DS,A} - V_{DS,A}^2] \qquad (4.22) \end{aligned}$$

[2] A step-by-step analysis of both CMOS NAND and NOR gates can be found in reference [6].

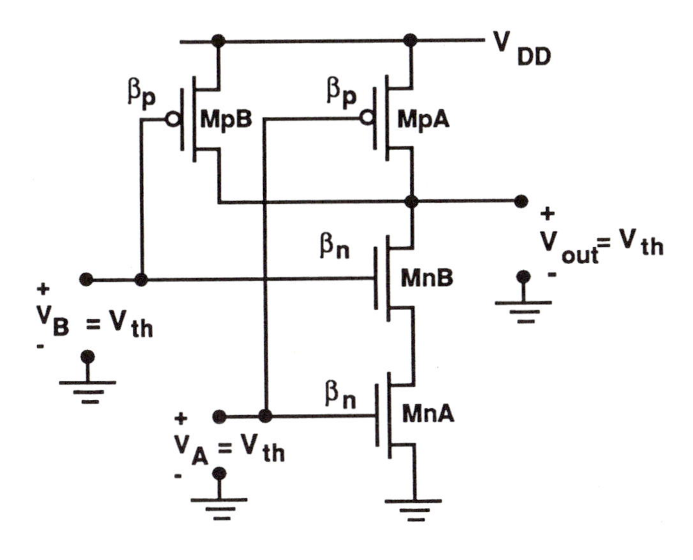

Figure 4.8: NAND2 V_{th} Circuit

which gives

$$V_{th} = V_{Tn} + 2\sqrt{\frac{I_D}{\beta_n}} \tag{4.23}$$

by eliminating $V_{DS,A}$. Both pMOS transistors are saturated and conduct equal currents. Using KCL we have

$$I_D = 2(\frac{\beta_p}{2})(V_{DD} - V_{th} - |V_{T0p}|)^2 \tag{4.24}$$

so combining with the nMOS expression and eliminating I_D gives

$$V_{th} = \frac{V_{T0n} + 2\sqrt{\beta_p/\beta_n}(V_{DD} - |V_{T0p}|)}{1 + 2\sqrt{\beta_p/\beta_n}}. \tag{4.25}$$

Comparing this equation with that derived for a simple inverter, it is seen that the only differences are in the factors of 2.

4.3.2 Transient Characteristics

Switching times can be estimated using the capacitances shown in Figure 4.9. We estimate

$$C_{out} = (C_{GDpA} + C_{GDpB}) + K(V_\ell)[C_{DBpA} + C_{DBpB}] + \sum C_n + C_{line} + C_G \tag{4.26}$$

where $\sum C_n$ accounts for the nMOS contributions which are important depending on the switching event. In addition, the capacitance between the two nMOS transistors (node X in Figure 4.7) may be included in some treatments. This may be used to find approximate equations to compute the transient performance; a computer verification is required if the switching times are particularly critical.

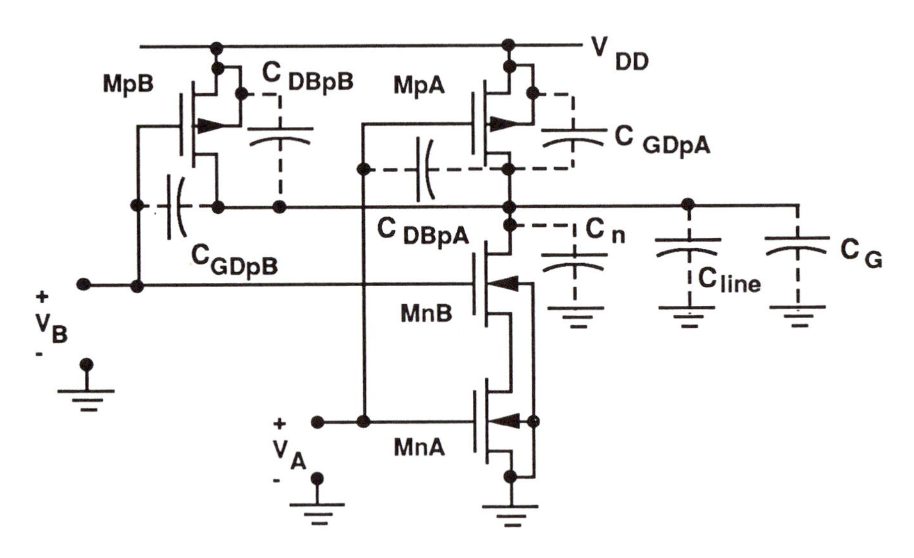

Figure 4.9: C_{out} Contributions

Output Charge Time

Consider first the low-to-high time t_{LH}. The initial condition on the output voltage is that $V_{out}(t=0) = 0$ [V] which implies that both $A = 1 = B$. If either A or B (or both) switch to logic 0 values, C_{out} charges through the appropriate pMOS transistors. The inverter analysis of Section 3.2.1 may be modified to give

$$t_{LH} = \tau_p \left[\frac{2(|V_{Tp}| - V_0)}{(V_{DD} - |V_{Tp}|)} + \ln\left(\frac{2(V_{DD} - |V_{Tp}|)}{V_0} - 1 \right) \right] \tag{4.27}$$

where $V_0 = 0.1 V_{DD}$ is the 10% point. The primary difference between the inverter and the present result is in the time constant structure and value. For the NAND2 gate, we have

$$\tau_p = \frac{C_{out}}{(\sum_j \beta_{pj})(V_{DD} - |V_{Tp}|)}, \tag{4.28}$$

where the summation is over the conducting p-channel transistors. If only A or B is 0, then $\sum_j \beta_{pj}$ is either β_{p1} or β_{p2}. The best-case situation is where $A = 0 = B$, since both pMOSFETs are conducting. This gives $\sum_j \beta_{pj} = (\beta_{p1} + \beta_{p2})$. The value of C_{out} varies with the switching combination; the largest value usually provides a reasonable first-design estimate.

Output Discharge Delay Time

To calculate the discharge delay time we assume an initial output voltage of $V_{out}(0) = V_{DD}$; this implies that at least one input is in a logic 0 state. Discharging occurs when both A and B go to logic 1 voltages. MOSFETs MnA and MnB are both active and provide a conducting path between C_{out} and ground. The can be modeled using the RC equivalent circuit in Figure 4.2. Applying the results of Section 4.2, the delay time is approximately

$$t_D \simeq R_{nA}C_{nA} + (R_{nA} + R_{nB})C_{out} \tag{4.29}$$

where R_{nA} and R_{nB} are the equivalent MOSFET resistances. This illustrates the dependence on the transistor aspect ratios.

Recall that t_D is an estimate of the time required for the voltage to fall from V_{DD} to a value of (V_{DD}/e). Approximating the output high-to-low time t_{HL} is more complicated. Following the inverter analysis we write

$$t_{HL} = \tau_n \left[\frac{2(V_{Tn} - V_0)}{(V_{DD} - V_{Tn})} + \ln\left(\frac{2(V_{DD} - V_{Tn})}{V_0} - 1 \right) \right], \tag{4.30}$$

and then modify the discharging time constant to

$$\tau_n \simeq \frac{C_{out}}{\beta_{n,eff}(V_{DD} - V_{Tn})}, \tag{4.31}$$

where $\beta_{n,eff}$ is an effective value for the series-connected pair. If both transistors have a channel width W, then the effective transconductance is given by

$$\beta_{n,eff} = k'_n \left(\frac{W}{L_{nA} + L_{nB}} \right) \tag{4.32}$$

where L_{nA}, L_{nB} are the respective channel lengths. This views the series transistors as a single device with a total channel length of $(L_{nA} + L_{nB})$, and neglects the stray capacitance between the two MOSFETs.

4.3.3 Design

Designing a NAND gate is straightforward. Since the logic function is a consequence of the circuit structure, the specific choices for MOSFET aspect ratios do not change the logic. Instead, the device sizes establish the DC critical voltages such as V_{th}, and directly determine the transient time intervals.

Some circuits require that the DC switching point be in a specific range. In this case, we first adjust the ratio of (β_n/β_p), and then choose the aspect ratios according to the technology limits and the desired transient response. If timing is critical, we may choose the aspect ratios needed to satisfy the switching response and only calculate the DC characteristics as an after thought.

The low-to-high time t_{LH} is controlled by the pMOS aspect ratios $(W/L)_{p1}$ and $(W/L)_{p2}$. Since Mp1 and Mp2 are in parallel, the worst-case situation occurs when only a single device is conducting. Thus, we can design both pMOSFETs to be the same size $(W/L)_p$ such that either transistor can individually meet the rise time specification. In terms of the charging time constant we have

$$\left(\frac{W}{L}\right)_p = \frac{C_{out}}{k'_p \tau_p (V_{DD} - |V_{Tp}|)} \tag{4.33}$$

which allows us to estimate the required aspect ratio.

The series-connected nMOS transistors limit the discharging response. The simplest design uses identical values of $(W/L)_n$ for both. This value may be computed from

$$\left(\frac{W}{L}\right)_n = \frac{2C_{out}}{k'_n \tau_n (V_{DD} - V_{Tn})}; \tag{4.34}$$

the factor of 2 arises from combining the channel lengths. This shows that the series-connected transistors must be larger to reduce the device resistance. A simple choice is to choose L to be the smallest value allowed in the processing, and then make both nMOSFETs two times larger than needed for an equivalent inverter.

4.3.4 N–Input NAND

The NAND2 structure can be extended to an N–input NAND gate by (a) using N nMOS input transistors in series, along with (b) N pMOSFETs in parallel. Analyzing the generic gate gives

$$V_{th} = \frac{V_{T0n} + N\sqrt{\beta_p/\beta_n}(V_{DD} - |V_{T0p}|)}{1 + N\sqrt{\beta_p/\beta_n}}. \tag{4.35}$$

where N is the number of inputs. Multiple-input NAND gates are easily implemented in circuit and logic designs. However, the output capacitance

C_{out} increases with N due to the pMOS parasitics C_{GDp} and C_{DBp}. The increased capacitance slows down the gate response. Also, the circuit is limited by the transient discharge through the nMOS chain. These considerations generally limit N to a maximum of 4 or 5 in realistic designs.

4.4 NOR Gate

A static 2-input NOR gate is illustrated in Figure 4.10. This performs the logical function

$$F = \overline{A + B} \tag{4.36}$$

where " $+$ " is used to denote the logical OR operation. Note again that each input is connected to both an nMOS and a pMOS transistor.

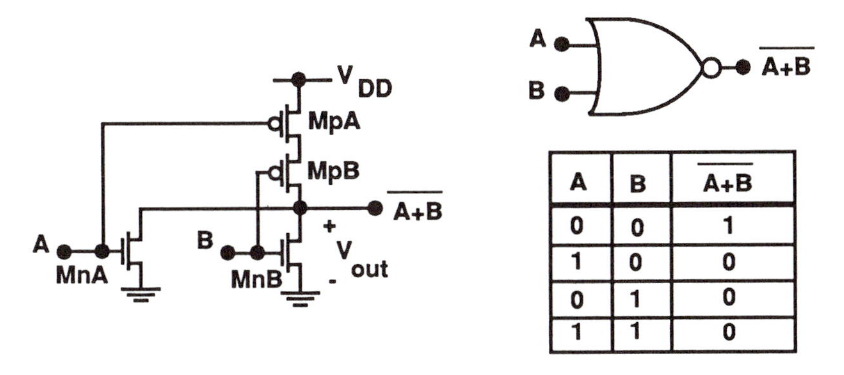

A	B	$\overline{A+B}$
0	0	1
1	0	0
0	1	0
1	1	0

Figure 4.10: 2-Input NOR Gate

Checking the circuit operation to see the NOR function is straightforward. Since the two n-channel transistors MnA and MnB are in parallel, having either $A = 1$ OR $B = 1$ will give current flow and pull the output voltage down to $V_{OL} = 0$ [V]. If the inputs are at logic values $A = 0$ and $B = 0$, then both MnA and MnB are off, while both of the p-channel devices are conducting; in this case, the pull-up characteristics of MpA and MpB give $V_{OH} = V_{DD}$. As with the NAND gate, the VTC depends on the input combinations. A typical plot is provided in Figure 4.11.

4.4.1 DC Characteristics

The switching characteristics of a NOR gate also varies with the input combination. Three different curves are shown in the VTC of Figure 4.12, each representing a different switching combination. Note that the behavior

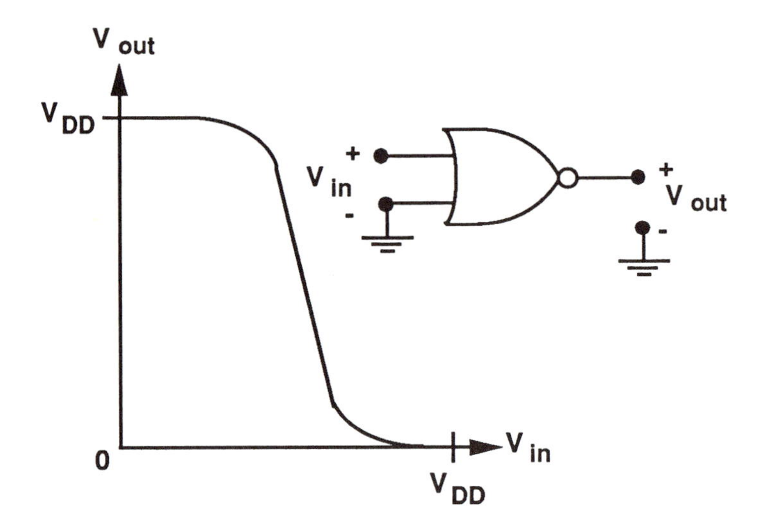

Figure 4.11: NOR Voltage Transfer Curve

is exactly opposite of that for a NAND gate. Analyzing the circuit for simultaneous input switching gives a gate threshold voltage of

$$V_{th} = \frac{V_{T0n} + (1/2)\sqrt{\beta_p/\beta_n}(V_{DD} - |V_{T0p}|)}{1 + (1/2)\sqrt{\beta_p/\beta_n}}; \tag{4.37}$$

body bias has been neglected in the derivation for simplicity. Comparing this with the results for both the inverter and the NAND gate show that the only difference is the factor of (1/2). The remaining two cases where ($A = 0, B$ switched) and (A switched, $B = 0$) can be analyzed using basic circuit techniques. The results show that the differences among the VTC switching points is due to the series connection of MpA and MpB.

4.4.2 Transient Times

Figure 4.13 provides the most important lumped-element contributions to C_{out}. This is estimated as

$$C_{out} = (C_{GDnA} + C_{GDnB}) + K(V_\ell)[C_{DBnA} + C_{DBnB}] + \sum C_p + C_{line} + C_G \tag{4.38}$$

with $\sum C_p$ used to include pMOS parasitics. As always, a computer simulation is mandatory for an accurate transient performance analysis.

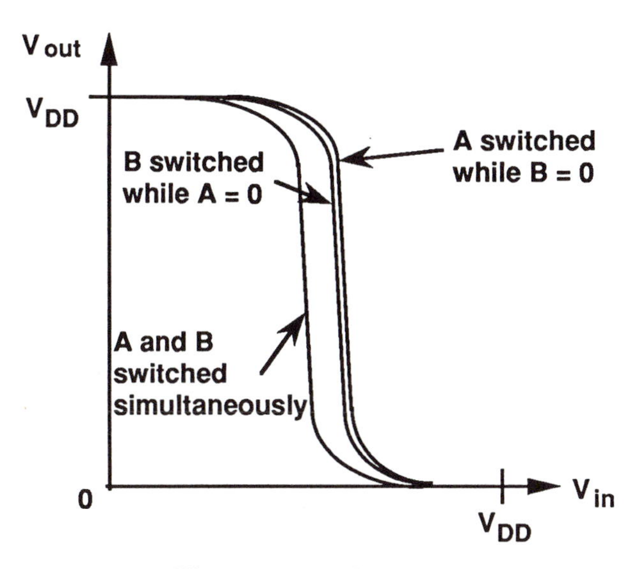

Figure 4.12: NOR2 VTC

Output Discharge Time

The high-to-low time t_{LH} is computed by noting that C_{out} discharges through the nMOS transistors MnA and MnB. Since these are in parallel, we estimate

$$t_{HL} = \tau_n \left[\frac{2(V_{Tn} - V_0)}{(V_{DD} - V_{Tn})} + \ln\left(\frac{2(V_{DD} - V_{Tn})}{V_0} - 1 \right) \right], \tag{4.39}$$

where the time constant is approximated by

$$\tau_n = \frac{C_{out}}{(\sum_j \beta_{nj})(V_{DD} - V_{Tn})}; \tag{4.40}$$

the summation is over all of the conducting nMOSFETs. The worst-case situation is when only a single transistor provides a path to ground. If both are active, then the discharge time is essentially cut in half.

Output Charge Delay Time

The charging delay time can be estimated using the techniques presented in Section 4.2. With both A and B at logic 0 levels, pMOS transistors MpA and MpB provide a conducting path between the output and the power supply. Using the RC model in Figure 4.3 gives

$$t_D \simeq R_{pA} C_{pA} + (R_{pA} + R_{pB}) C_{out} \tag{4.41}$$

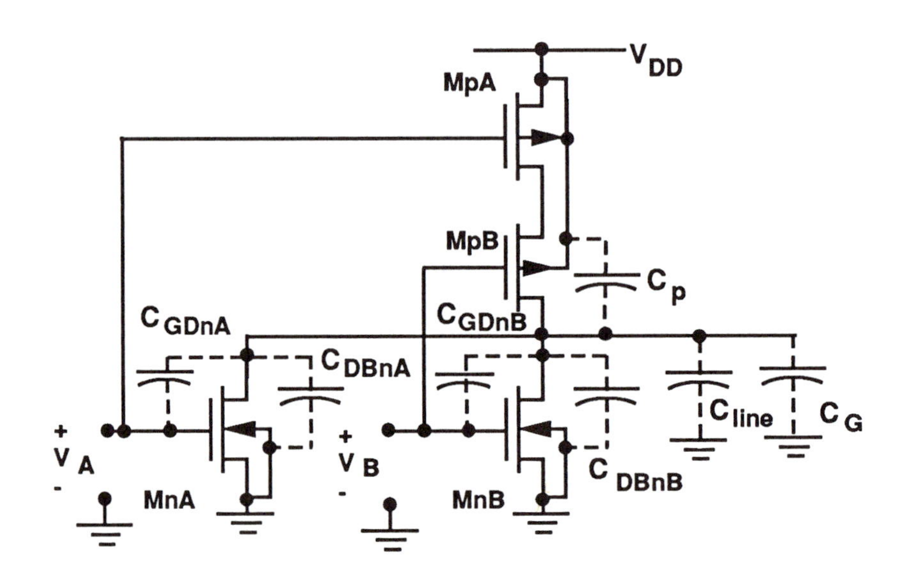

Figure 4.13: Circuit for NOR2 C_{out}

where R_{pA} and R_{pB} are the pMOS resistances.

To estimate the low-to-high time t_{LH} we use

$$t_{LH} = \tau_p \left[\frac{2(|V_{Tp}| - V_0)}{(V_{DD} - |V_{Tp}|)} + \ln\left(\frac{2(V_{DD} - |V_{Tp}|)}{V_0} - 1 \right) \right] \tag{4.42}$$

with a time constant of

$$\tau_p \simeq \frac{C_{out}}{\beta_{p,eff}(V_{DD} - |V_{Tp}|)}. \tag{4.43}$$

The effective device transconductance $\beta_{p,eff}$ models the series pair of pMOS transistors. Assuming that both have the same width W gives

$$\beta_{p,eff} \simeq k_p' \left(\frac{W}{L_{p,eff}} \right), \tag{4.44}$$

where

$$L_{p,eff} = L_{pA} + L_{nB} \tag{4.45}$$

is the total effective channel length.

4.4.3 Design

Functional logic design for a NOR gate is automatic with the placement of the transistors in the circuit: nMOSFETs are in parallel and pMOSFETs are in series. The choice of aspect ratios affects the gate threshold voltage V_{th} and the transient characteristics, as is the case for all static CMOS gates.

NOR gate design is similar to that discussed for NAND circuits. To set the value of V_{th}, we adjust the ratio (β_n/β_p); transient specifications then dictate the specific aspect ratios. The high-to-low time t_{HL} is set by the values of $(W/L)_{nA}$ and $(W/L)_{nB}$. Since the two are in parallel, we may choose both nMOS transistors to have the same $(W/L)_n$ with

$$\left(\frac{W}{L}\right)_n = \frac{C_{out}}{k'_n \tau_p (V_{DD} - V_{Tn})} \tag{4.46}$$

providing a reasonable design value.

Charging is limited by the p-channel MOSFET chain. Choosing a value of $(W/L)_p$ by means of

$$\left(\frac{W}{L}\right)_p = \frac{2C_{out}}{k'_p \tau_p (V_{DD} - |V_{Tp}|)} \tag{4.47}$$

gives the simplest design.

4.4.4 N–Input NOR

An N–input NOR gate can be constructed by paralleling N nMOS transistors and series connecting N pMOS transistors such that each input is connected to one of each type. Analyzing this configuration gives a gate threshold voltage of

$$V_{th} = \frac{V_{T0n} + (1/N)\sqrt{\beta_p/\beta_n}(V_{DD} - |V_{T0p}|)}{1 + (1/N)\sqrt{\beta_p/\beta_n}} \tag{4.48}$$

where N is the number of inputs. NOR gate logic is very straightforward to implement, and is also very popular. However, it also has an output capacitance problem due to the parallel nMOS devices, and requires charging through a pMOS chain, so that N is usually kept to a maximum of 4 or 5 (unless speed is not an issue).

4.5 Comparison of NAND and NOR Gates

Both NAND and NOR gates are easy to implement in CMOS logic. However, for equal numbers of inputs and device sizes, NAND gates have better transient response than NOR gates, making them more popular in high-performance design. The reasoning behind this statement can be understood by recalling the delay times. The series-connected transistors are the limiting factor[3]. In a NAND gate, the discharge delay time $t_{D,NAND}$ is determined by a chain of n-channel MOSFETs. A NOR gate, on the other hand, has a charging time $t_{D,NOR}$ which is due to charging through a string of p-channel transistors. Since $R \propto (1/\beta)$ and $k'_n > k'_p$ is always true, $R_n < R_p$. Thus, given nMOS and pMOS transistor chains of the same size, the nMOS chain always discharges faster than the pMOS chain can charge.

Even though NOR gates are slower than NAND gates, both are widely used in practice. They provide the main logic components in full-custom, ASIC[4], and gate arrays.

4.6 OR and AND Gates

Static CMOS directly provides NAND, NOR, and NOT (invert) functions which is more than sufficient to implement arbitrary logic functions. However, sometimes basic functions OR and AND are useful. The obvious solution is to produce the OR by a NOR-NOT cascade, and implement the AND using a NAND-NOT cascade.

A more efficient approach is to combine NOR and NAND gates using standard Boolean algebraic identities. Applying the DeMorgan rules allows us to write the OR function as

$$\begin{aligned} F &= \overline{\overline{A} \cdot \overline{B}} \\ &= A + B. \end{aligned} \tag{4.49}$$

Similarly, the AND function can be obtained using

$$\begin{aligned} G &= \overline{\overline{A} + \overline{B}} \\ &= A \cdot B. \end{aligned} \tag{4.50}$$

The drawings in Figure 4.14 illustrate the use of these basic transformation.

Often times we can construct high-performance AND or OR gates using smaller NAND and NOR circuits. As an example let us examine the 4-input

[3] The parallel-connected transistors (pMOS in a NAND or nMOS in a NOR) provide fast transient response since multiple conduction paths exists.

[4] Application-specific integrated circuits.

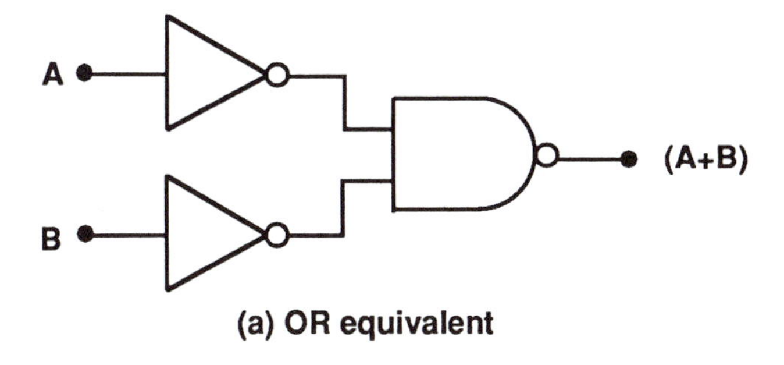

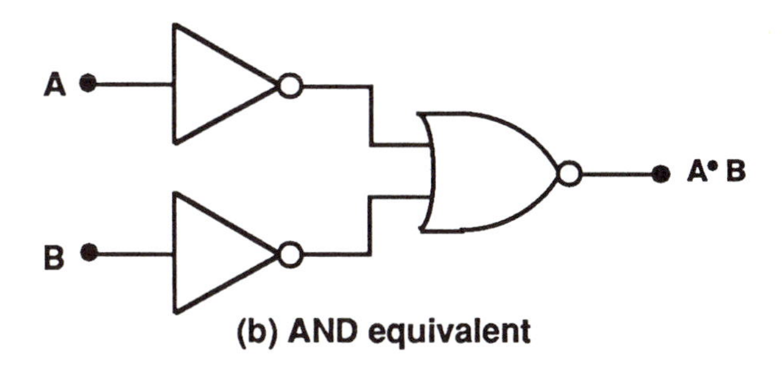

Figure 4.14: DeMorgan Reductions

AND gate shown in Figure 4.15 which uses 2 NAND2 gates cascaded into a NOR2 to form

$$\begin{aligned} F &= \overline{\overline{AB} + \overline{BC}} \\ &= ABCD. \end{aligned} \tag{4.51}$$

At first glance it may appear that this circuit automatically has a slow response due to the number of stages. However, as discussed in Chapter 8, the switching speed of a static logic chain depends on both the transistor sizing and the number of gates in the chain. Optimum sizing criteria can be applied to make this as fast (or faster) than a single NAND4.

4.7 Combinational Logic

The NAND and NOR circuits can easily be extended to build more complex logic gates. The approach presented in this section yields what we will call **standard CMOS** logic circuits. This class of static logic gates starts with the CMOS inverter. Logic functions are synthesized by adding series- and parallel-connected transistor combinations according to a simple set of

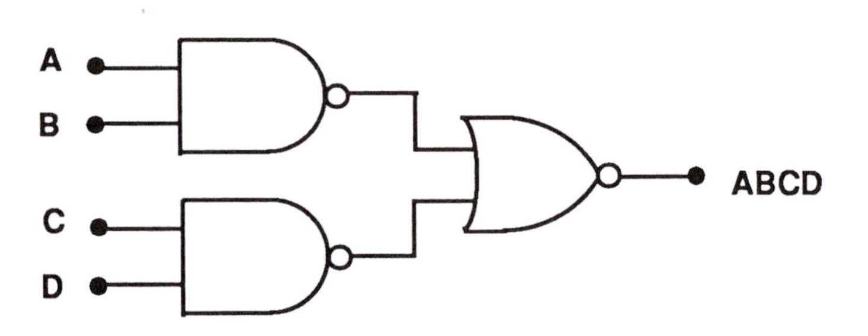

Figure 4.15: AND4 Composite Gate

rules. Static gates of this type are very common in practice due to their simplicity, straightforward design, and ease-of-use.

To implement an arbitrary combinational logic function in standard CMOS, we adhere to the following basic rules:

- Each input is connected to the gate of both an nMOS and a pMOS transistor;
- The pMOS and nMOS circuits are complements of each other.

A gate with N input variables requires a total of $2N$ MOSFETs. Although a single polarity design (i.e., all nMOS) is sufficient to generate the logic function, the complementary structure insures the low-power dissipation property. The generic gate in Section 4.1 is redrawn in Figure 4.16 for reference. We have chosen the logic blocks to be labeled as $\overline{F}$ (nMOS) and F (pMOS); the output of the logic gate is $\overline{F}$. This labelling of nMOS and pMOS blocks is arbitrary. The important property to remember is that when one logic block is conducting (a "1" state), then the complementary block is open (a "0" state).

4.7.1 Logic Formation

Logic structuring is most easily understood using nMOSFETs as a basis since they directly implement positive logic. Primitive logic functions are obtained using two simple nMOS rules:

Rule 1: nMOS transistors in series implement the AND operation;

Rule 2: nMOS transistors in parallel give the OR operation.

An additional level of design is provided by the third and fourth rules:

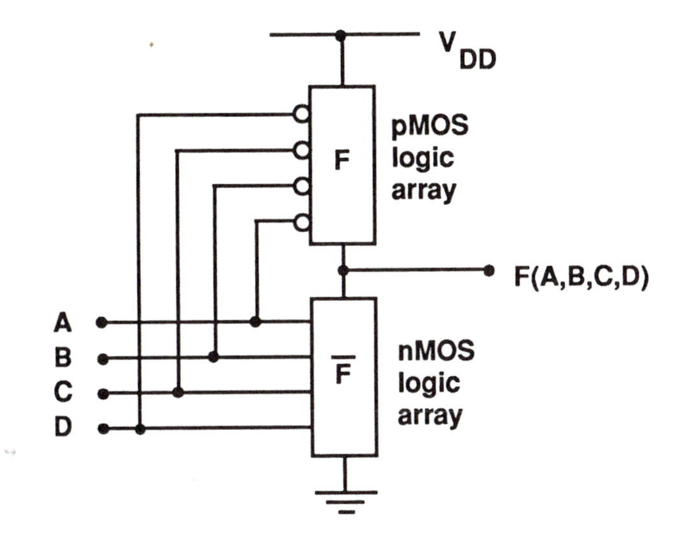

Figure 4.16: General CMOS Static Logic Gate

Rule 3: Parallel nMOS branches OR the individual branch functions;

Rule 4: Logic functions in series are ANDed together.

These rules are summarized in Figure 4.17.

To build a logic gate in standard static CMOS, we first structure the nMOS logic transistors according to the rules above. This forms the lower half of the circuit, so that

Rule 5: The output is the complement of the nMOS logic.

Finally, we add the p-channel transistors according to the rule

Rule 6: The pMOS circuit is the exact **dual** of the nMOS circuit

to complete the design. The last rule is very easy to interpret: inputs to nMOS which are in series become parallel-connected in the pMOS section, while nMOS parallel logic is transformed to series-connected pMOS circuits[5]. Duality is also applied to any logic sub-blocks. This general procedure can be used to implement arbitrary logic functions. It is classified as **Series-Parallel** CMOS logic due to the structure used to implement logic functions.

[5]The DeMorgan relations provide the basis for the transformations.

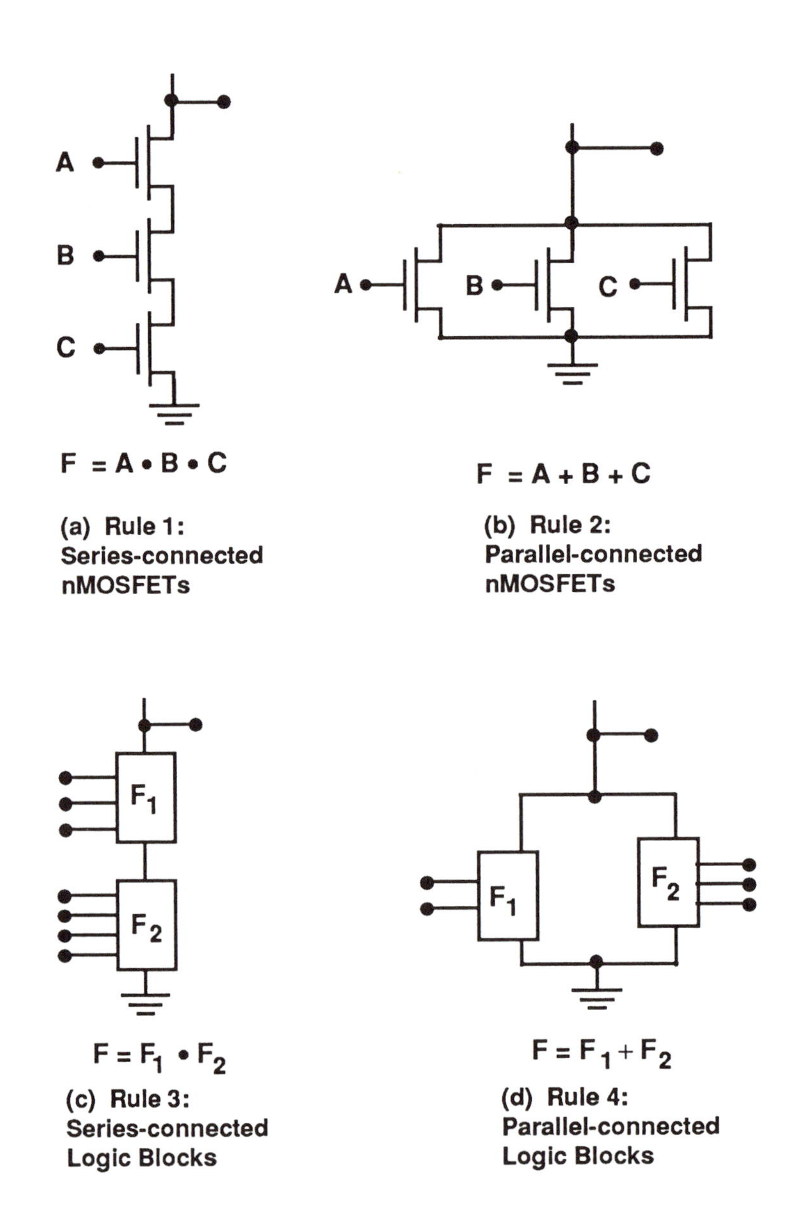

Figure 4.17: nMOS Logic Formation

EXAMPLE 4.1: Basic AOI Logic

Suppose that we want to implement the logic function

$$F = \overline{ABC + AD}$$

in static combinational CMOS. First, we factor A to write

$$F = \overline{A(BC + D)}$$

as a simpler expression. The first term BC is obtained by series connecting two nMOS inputs; the corresponding pMOS devices are in parallel. To OR BC with D requires that the series BC nMOS branch be in parallel with a single D nMOSFET; the pMOS section is formed in a complementary manner. Finally, ANDing A with $(BC+D)$ is achieved by series connecting an A nMOSFET with the nMOS logic for the OR term, and structuring the pMOS complement. This results in the circuit below.

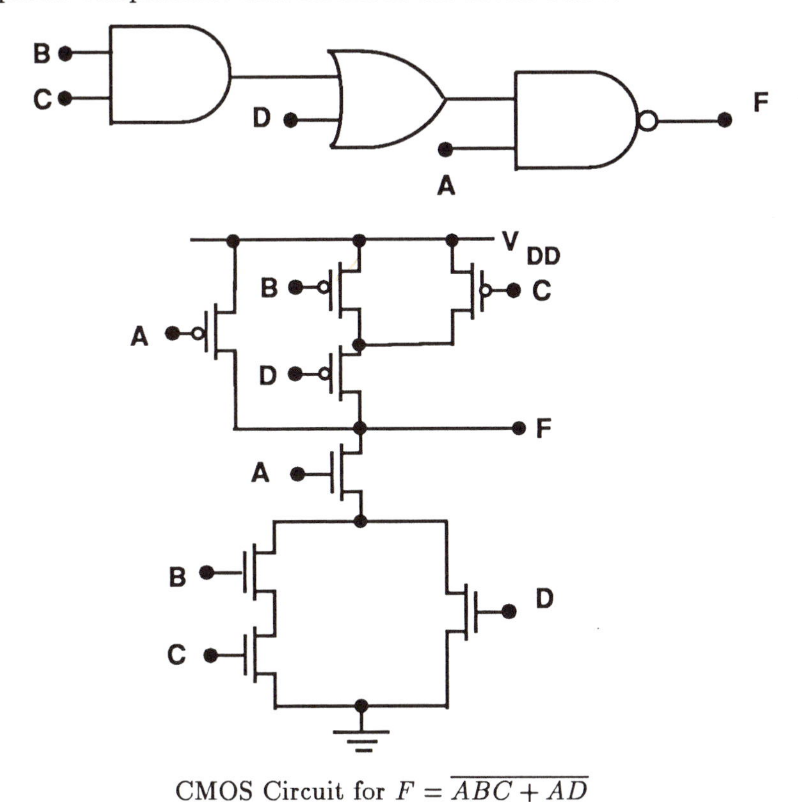

CMOS Circuit for $F = \overline{ABC + AD}$

EXAMPLE 4.2: 4-to-1 Multiplexer (MUX)

A 4-to-1 multiplexer selects one of four possible inputs according to the control bits (S_0, S_1). Denote the input data lines by D_0, D_1, D_2, D_3. We will design a MUX unit with an output F of

$$F = D_0\overline{S}_0\overline{S}_1 + D_1\overline{S}_0S_1 + D_2S_0\overline{S}_1 + D_3S_0S_1 \tag{4.52}$$

using three basic AOI gates. The logic is shown as below.

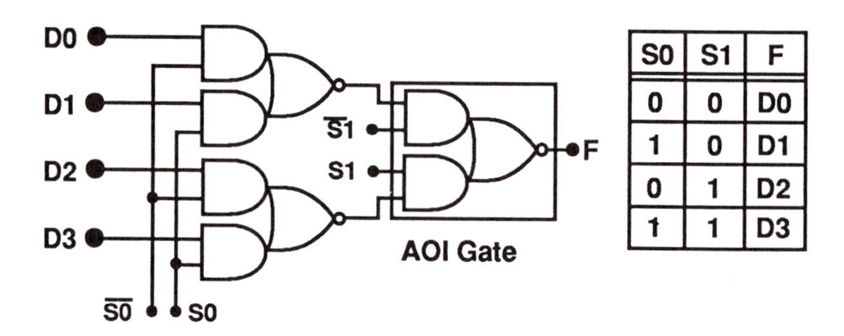

S0	S1	F
0	0	D0
1	0	D1
0	1	D2
1	1	D3

All three gates are identical, and the rules allow direct construction of the circuit.

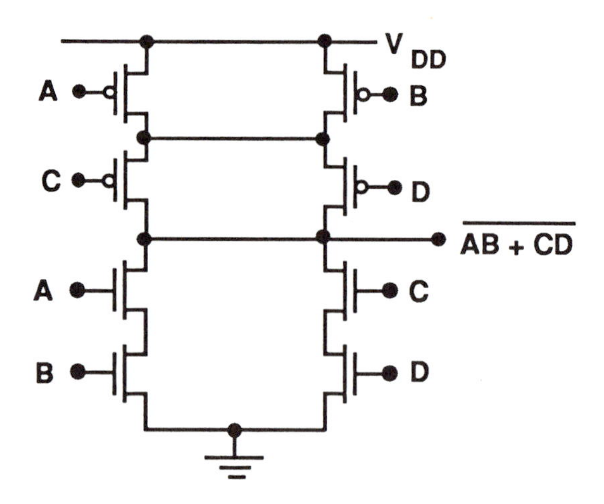

AOI Circuit for MUX Implementation

4.7.2 Canonical Logic Forms

CMOS directly provides both **AOI** (AND-OR-Invert) and **OAI** (OR-AND-Invert) logic functions (these simply give the order of logic structuring from the input to the output). An example of an AOI structure is given by

$$F1 = \overline{AB + BC + AC}; \tag{4.53}$$

in classical terminology, the term $(AB+BC+AC)$ is in **sum-of-products** (SOP) form. An OAI form is illustrated in the function

$$F2 = \overline{(A + B)(C + D)}; \tag{4.54}$$

the OA expression $(A+B)(C+D)$ is in classical **product-of-sums** (POS) form. Both approaches can be nested to arbitrarily deeper levels. For example, we can perform an OR operation inside of an AOI function:

$$F3 = \overline{(A + B)\overline{C} + A\overline{C}}\ . \tag{4.55}$$

The CMOS circuit implementation still follows the rules listed in the previous section, regardless of the logic form. However, the transient response times are sensitive to the transistor placement, layout, and switching order, so that care must be exercised in high-performance design.

4.7.3 Circuit Design

Gates can be structured to perform arbitrary logic functions using the simple rules developed at the beginning of this section. The electrical characteristics are a direct result of the circuit topology, the device geometries, and the layout. At the schematic level, the aspect ratios and estimated capacitances are the most critical to determining the performance.

DC Characteristics

The voltage transfer characteristics described in the plot of V_{out} vs. V_{in} depend on the input switching combination. The shape of the curve and the critical voltages vary with each possibility. Computer simulation is the easiest way to study the family of curves for a particular circuit.

Transient Response

Estimating the rise and fall times for a complex gate is accomplished by using RC-equivalent models for the charge and discharge paths. Each MOSFET is replaced by a resistor of value

$$R = \frac{1}{\beta(V_{DD} - V_T)} \tag{4.56}$$

EXAMPLE 4.3: SOP/AOI Logic Circuit

Consider the canonical AOI form

$$\begin{aligned} F &= \overline{AB + AC + BC} \\ &= \overline{AB + (A + B)C}. \end{aligned}$$

The CMOS gate obtained by applying the logic formation rules is shown below.

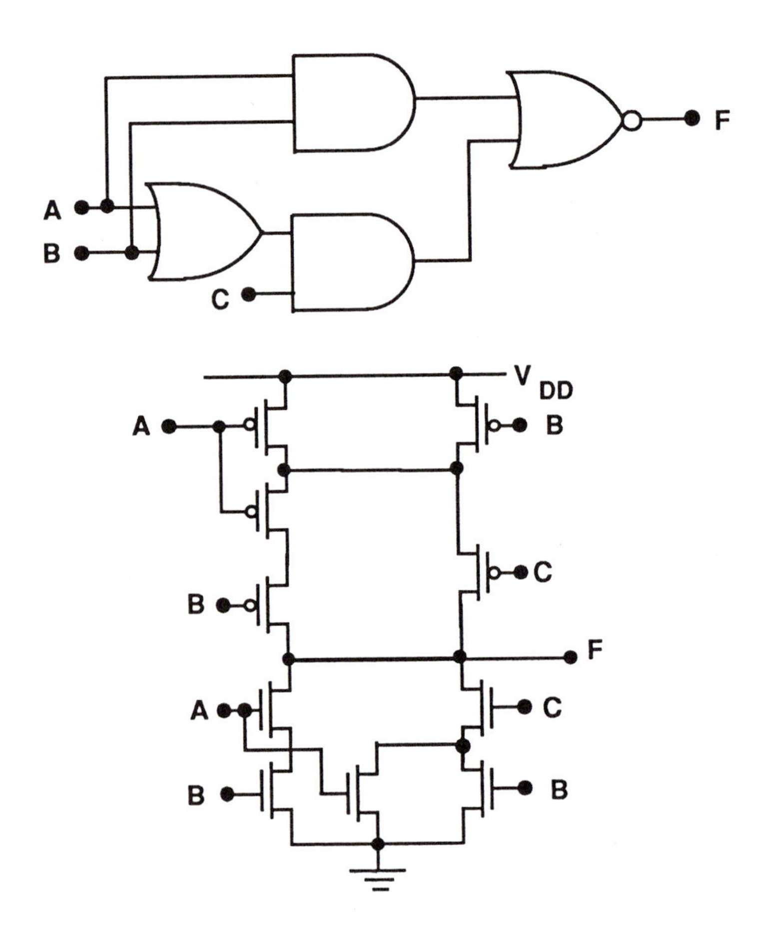

AOI CMOS Circuit for $F = \overline{AB + AC + BC}$

where $\beta = k'(W/L)$ is set by the design. Parallel and series-connected MOSFETs are combined using standard resistor reductions. The most critical value is the largest resistance path between the output node and V_{DD} (for t_{LH}) or ground (for t_{HL}). Capacitance estimates are based on the device size and process parameters.

Procedure

Transient response is often the most important aspect of the circuit design. Although the design methodology presented here only provides first estimates, it is useful for initial performance checks and sizing.

The following sequence provides a straightforward approach to designing complex logic gates.

1. Estimate C_{out} for the circuit. Then design an **inverter** with the desired transient response by computing $(W/L)_{n,inv}$ and $(W/L)_{p,inv}$.

2. Construct the nMOS logic block and count the largest possible number m of series-connected transistors. Choose each device to be identical with $(W/L)_n = m(W/L)_{n,inv}$.

3. Construct the pMOS logic block and count the largest possible number ℓ of series-connected transistors. Choose each device to be identical with $(W/L)_p = \ell(W/L)_{p,inv}$.

The resulting circuit will have response times which are longer than the inverter designed in Step 1 since the inter-transistor capacitance is ignored in the procedure. This also illustrates the problem with maintaining speed in a complex CMOS gate. Since the nMOS and pMOS logic blocks are complementary, series-connected MOSFETs are unavoidable; real-estate requirements increase accordingly. As a rule of thumb, circuits with series-connected nMOSFETs switch faster than those which use the same number of pMOSFETs.

4.8 Exclusive-OR and Equivalence

The 2-input **exclusive-OR** (XOR) function takes A and B and generates the function

$$\begin{aligned} F &= A \oplus B \\ &= A\overline{B} + \overline{A}B. \end{aligned} \tag{4.57}$$

This produces a logic 1 output when either $A = 1$ or $B = 1$ exclusively, i.e., the input combinations where $A = B$ give a logic 0 output. Figure 4.18

EXAMPLE 4.4: Complex Circuit Design

Consider again the circuit for

$$F = \overline{AB + AC + BC}.$$

discussed in the Example 4.3. To choose the aspect ratios, we will estimate an output capacitance of $C_{out} = .20$ [pF]. Assuming $V_{DD} = 5$ [V], $V_{Tn} = +0.7$ [V] $= |V_{Tp}|$, $k_n' = 55$ [μA/V^2] and $k'p = 25$ [μA/V^2], we will set the time constants at $\tau_n = \tau_p = 0.5$ [ns] corresponding to $t_{LH} = t_{HL} \simeq 1.5$ [ns] with a 5 [V] power supply. Using the design formulas in Section 3.4 gives

$$\left(\frac{W}{L}\right)_{n,inv} = \frac{0.2 \times 10^{-12}}{(55 \times 10^{-6})(0.5 \times 10^{-9})(5 - .7)} \simeq 1.7,$$
$$\left(\frac{W}{L}\right)_{p,inv} = \frac{0.2 \times 10^{-12}}{(25 \times 10^{-6})(0.2 \times 10^{-9})(5 - .7)} \simeq 3.7.$$

Both the nMOS and pMOS logic blocks have a worst-case current path of 2 series MOSFETs. Thus,

$$\left(\frac{W}{L}\right)_n \simeq 3.4,$$
$$\left(\frac{W}{L}\right)_p \simeq 7.4,$$

which can be rounded to 4 and 8, respectively, to compensate for some of the stray capacitance.

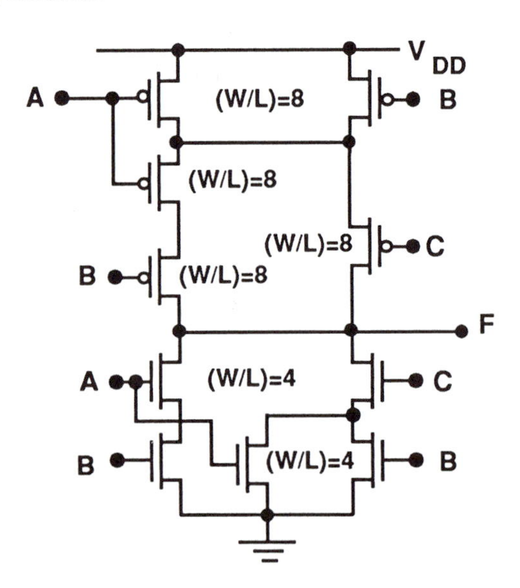

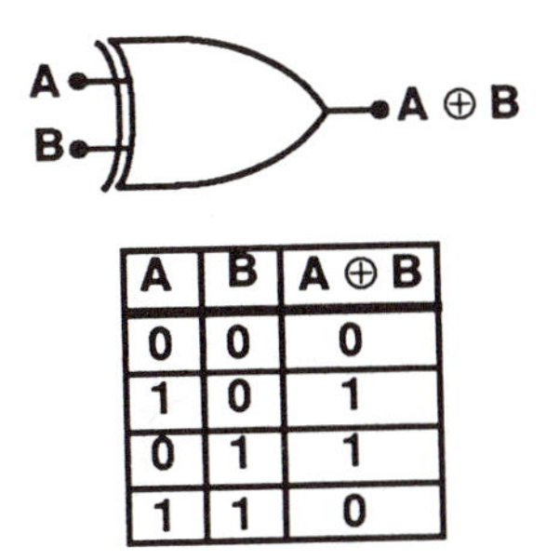

A	B	A ⊕ B
0	0	0
1	0	1
0	1	1
1	1	0

Figure 4.18: Exclusive-OR

shows the shape-specific XOR gate symbol and truth table. Combinational logic can be used to implement the XOR function directly. Since the defining equation is already in AOI form, the CMOS gate may be constructed as shown in Figure 4.19.

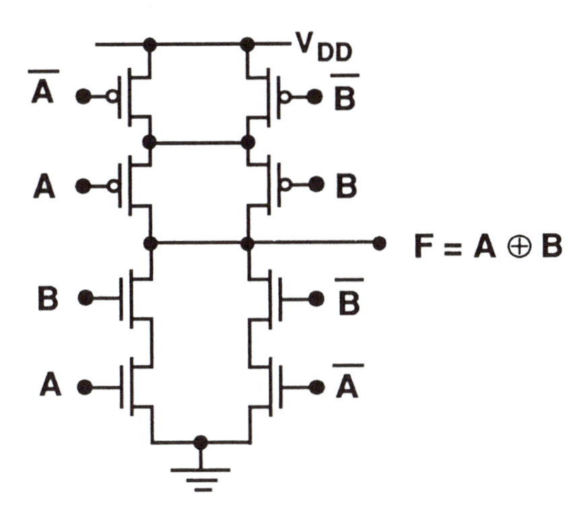

Figure 4.19: AOI XOR Circuit

The **equivalence function** is designed to produce a logical 1 output when the inputs are equal. For a 2-input gate with inputs A and B, the equivalence is defined by

$$\begin{aligned} F &= A \odot B \\ &= AB + \overline{A}\,\overline{B} \\ &= \overline{A \oplus B}. \end{aligned} \tag{4.58}$$

As seen by the truth table in Figure 4.20, the 2-input equivalence function is the complement of the XOR and is sometimes referred to as the exclusive-NOR (XNOR) function. AOI structuring is again appropriate for building the CMOS gate and results in the circuit illustrated in the same figure.

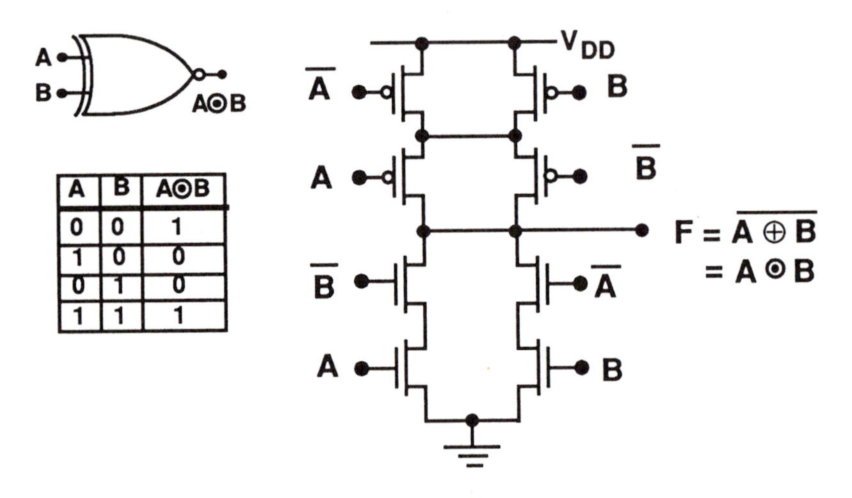

A	B	A⊙B
0	0	1
1	0	0
0	1	0
1	1	1

Figure 4.20: Equivalence Function

A useful application of the XOR function is in performing binary addition. A **half-adder** has two inputs (A_0, B_0) and generates the sum S_0 and carry C_0 bits by

$$\begin{aligned} S_0 &= A_0 \oplus B_0 \\ C_0 &= A_0 B_0. \end{aligned} \tag{4.59}$$

Both terms may be directly implemented in static logic using the gates above.

Designing a **full adder** is also straightforward. Denoting the input bits by (A_n, B_n), and the carry input by C_{n-1}, the full adder generates sum and carry bits of

$$\begin{aligned} S_n &= A_n \oplus B_n \oplus C_{n-1}, \\ C_n &= A_n B_n + C_{n-1}(A_n + B_n), \end{aligned} \tag{4.60}$$

respectively. To implement this function in AOI logic, we expand the expressions in the form

$$\begin{aligned} S_n &= (A_n + B_n + C_{n-1})\overline{C}_n + (A_n B_n C_{n-1}), \\ C_n &= A_n B_n + A_n C_{n-1} + B_n C_{n-1}, \end{aligned} \tag{4.61}$$

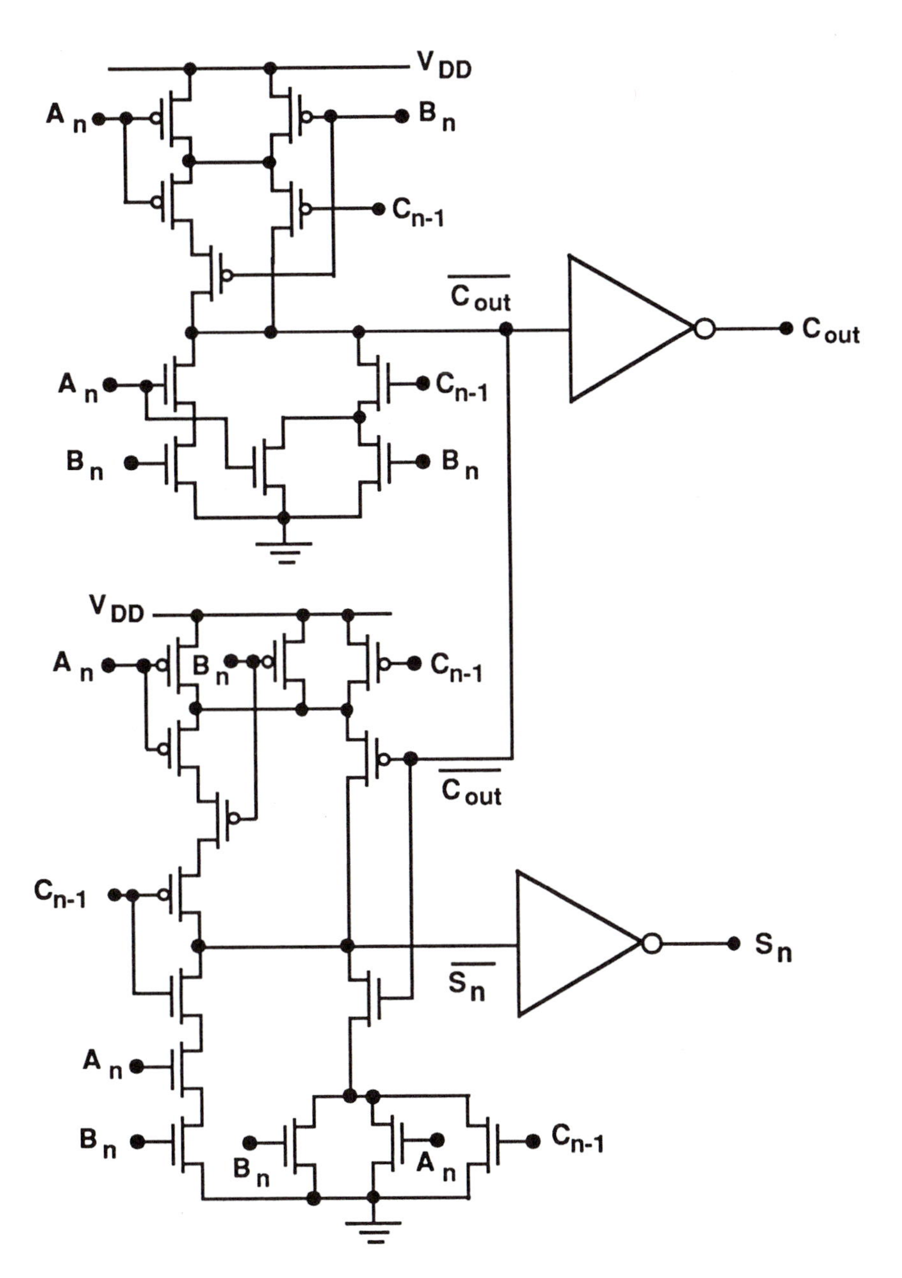

Figure 4.21: AOI Full Adder

to obtain the logic diagram and circuit shown in Figure 4.21.

An N-bit adder can be constructed using a half-adder and (N-1) full adders in the usual manner; an 8-bit adder is shown in Figure 4.22, Although this is a straightforward design, it uses a ripple carry architecture where the $n-$th sum bit S_n is not valid until the input carry bit C_{n-1} from the adjacent adder stabilizes. Faster carry algorithms and circuits, such as the "carry look-ahead" (CLA) network, can be used if the carry propagation time delay is excessive[6].

4.9 Structural Variations

The approach to logic synthesis discussed above is based on series and parallel branches of MOSFETs. While these standard arrangements are the easiest to work with, other switching topologies are easily created. An example is shown in Figure 4.23 where MOSFETs Mn and Mp are used to bridge across the branches. The output function for this particular example is given by

$$F = \overline{A(B + CE) + D(E + BC)} \tag{4.62}$$

as verified by counting nMOS conduction paths from the output node to ground. Cutset analysis may be used for this type of logic, and the reader is referred to the literature for more in-depth discussions [3].

4.10 Tri-State Output

Tri-state circuits provide three distinct output values: Logic 0, Logic 1, and **infinite impedance**. The infinite impedance state is used to isolate a logic gate from the rest of the system. Bidirectional communication and other important structural features can be implemented using tri-state properties.

A simple tri-state buffer is shown in Figure 4.24. Data bit A is the input, while the enable signal EN controls the output. When $EN = 1$, the circuit functions as a normal buffer and provides low and high voltages. If $EN = 0$, both the nMOS and pMOS output transistors are in cutoff, isolating the output node. Ideally, the transistors would have zero drain current giving an impedance of $Z \to \infty$. In practice, only leakage currents exist so that the output impedance is extremely large.

An inverting tri-state buffer is shown in Figure 4.25. This uses an extra transistor (MnL) as the nMOS logic transistor. Pull-up or pull-down resistors or transistors can be added to either circuit if needed.

[6]The CLA algorithm is developed in Chapter 7 for a dynamic NORA circuit.

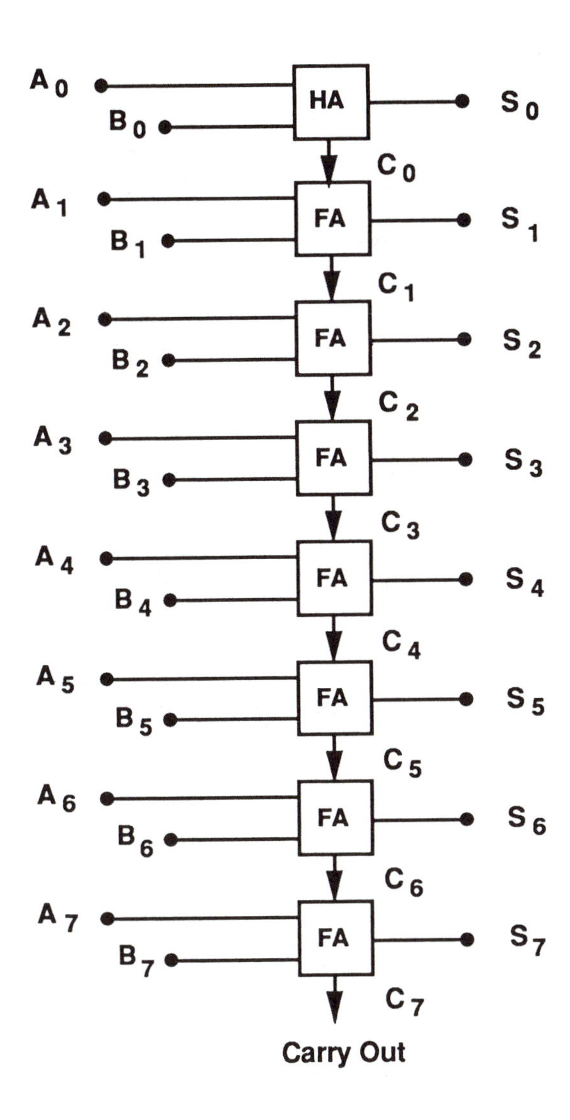

Figure 4.22: 8-Bit Adder

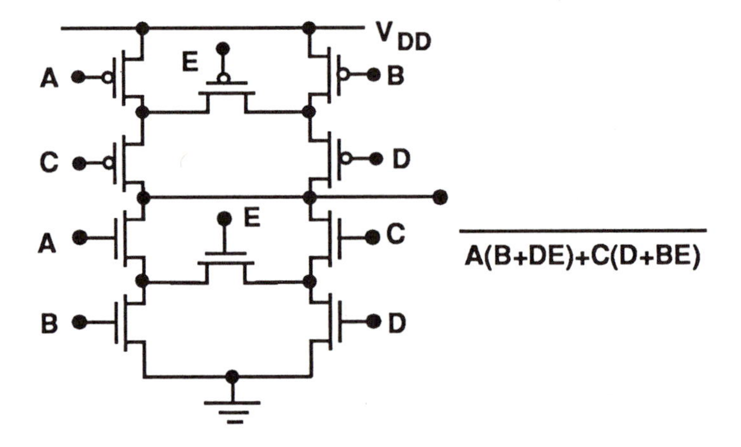

Figure 4.23: Logic Variation Example

4.11 Pseudo-nMOS/pMOS Logic

Designing CMOS gates to have negligible static power dissipation requires that the logic be implemented using both nMOS and pMOS transistors. An N-input gate requires $2N$ MOSFETs, which is twice the number needed to perform the logical operation. If the transistor count becomes excessive, or if layout problems start to overwhelm the need to maintain low power dissipation, then **pseudo-nMOS** or **pseudo-pMOS** structuring may be useful to consider. As discussed below, an N input gate only requires $N+1$ transistors in this approach, but the gates dissipate static power. Other drawbacks of pseudo circuits are that the logic swing $V_{ell} = (V_{OH} - V_{OL})$ is always less than V_{DD}, and that large transistors are usually needed to achieve acceptable output voltages.

4.11.1 Pseudo-nMOS

Pseudo-nMOS logic uses nMOS transistors to form the logic function, and a single pMOSFET as a load. An example is the inverter shown in Figure 4.26. Logically this is identical to the standard CMOS approach. The main difference is that the gate of pMOS transistor Mp is grounded giving

$$V_{SGp} = V_{DD} \tag{4.63}$$

so that it is always in a conducting mode.

To understand the DC characteristics, suppose that we use a logic 0

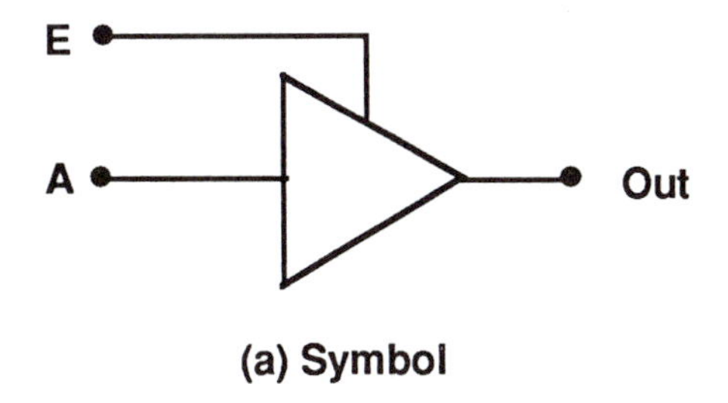

(a) Symbol

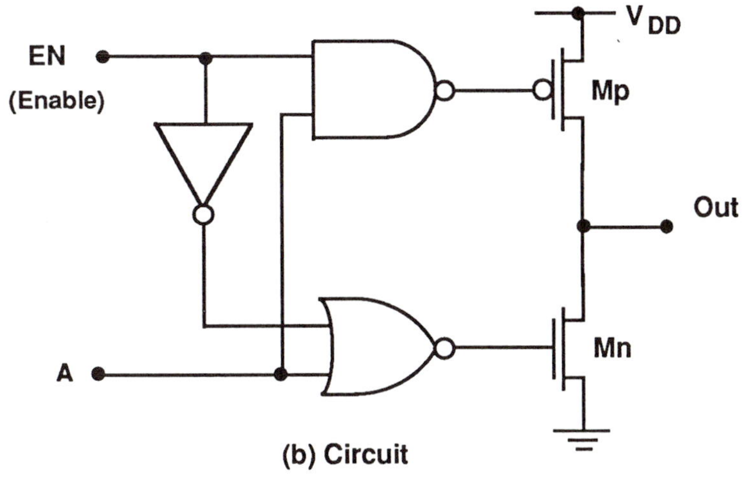

(b) Circuit

Figure 4.24: Tri-State Buffer

input with $V_{in} < V_{Tn}$; this places Mn in cutoff. Since Mp is biased on,

$$V_{OH} = V_{DD} \tag{4.64}$$

is achieved. Analyzing a logic 1 input state is more complicated. If $V_{in} = V_{DD}$, Mn is switched on and the output node has a conducting path to ground. Unlike standard CMOS, however, Mp is still in a conducting mode. This prohibits V_{out} from ever reaching 0 [V] and is typical of **ratioed** nMOS logic circuit [7] where the relative values of (W/L) set V_{OL}.

Let us analyze the circuit to find V_{OL}. With $V_{in} = V_{DD}$, Mn is non-saturated while Mp is saturated (for a reasonable design with a low V_{OL}).

[7] See reference [6] for a general discussion of nMOS circuit problems.

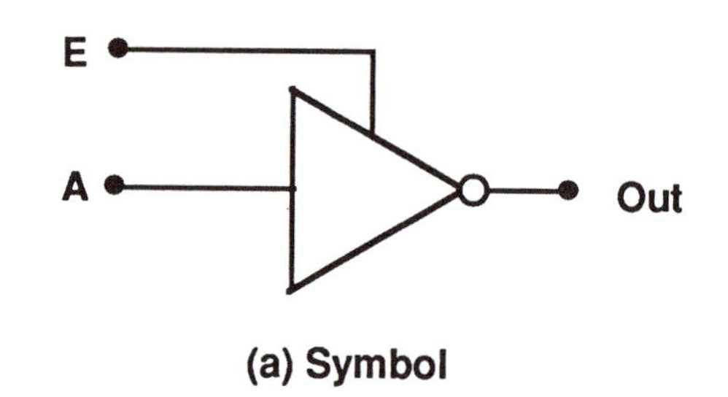

(a) Symbol

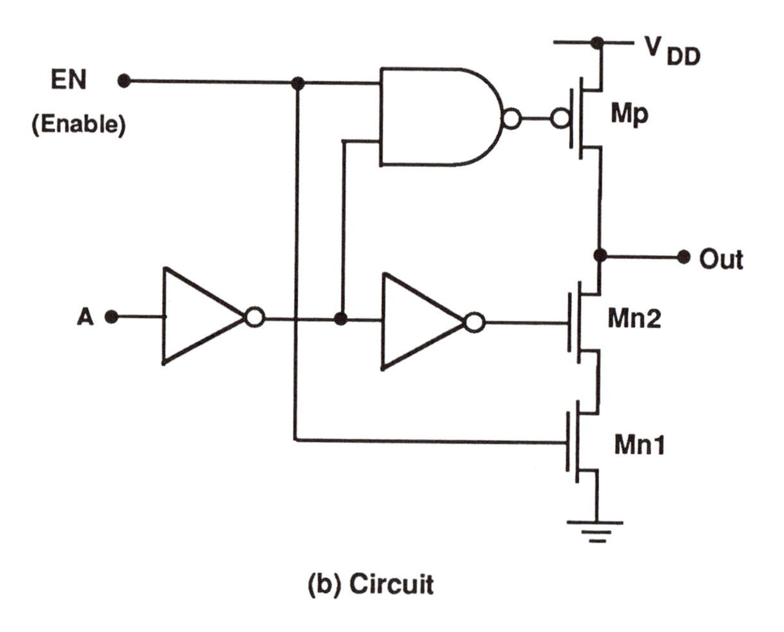

(b) Circuit

Figure 4.25: Inverting Tri-State Buffer

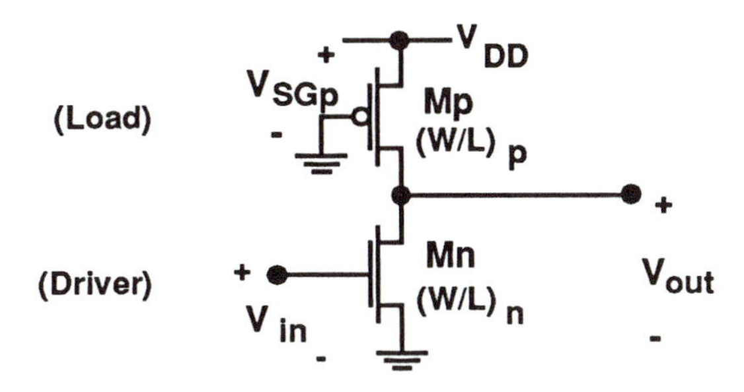

Figure 4.26: Pseudo nMOS Inverter

Equating currents gives

$$\frac{\beta_n}{2}[2(V_{DD} - V_{Tn})V_{OL} - V_{OL}^2] = \frac{\beta_p}{2}(V_{DD} - |V_{Tp}|)^2. \tag{4.65}$$

This is a quadratic equation for V_{OL} with a physical solution of

$$V_{OL} = (V_{DD} - V_{Tn}) - \sqrt{(V_{DD} - V_{Tn})^2 - \frac{\beta_p}{\beta_n}(V_{DD} - |V_{Tp}|)^2}. \tag{4.66}$$

The value of V_{OL} is thus set by the **driver-to-load ratio**

$$\frac{\beta_n}{\beta_p} = \frac{k'_n(W/L)_n}{k'_p(W/L)_p}, \tag{4.67}$$

where the nMOS transistor acts as the logic driver with β_n, while the pMOSFET is being used as an active load with β_p.

Ratioed logic places constraints on the layout and design. For an operational inverter, we must have

$$\frac{\beta_n}{\beta_p} > \left(\frac{V_{DD} - |V_{Tp}|}{V_{DD} - V_{Tn}}\right)^2 \tag{4.68}$$

to keep the square root term real; if this is not satisfied, then our original assumptions on the MOSFET conduction modes are not valid. In a process where $V_{Tn} = |V_{Tp}|$, this requires that

$$\frac{\beta_n}{\beta_p} > 1, \tag{4.69}$$

showing that relative device size is critical for setting DC characteristics. Moreover, a small V_{OL} requires that $(\beta_n/\beta_p) >> 1$. The design equation is obtained by rearranging the current expression to read

$$\frac{\beta_n}{\beta_p} = \frac{(V_{DD} - |V_{Tp}|)^2}{2(V_{DD} - V_{Tn})V_{OL} - V_{OL}^2}. \tag{4.70}$$

This provides the minimum driver-to-load ratio needed for a desired V_{OL}.

The pseudo-nMOS inverter provides the basis for implementing arbitrary logic functions using standard logic structuring. However, real estate and power tradeoffs should be examined before opting to use ratioed logic gates in CMOS. To understand this, consider a NAND2 gate which is constructed by adding another driver nMOS transistor in series with the inverter transistors. Figure 4.27 provides the circuit diagram. Denoting

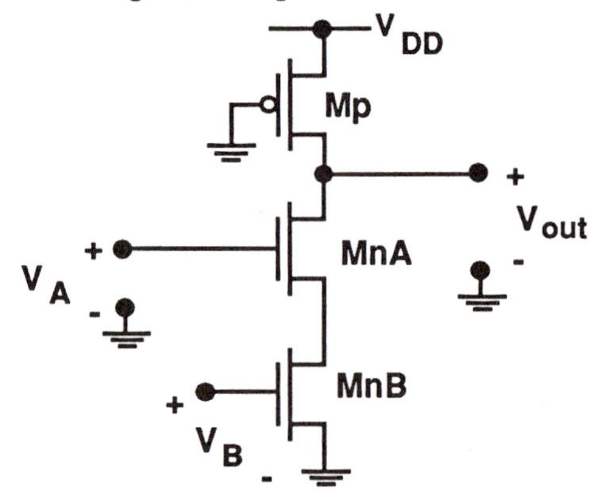

Figure 4.27: Pseudo-nMOS NAND2 Circuit

the devices by MnA and MnB and assuming that W is the same for both gives an effective β for the series devices of

$$\beta_{n,eff} = k'_n\left(\frac{W}{L_A + L_B}\right). \tag{4.71}$$

Thus, the individual aspect ratios $(W/L)_A$ and $(W/L)_B$ must be twice the size of the inverter for the same V_{OL} design. A NAND3 gate requires triple-area inverters, and so on, so that NAND logic should be avoided in this type of circuit.

Ratioed nMOS logic does allow efficient NOR gate design. In this case, an N-input NOR gate is constructed by placing N-nMOS logic transistors in parallel. Every nMOS transistor can be designed with the same (W/L) value determined for an inverter. The real estate tradeoff depends on whether this design with (N+1) total transistors in a ratioed configuration will result in a smaller area than an equivalent standard CMOS with 2N unratioed devices. Since ratioed logic always results in DC power dissipation, the tradeoff becomes more complex.

Another situation where pseudo-nMOS logic is a viable alternative is in structured arrays. Figure 4.28 illustrates a circuit which uses nMOS logic transistors in a NOR-based arrangement. Layout is simplified from both the use of arrayed logic and the elimination of the pMOS interconnects found in standard CMOS.

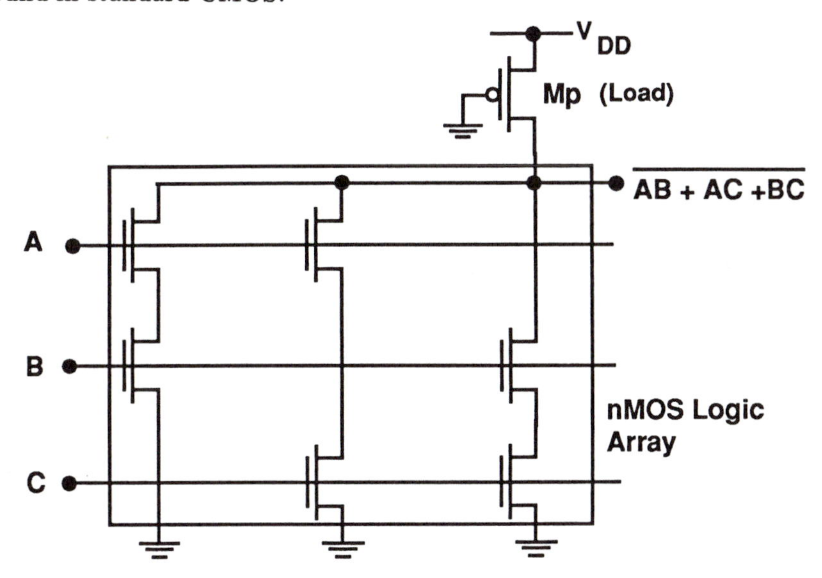

Figure 4.28: Pseudo-nMOS Logic Array

4.11.2 Pseudo-pMOS

A pseudo-pMOS design is obtained using pMOS transistors to form the logic and a single nMOSFET as a load. Figure 4.29 shows a basic inverter in pseudo- pMOS. Mp is now the logic driver, while Mn is used as a load device; Mn is always biased into conduction since its gate is connected with

$$V_{GSn} = V_{DD} \tag{4.72}$$

maintained.

Operation of the circuit is straightforward. Consider a logic 1 input voltage with V_{in} high. Logic transistor Mp is off, so that the output node has a direct conduction path to ground through the load transistor Mn giving

$$V_{OL} = 0 \text{ [V]}. \tag{4.73}$$

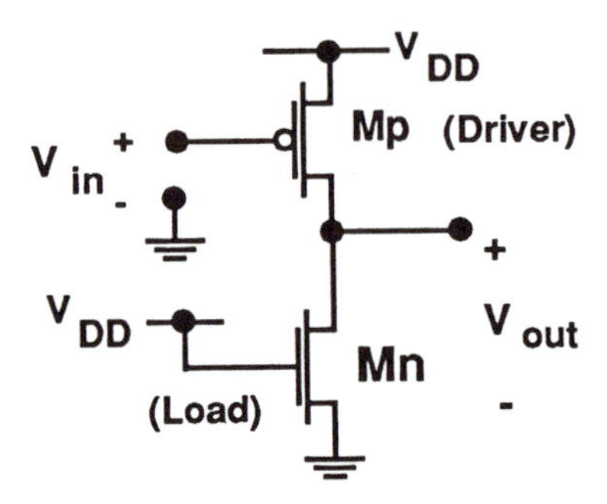

Figure 4.29: Pseudo-pMOS Inverter

If instead we use a logic 0 input voltage $V_{in} = 0$ [V], the output becomes ratioed and $V_{OH} < V_{DD}$.

The ratioed analysis proceeds as follows. An input voltage of $V_{in} = 0$ [V] places Mp into non-saturation; Mn will be saturated at this point. Equating currents and noting that $V_{SDp} = (V_{DD} - V_{out})$ gives

$$\frac{\beta_p}{2}[2(V_{DD}-|V_{Tp}|)(V_{DD}-V_{OH})-(V_{DD}-V_{OH})^2] = \frac{\beta_n}{2}(V_{DD}-V_{Tn})^2; \quad (4.74)$$

solving,

$$V_{OH} = |V_{Tp}| + \sqrt{(V_{DD} - |V_{Tp}|)^2 - \frac{\beta_n}{\beta_p}(V_{DD} - V_{Tn})^2}. \quad (4.75)$$

For this circuit, the driver-to-load ratio must satisfy

$$\frac{\beta_p}{\beta_n} >> 1 \quad (4.76)$$

to achieve V_{OH} close to V_{DD}. The design equation is

$$\frac{\beta_p}{\beta_n} = \frac{(V_{DD} - V_{Tn})^2}{2(V_{DD} - |V_{Tp}|)(V_{DD} - V_{OH}) - (V_{DD} - V_{OH})^2} \quad (4.77)$$

which establishes the minimum driver-to-load ratio for a desired V_{OH}.

Complex pseudo-pMOS gates are subject to the complementary limitations discussed for pseudo-nMOS circuits. In particular, positive-logic NAND gates are efficient with this approach since each pMOS logic transistor can be designed with the minimum inverter aspect ratio. NOR gates require excessive logic transistor sizes and should be avoided. Arrays are possible, but the slower response of pMOS logic usually precludes applying pseudo-pMOS logic to any great extent.

4.12 Flip-Flops

Static CMOS logic gates can be used to construct standard flip-flops. A NOR- based SR flip-flop is shown in Figure 4.30. The inputs are denoted by S (Set) and R (Reset), while the outputs are Q and $\overline{Q}$. The operation is straightforward. If both S and R are 0, then the output state is unchanged. Setting the circuit with $S = 1$ gives $(Q, \overline{Q}) = (1, 0)$. Resetting the output to $(Q, \overline{Q}) = (0, 1)$ is accomplished by an input $R = 1$. The input state $R = 1, S = 1$ is not used since it gives a contradiction at the output (they are not complements).

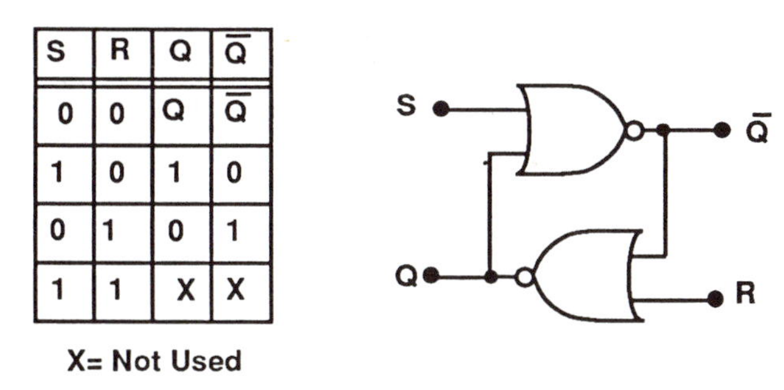

S	R	Q	$\overline{Q}$
0	0	Q	$\overline{Q}$
1	0	1	0
0	1	0	1
1	1	X	X

X= Not Used

Figure 4.30: NOR-Based SR Flip-Flop

Implementing the circuit in CMOS is accomplished by using two cross-coupled NOR gates as shown in Figure 4.31. The bistable properties are obtained using feedback.

An alternate implementation of an SR flip-flop is the NAND-based arrangement shown in Figure 4.32. In this case, an input condition of $S = 1, R = 1$ holds the output state while $S = 0, R = 0$ is not used. Since this uses series-connected nMOS transistors while the NOR circuit requires series-connected pMOSFETs, the NAND-based circuit will switch faster in equal real-estate designs. Aside from this difference, the performance of the two can be made to be approximately equal.

A D flip-flop (where **D** stands for *data* or *delay*) has a single input and is used to store a data bit. Adding an inverter and a clocking signal ϕ to an SR flip-flop give a D flip-flop as shown in Figure 4.33. The data state D is latched when $\phi = 1$; the circuit also eliminates the possibility of having an input state where both S and R are 1's.

It is possible to construct any flip-flop type (toggle, JK, etc.) using static logic circuits. In CMOS design, however, flip-flops which are based on the properties of bidirectional transmission-gates (TGs) provide superior performance. Transmission-gate logic is detailed in the next chapter, with

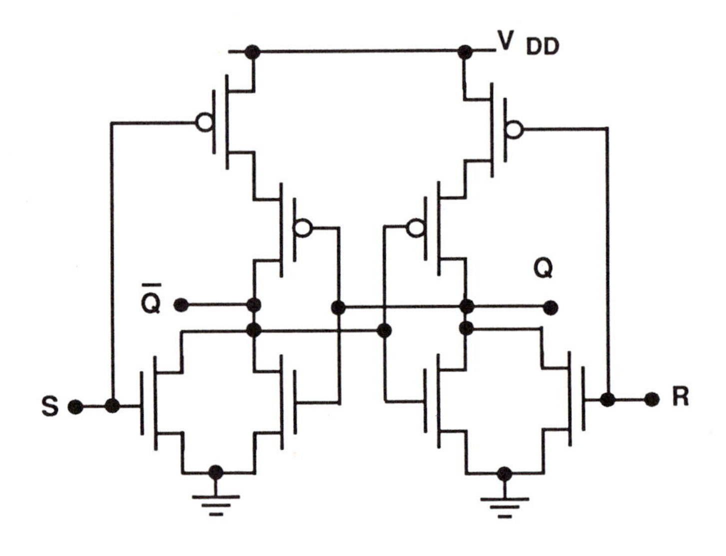

Figure 4.31: CMOS NOR SR Flip-Flop

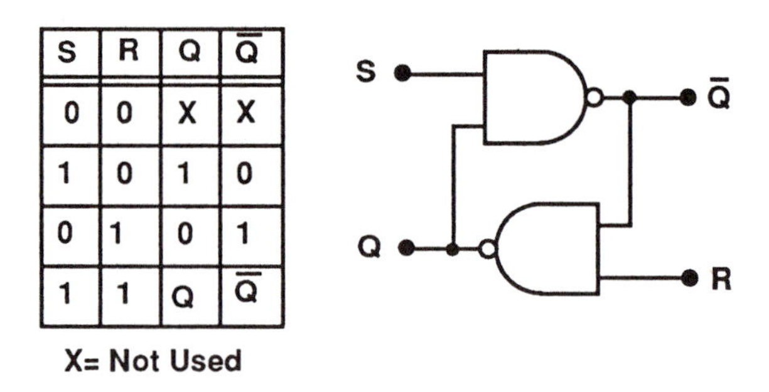

S	R	Q	$\overline{Q}$
0	0	X	X
1	0	1	0
0	1	0	1
1	1	Q	$\overline{Q}$

X= Not Used

Figure 4.32: NAND-Based SR Flip-Flop

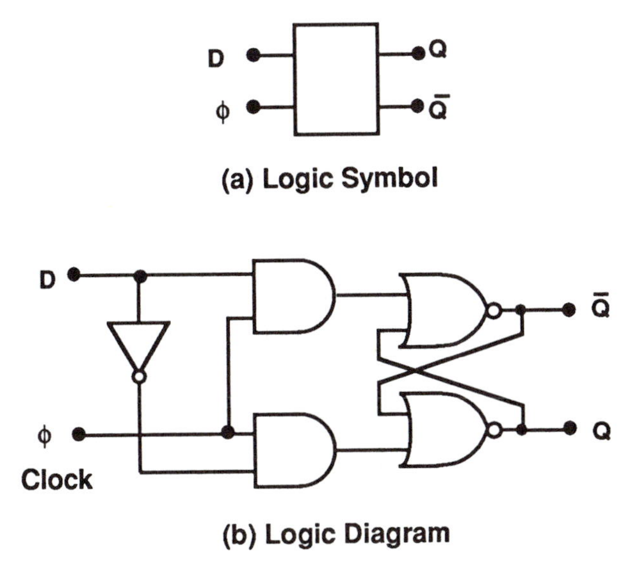

Figure 4.33: D Flip-flop

flip-flops examined in Section 5.5.

4.13 Schmitt Trigger

Schmitt trigger circuits are threshold switches. They are designed to be resistant to small variations of the input voltage, and require that the input voltage pass through trigger values to induce a change. Figure 4.34 illustrates the voltage transfer characteristic of an ideal inverting Schmitt trigger. Hysteresis is evident from the presence of the input trigger voltages: V^+ for increasing V_{In}, and V^- for decreasing V_{in}. The **hysteresis voltage** is defined by

$$V_H = V^+ - V^- \tag{4.78}$$

and gives a measure of the gate characteristics. Schmitt triggers are particularly nice for design start-up (delay) circuits, and can also be useful in noisy environments. A symmetric CMOS implementation is shown in the Figure 4.35. This is split into upper (pMOS) and lower (nMOS) segments which are functional complements of one another. The design of the circuit is accomplished by adjusting transistor aspect ratios as discussed below.

Circuit operation is most easily understood by examining the nMOS switching segment shown in Figure 4.36. MOSFET Mn2 is the main switching device, while Mn1 and Mn3 act as a feedback network which controls

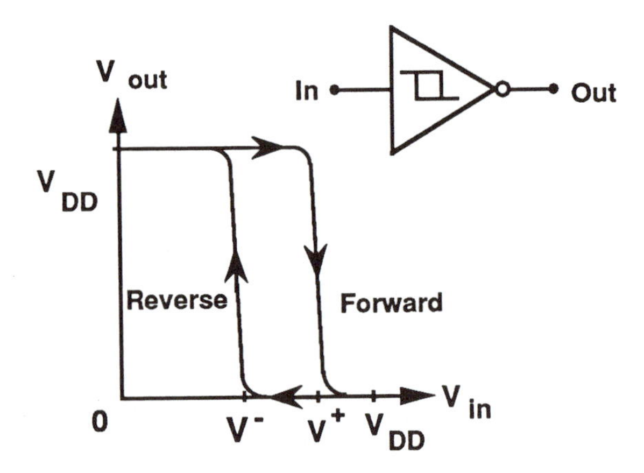

Figure 4.34: Schmitt Trigger

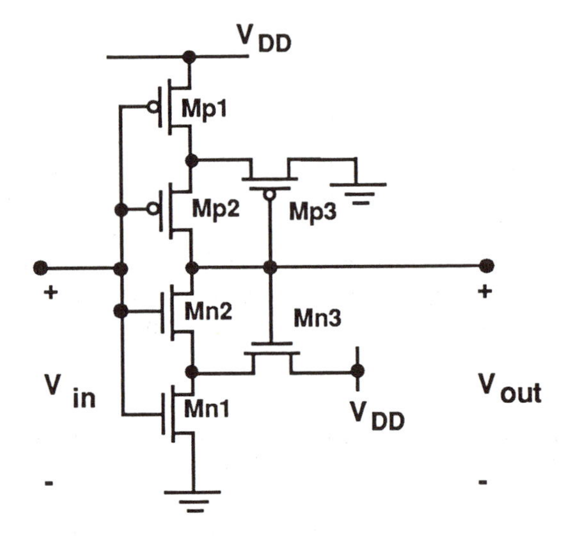

Figure 4.35: Symmetric Schmitt Trigger Circuit

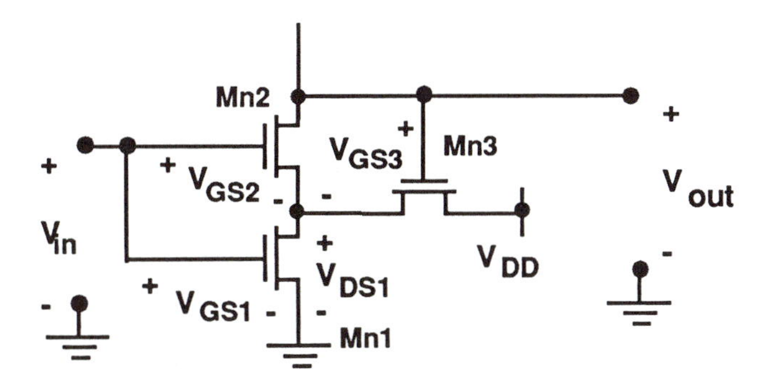

Figure 4.36: nMOS Trigger Circuit

the value of V^+. Assume that the input is set to $V_{in} = 0$ and then increased; all of the n-channel transistors are initially in cutoff. Conduction depends on the gate-source voltages

$$\begin{aligned} V_{GS1} &= V_{in} \\ V_{GS2} &= V_{in} - V_{DS1} \\ V_{GS3} &= V_{out} - V_{DS1}. \end{aligned} \tag{4.79}$$

Mn1 turns on when $V_{GS1} = V_{Tn1}$; however, Mn2 requires an input voltage of

$$V_{in} = V_{Tn2} + V_{DS1} \tag{4.80}$$

$$\equiv V^+ \tag{4.81}$$

to enter active conduction. The value of V_{DS1} is controlled by the MOSFET pair (Mn1,Mn3) acting as a feedback network. Increasing V_{in} decreases V_{DS1} until the critical switch condition is met to turn Mn2 on. When this point is reached, the output node has a discharge path to ground through Mn1 and Mn2, and the output voltage falls to zero.

The forward trigger voltage V^+ can be estimated by ignoring body bias effects. To turn on Mn2 requires a drain-source voltage of

$$V_{DS1} = V^+ - V_{Tn} \tag{4.82}$$

on MOSFET Mn1. This place Mn1 on the edge of saturation with a current of

$$I_1 = \frac{\beta_1}{2}(V^+ - V_{Tn})^2. \tag{4.83}$$

Mn3 is also saturated (since $V_{DS3} = V_{GS3}$) with a current

$$I_3 = \frac{\beta_3}{2}(V_{DD} - V^+)^2. \tag{4.84}$$

Equating currents $I_1 = I_3$ and rearranging gives

$$V^+ = \frac{V_{DD} + \sqrt{\beta_1/\beta_3}V_{Tn}}{1 + \sqrt{\beta_1/\beta_3}} \tag{4.85}$$

as the forward trigger voltage. Design is accomplished by the adjusting the ratio

$$\frac{\beta_1}{\beta_3} = \frac{(W/L)_1}{(W/L)_3} \tag{4.86}$$

for Mn1 and Mn3. The reverse trigger voltage V^- is obtained by a complementary analysis of the pMOS half of the circuit. The analysis gives

$$V^- = \frac{\sqrt{\beta_4/\beta_6}\,(V_{DD} - V_{Tn})}{1 + \sqrt{\beta_4/\beta_6}} \tag{4.87}$$

illustrating that the reverse trigger voltage is controlled by Mp4 and Mp6.

This CMOS circuit allows for designing a symmetric trigger where

$$V^+ = \frac{1}{2}V_{DD} + \Delta V \tag{4.88}$$

$$V^- = \frac{1}{2}V_{DD} - \Delta V; \tag{4.89}$$

in this case, the hysteresis voltage is given by

$$V_H = 2(\Delta V). \tag{4.90}$$

Defining the relative β value

$$\beta_r = \frac{\beta_1}{\beta_3} = \frac{\beta_4}{\beta_6} \tag{4.91}$$

and assuming that $V_{T0n} = |V_{T0p}| = V_T$ gives

$$\Delta V = \frac{V_{DD}(1 - \sqrt{\beta_r}\,) + 2\sqrt{\beta_r}\,V_T}{2(1 + \sqrt{\beta_r}\,)}. \tag{4.92}$$

Rearranging gives

$$\sqrt{\beta_r} = \frac{V_{DD} - 2(\Delta V)}{V_{DD} + 2(\Delta V) - 2V_T} \tag{4.93}$$

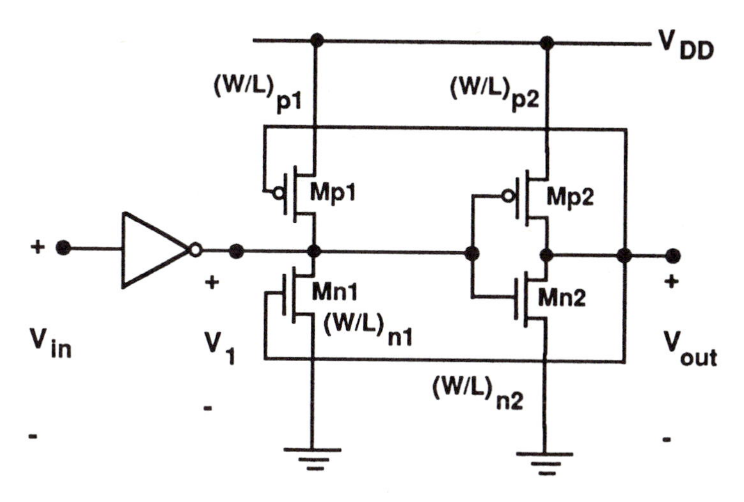

Figure 4.37: Alternate Schmitt Trigger

for a simple design equation.

A non-inverting CMOS Schmitt trigger circuit is shown in Figure 4.37. The switching voltages of this circuit are set by the transistor aspect ratios by using feedback to control MOSFETs Mp1 and Mn1.

The operation of the network is based on the VTC of a simple inverter. Recall that the ratio (β_n/β_p) determines V_{th}, i.e., the switching point. Increasing this ratio gives smaller V_{th}, while decreasing the ratio gives larger V_{th}. Schmitt trigger action is obtained by electronically altering the value of (β_n/β_p) depending on whether the voltage V_1 is increasing or decreasing.

Suppose that initially $V_{in} = 0$ so that $V_1 = V_{DD}$. The feedback loop biases Mp1 into conduction while Mn1 is in cutoff. The output voltage V_{out} is determined by the inverter made up of parallel-connected pMOSFETs Mp1 and Mp2, and the nMOS transistor Mn2. Thus, the effective transconductance ratio is

$$\beta_A = \frac{\beta_{n1}}{\beta_{p1} + \beta_{p2}}. \tag{4.94}$$

This is shown in Figure 4.38, and indicates an increased value of V_{th}. The value of β_A controls forward switching and, therefore, the value of V_a. If, on the other hand, the circuit is initially at $V_{in} = V_{DD}$ and $V_1 = 0$, then Mp1 is in cutoff while Mn1 is active. Now the inverter consists of parallel-connected nMOSFETs Mn1 and Mn2, but only a single pMOSFET Mp2.

The effective circuit ratio for this case is

$$\beta_B = \frac{\beta_{n1} + \beta_{n2}}{\beta_{p2}} \tag{4.95}$$

which gives a smaller V_b than in the forward switching case. Switching is

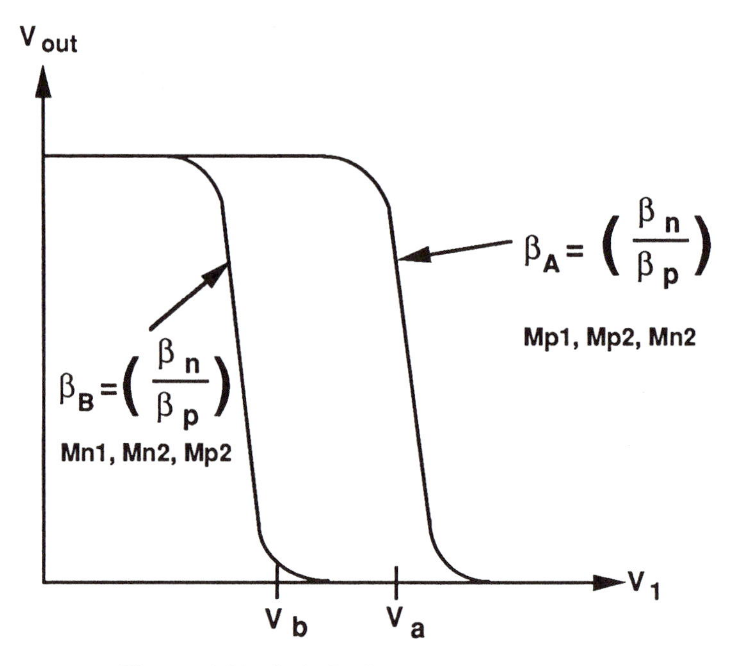

Figure 4.38: Switched Inverter Characteristics

used to set the values of V^+ and V^- in the Schmitt trigger circuit.

4.14 References

Static logic circuits are discussed in the books listed below.

[1] M. Annaratone, **Digital CMOS Circuit Design**, Kluwer Academic Publishers, Boston, 1986.

[2] E. D. Fabricius, **Introduction to VLSI Design**, McGraw-Hill, New York: 1990.

[3] N.K. Jha and S. Kundu, **Testing and Reliable Design of CMOS Circuits**, Kluwer Academic Publishers, Norwell, MA: 1990.

[4] A. Mukherjee, **Introduction to nMOS and CMOS VLSI Systems Design**, Prentice-Hall, Englewood Cliffs, NJ: 1986.

[5] M. Shoji, **CMOS Digital Circuit Technology**, Prentice-Hall, Englewood Cliffs, NJ, 1988.

[6] J.P. Uyemura, **Fundamentals of MOS Digital Integrated Circuits**, Addison-Wesley, Reading, MA: 1988.

[7] N. Weste and K. Eshraghian, **CMOS VLSI Design**, Addison-Wesley, Reading, MA: 1985.

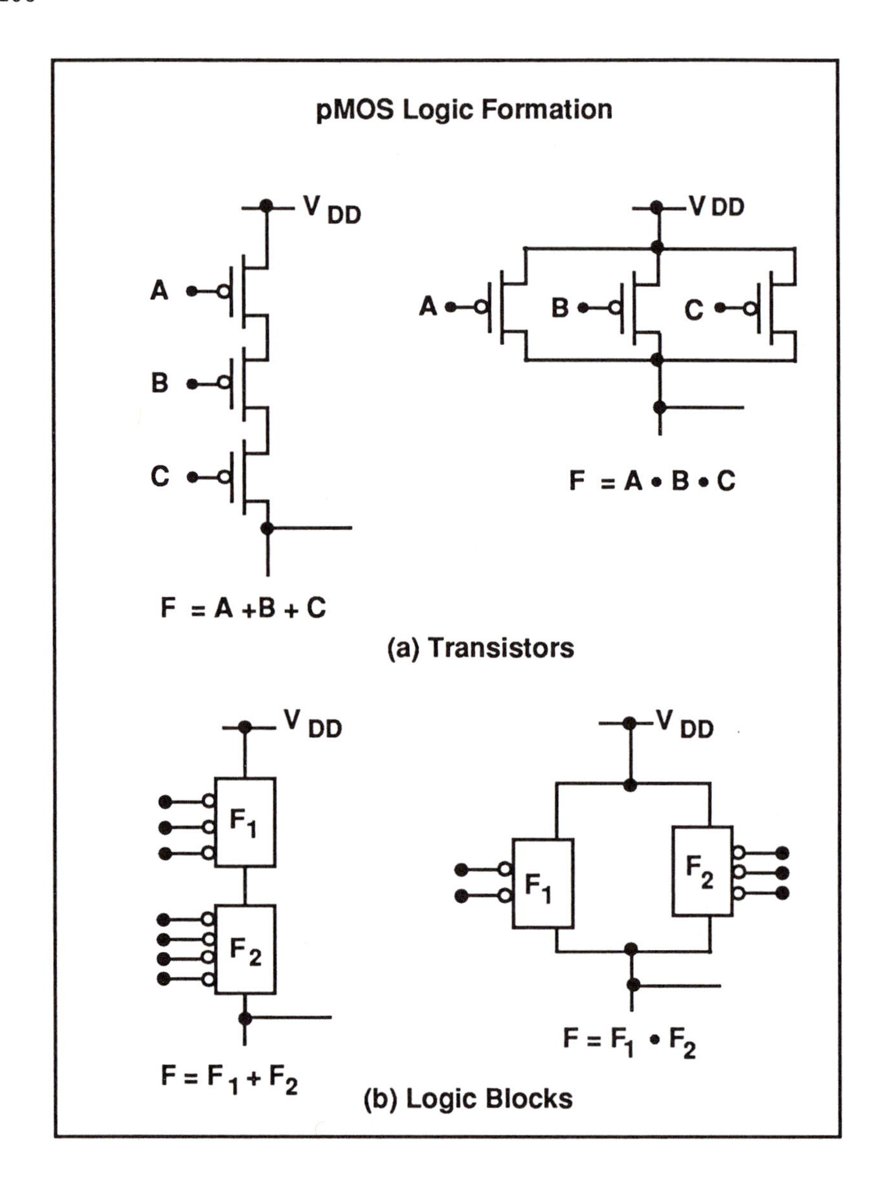
pMOS Logic Formation
V DD
A
B
C
F = A +B + C
V DD
A
B
C
F = A • B • C
(a) Transistors
V DD
F1
F2
F = F1 + F2
V DD
F1
F2
F = F1 • F2
(b) Logic Blocks

Chapter 5

CMOS Switch Logic

Conventional static CMOS logic circuits provide the foundation for many system designs. However, other circuit variations are possible which often allow greater flexibility or give better performance than that offered by standard CMOS. In this chapter we will examine the more popular alternatives to static circuits. All can be discussed (with varying degrees of success) using the concept of switched logic circuits, so we have chosen to group them together here.

This first part of this chapter covers the behavior, characterization, and application of CMOS transmission gates to digital logic networks. The latter sections discuss the advanced techniques of differential cascode voltage switch logic (DCVS logic or CVSL), complementary pass logic (CPL), and differential split-level (DSL) logic.

5.1 CMOS Transmission Gates

Transmission gates (TG) are used as primitive switching or logic elements in CMOS. They are used to implement basic switching schemes, and can be extended to provide advanced logic functions. Figure 5.1 shows two common TG symbols. The conduction path through the element is controlled by the complementary signal pair $(S, \overline{S})$ such that $S = 1$ gives $B = A$; when $S = 0$, a high-impedance state exists which blocks the logic flow. A transmission gate can thus be viewed as a voltage-controlled (or, logic-controlled) switch. CMOS transmission gates are made using parallel-connected nMOS and pMOS transistors as shown in Figure 5.2. The electrical behavior of a CMOS transmission gate is most easily understood by first studying the characteristics of individual MOSFET pass transistors, and then construct-

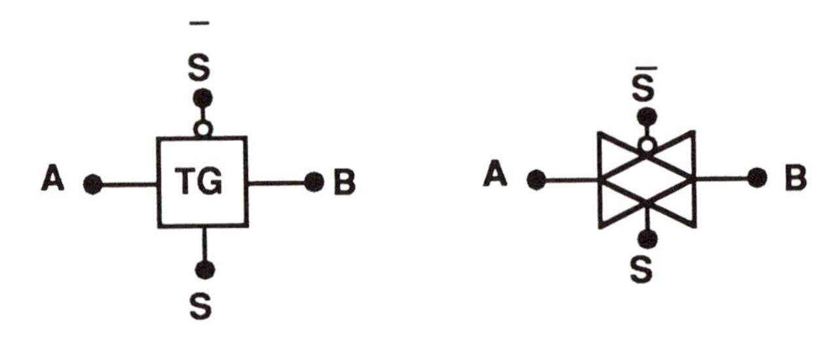

Figure 5.1: CMOS Transmission Gate Symbols

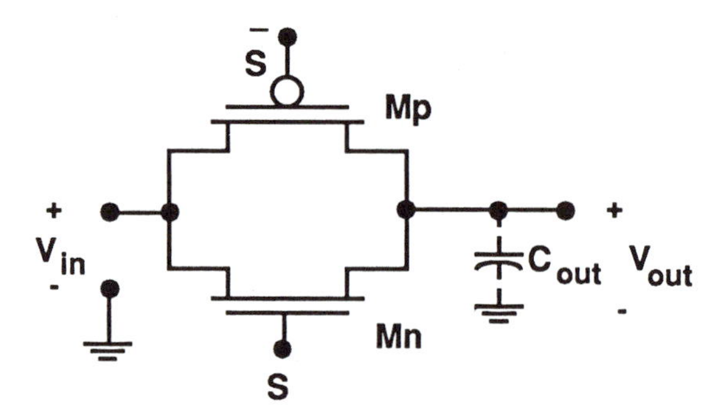

Figure 5.2: CMOS Transmission Gate

ing the parallel circuit.

5.1.1 nMOS Transmission Properties

Consider a single nMOSFET as shown in Figure 5.3. To analyze the pass

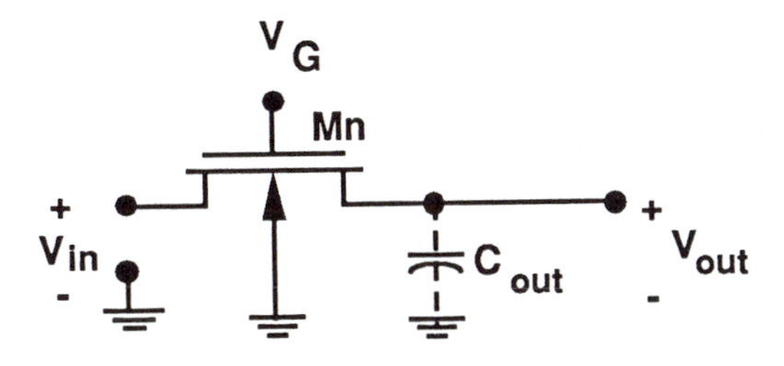

Figure 5.3: nMOS Pass Transistor

characteristics, we place a high voltage $V_G = V_{DD}$ onto the gate. Logic 1 transferral corresponds to having $V_{in} = V_{DD}$. Assuming that C_{out} is initially uncharged, the analysis of the current flow gives an output voltage $V_{out}(t)$ which increases according to[1]

$$V_{out}(t) = (V_{DD} - V_{Tn})\left[\frac{(t/\tau_{ch})}{1 + (t/\tau_{ch})}\right], \tag{5.1}$$

where

$$\tau_{ch} = \frac{2C_{out}}{\beta_n(V_{DD} - V_{Tn})} \tag{5.2}$$

is the **charging time constant**.

The limitation on using a single nMOS pass transistor is obvious. For time $t \to \infty$,

$$V_{out}(t) \to V_{max} = V_{DD} - V_{Tn}, \tag{5.3}$$

where

$$V_{Tn} = V_{T0n} + \gamma_n(\sqrt{2|\phi_{Fp}| + V_{max}} - \sqrt{2|\phi_{Fp}|}). \tag{5.4}$$

This expression illustrates what is called a **threshold voltage drop** from the input to the output. The problem originates from the fact that maintaining an inversion layer requires that V_{GS} be a minimum of one threshold voltage. An isolated nMOS pass transistor is said to only transmit a "weak" logic 1 state due to this voltage reduction.

[1] See reference [10] for details

Transferring a logic 0 through the transistor corresponds to the case where we set $V_{in} = 0$ and $V_{out}(t = 0) = V_{max}$. Analyzing the current flow gives the capacitor voltage as

$$V_{out}(t) = (V_{DD} - V_{Tn})\left[\frac{2e^{-(t/\tau_{dis})}}{1 + e^{-(t/\tau_{dis})}}\right], \tag{5.5}$$

where the **discharge time constant** is given by

$$\tau_{dis} = \frac{C_{out}}{\beta_n(V_{DD} - V_{Tn})}. \tag{5.6}$$

Since $V_{out}(t \to \infty) = 0$, the n-channel MOSFET does not have any problem passing a low voltage (logic 0 state).

Figure 5.4 shows the input-output characteristics of an nMOS pass transistor. Both the DC and transient properties affect the performance of

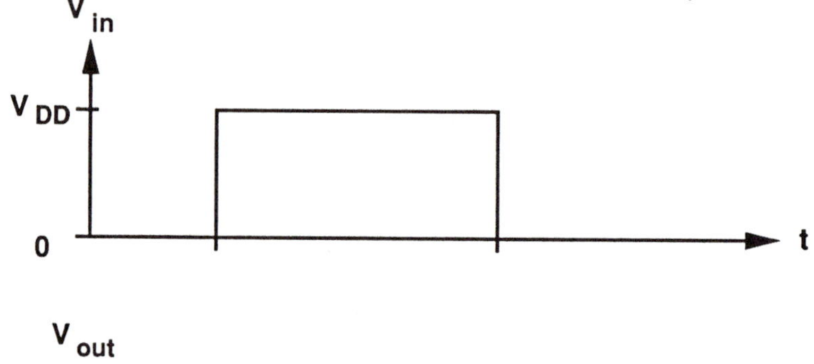

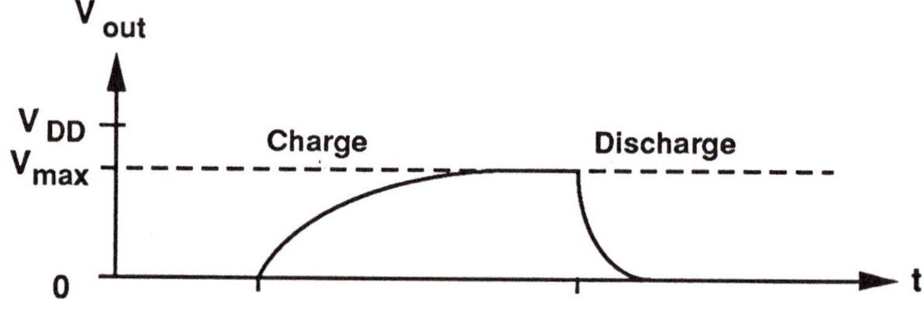

Figure 5.4: nMOS Pass Characteristics

nMOS-only switching.

5.1.2 pMOS Transmission Characteristics

Figure 5.5 shows a pMOSFET used as a controlled switching device. Logic 1 transmission is described by setting $V_{in} = V_{DD}$ with the gate voltage at

$V_G = 0$ [V]. A moments reflection shows that this situation is identical to the problem of finding t_{LH} in a CMOS inverter. Assuming that the output voltage is initially at a value $V_{out}(t = 0) = |V_{Tp}|$ (as justified below) gives the charging as

$$V_{out}(t) = V_{DD} - (V_{DD} - |V_{Tp}|)\left[\frac{2e^{(-t/\tau_{ch})}}{1 + e^{(-t/\tau_{ch})}}\right], \tag{5.7}$$

where

$$\tau_{ch} = \frac{C_{out}}{\beta_p(V_{DD} - |V_{Tp}|)} \tag{5.8}$$

is the charging time constant for the pMOS switch. This shows that $V_{out} \to V_{DD}$ implying that an isolated pMOS transistor passes a "strong" logic 1 voltage without any problems.

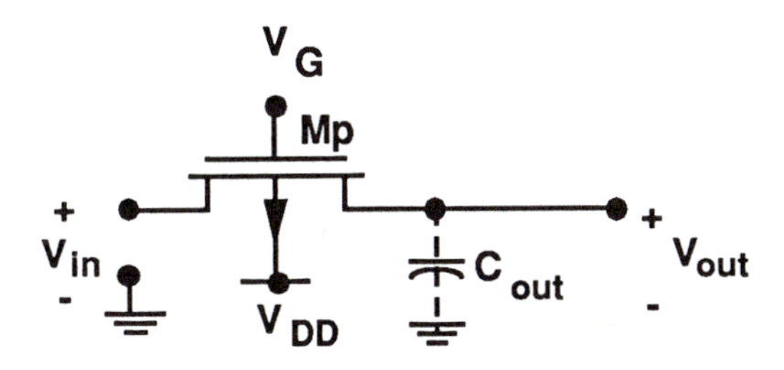

Figure 5.5: pMOS Pass Transistor

Transmitting a logic 0 state, on the other hand, does lead to a threshold voltage loss problem. Setting $V_{in} = 0$ and $V_{out}(t = 0) = V_{DD}$ in the circuit shows that

$$V_{SGp} = V_{out} = V_{SDp}, \tag{5.9}$$

so that Mp remains saturated during the discharge event. Solving the current flow equation yields an output voltage of the form

$$V_{out}(t) = |V_{Tp}| + \frac{(V_{DD} - |V_{Tp}|)}{1 + (V_{DD} - |V_{Tp}|)(t/2\tau_{ch})}. \tag{5.10}$$

This shows that for large times t,

$$V_{out}(t \to \infty) = |V_{Tp}| = V_{min}, \tag{5.11}$$

so that it is not possible to discharge the capacitor to 0 volts through the p-channel transistor. Including body-bias effects gives the complete expression

$$V_{min} = |V_{T0p} - \gamma_p(\sqrt{2\phi_{Fn} + (V_{DD} - V_{min})} - \sqrt{2\phi_{Fn}})|; \tag{5.12}$$

the pMOS transistor is thus said to pass a "weak" logic 0.

The curves in Figure 5.6 shows the transfer characteristics of the pMOS transistor. Once again note that the pMOSFET is the voltage complement of the nMOSFET.

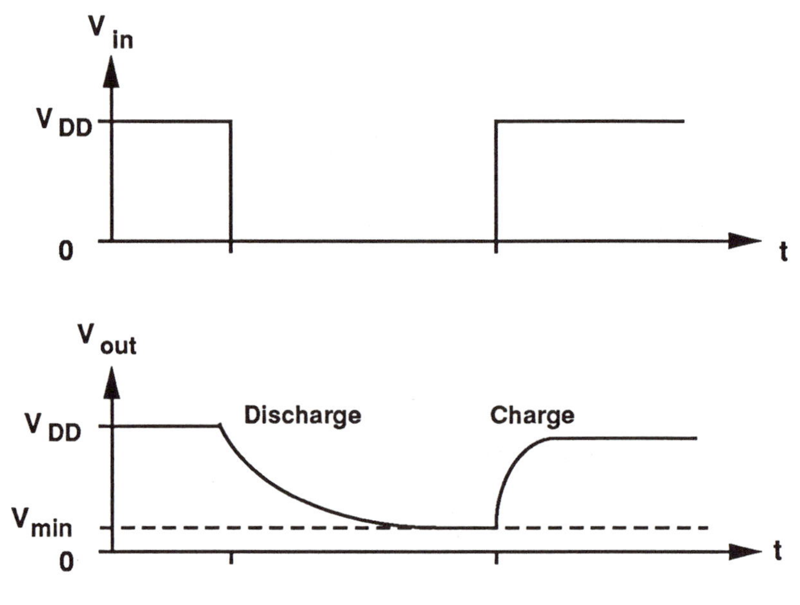

Figure 5.6: pMOS Pass Characteristics

5.2 Transmission Gate Model

A CMOS TG uses both an nMOS and a pMOS transistor in parallel to achieve complementary transmission. Both logic 0 and logic 1 transmitted states are strong, i.e., the CMOS TG can pass the entire range 0 to V_{DD} since at least one MOSFET is always conducting. The structure eliminates the problem of threshold voltage losses which occur when only a single transistor is used making it compatible with the rail-to-rail logic swings which characterize CMOS circuits.

Analysis of transmission gate properties is straightforward. Consider the basic circuit in Figure 5.7. V_{in} represents the input voltage, while V_{out} is the output across the load capacitor C_{out}. When $S = 1$, the TG acts like a closed bidirectional switch. If $V_{in} = V_{DD}$, capacitor C_{out} will charge through the transmission gates. On the other hand, an input voltage of $V_{in} = 0$ will allow the capacitor to discharge to a zero- voltage state. Table

5.1 summarizes the voltage transmission properties for single-polarity pass gates and the CMOS TG.

Logic Level	nMOS	pMOS	CMOS
Logic 0	0	$\|V_{Tp}\|$	0
Logic 1	$(V_{DD} - V_{Tn})$	V_{DD}	V_{DD}

Table 5.1: Transmission Characteristics

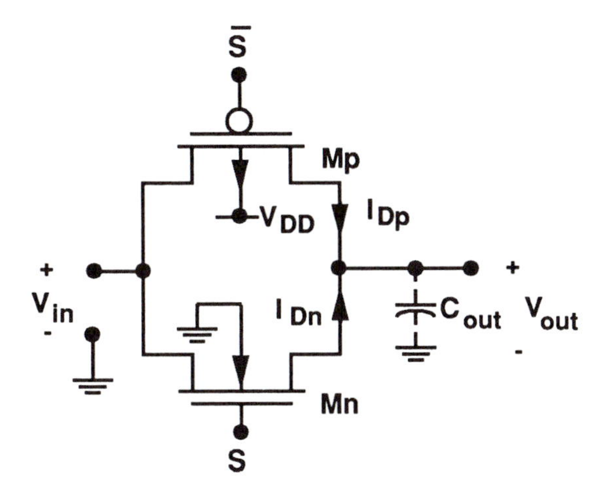

Figure 5.7: CMOS Transmission Gate

A detailed treatment of the charging and discharge problem requires that we sum the currents through both transistors. By inspection,

$$I_{Dn} + I_{Dp} = C_{out}\frac{dV_{out}}{dt}, \tag{5.13}$$

where the equations for I_{Dn} and I_{Dp} depend on the voltages. As indicated by the individual analyses, both saturated and non-saturated modes can occur. Figure 5.8 shows the MOSFET states for a logic 1 transfer. Although we can study the circuit analytically, it is easier to either use a simple model or implement a full computer simulation. Modelling is very useful for understanding the important characteristics and for providing initial design considerations.

The simplest model for a TG is the resistor-switch combination illustrated in Figure 5.9. Logic transfer is controlled by $(S, \overline{S})$ such that (1,0)

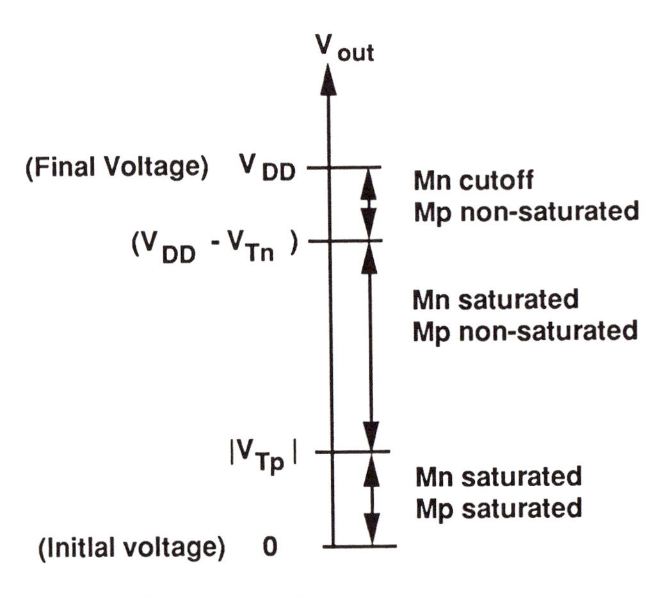

Figure 5.8: MOSFET Operational States

gives transmission, while $(0,1)$ blocks the data path. Let us denote the equivalent transmission gate resistance by R_{TG}. Transfer of a logic 1 state is equivalent to the capacitor charging through a resistor as described by

$$V_{out}(t) = V_{DD}[1 - e^{-(t/\tau_{TG})}], \tag{5.14}$$

where

$$\tau_{TG} = R_{TG}C_{out} \tag{5.15}$$

is the time constant. Similarly, a logic 0 transfer corresponds to the discharging the capacitor such that the voltage is given by

$$V_{out}(t) = V_{DD}e^{-(t/\tau_{TG})}. \tag{5.16}$$

The input and output voltage plots are shown in Figure 5.10. The transfer time is limited by the time constant, which is in turn set by the geometry and layout.

5.2.1 Equivalent Resistance

The equivalent transmission gate resistance R_{TG} can be defined by

$$R_{TG} = \frac{V_{TG}}{I_{Dn} + I_{Dp}} \tag{5.17}$$

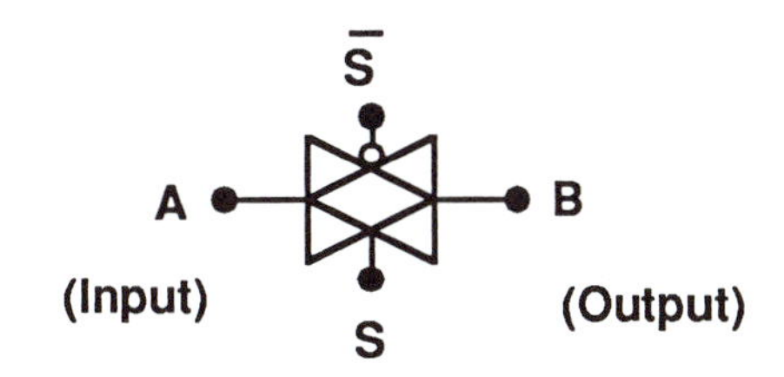

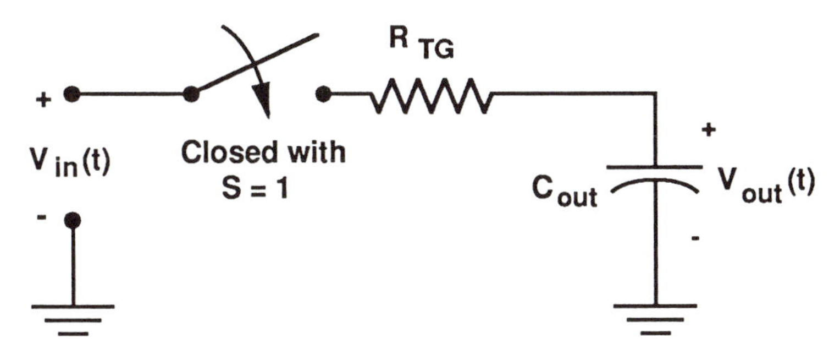

Figure 5.9: TG Resistor-Switch Model

where V_{TG} is the voltage. Since the MOSFETs are nonlinear, R_{TG} is itself a nonlinear function of the voltages. Although the functional dependence can be determined, it is easier to approximate R_{TG} as a constant. This is sufficient for initial design or analysis estimates; computer simulations may be used for greater accuracy.

Let R_n and R_p respectively be the equivalent nMOS and pMOS drain-source resistances. The total TG resistance may be viewed as R_n and R_p in parallel: $R_{TG} = (R_n \parallel R_p)$. When a MOSFET is in cutoff, its equivalent resistance is infinite. Figure 5.11 illustrates the behavior of R_n, R_p, and R_{TG} as a function of the TG voltage during a logic 1 transfer event. Although R_{TG} exhibits large variations, we can use an average or a maximum value for initial circuit modeling.

Different approaches may be used to estimate R_{TG}. Average MOSFET resistances can be calculated using

$$\begin{aligned} R_n &= \frac{1}{\beta_n(V_{DD} - V_{Tn})}, \\ R_p &= \frac{1}{\beta_p(V_{DD} - |V_{Tp}|)}, \end{aligned} \tag{5.18}$$

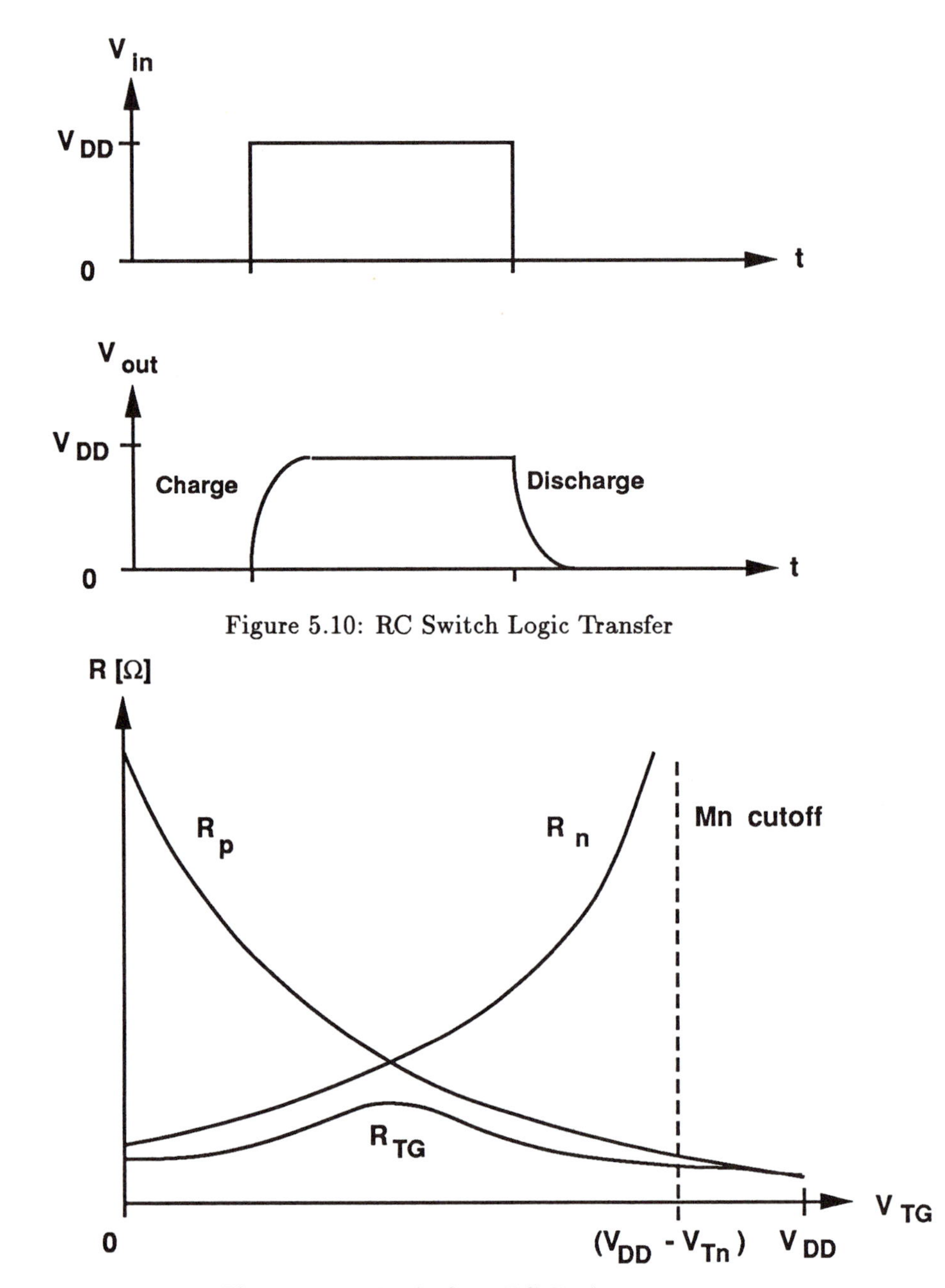

Figure 5.10: RC Switch Logic Transfer

Figure 5.11: Equivalent TG Resistances

so that the transmission gate conductance $G_{TG} = (1/R_{TG})$ is given by

$$G_{TG} = k_n'(\frac{W}{L})_n(V_{DD} - V_{Tn}) + k_p'(\frac{W}{L})_p(V_{DD} - |V_{Tp}|). \tag{5.19}$$

This is generally adequate for a first-order hand calculation. We see that a small R_{TG} requires large β's, i.e., the resistance is controlled by the aspect ratios $(W/L)_n$ and $(W/L)_p$. This is, of course, expected from the DC characteristics. The simplest layouts will employ $(W/L)_n = (W/L)_p$, giving unequal values of R_n and R_p.

5.2.2 Load Capacitance

The value of C_{out} is sensitive to the TG layout, the interconnect geometry, and the fanout circuitry. Since $\tau_{TG} = R_{TG}C_{out}$, reducing the output capacitance increases the switching speed of the circuit. The contributions illustrated in Figure 5.12 are combined to give the approximate value

$$C_{out} \simeq (C_{GDn} + C_{GSp}) + K(V_\ell)[C_{DBn} + C_{SBp}] + C_{line} + C_{in} \tag{5.20}$$

where C_{in} represents the input capacitance of the following stage. Since $(C_{line} + C_{in})$ is usually a constant MOS-type capacitance, the important aspects of TG design centers around an examination of the MOSFET parasitics.

5.3 Layout Considerations

Design tradeoffs are not apparent until the layout is considered. Figure 5.13 shows a basic transmission gate with the important dimensions labeled. To see the overall problem, recall that large values of (W/L) reduce the resistance R_{TG}. However, this implies that W_n and W_p are large. Using the basic capacitance analysis shows that the MOSFET parasitic capacitances C_o, C_{GS}, C_{GD}, C_{DB}, and C_{SB} are all proportional to W. Increasing W_n and W_p decreases the resistance, but increases the capacitance.

The transient performance is strongly dependent on the line and input capacitance $(C_{line} + C_{in})$. If

$$C_{\mathrm{MOSFET}} >> (C_{line} + C_{in}), \tag{5.21}$$

then the transistors are basically charging and discharging their own parasitic capacitances. On the other hand, if the layout gives

$$(C_{line} + C_{in}) >> C_{MOSFET} \tag{5.22}$$

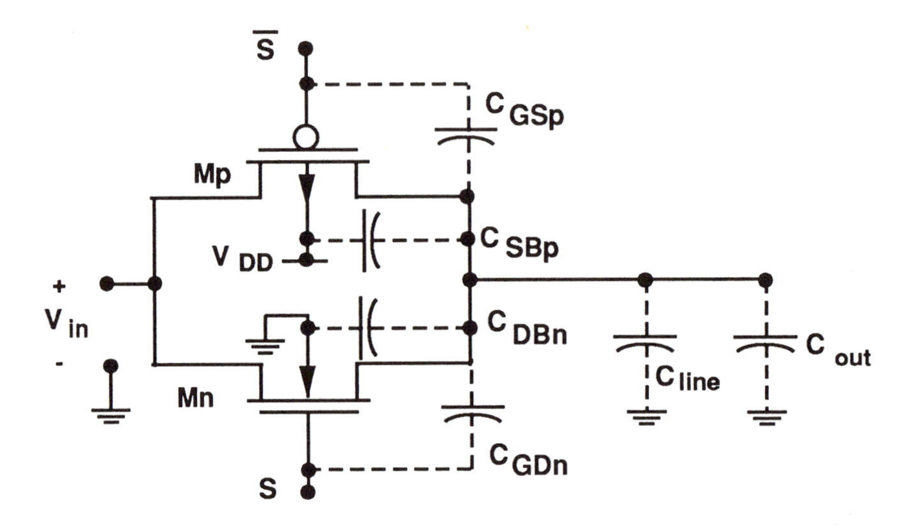

Figure 5.12: TG Output Capacitance

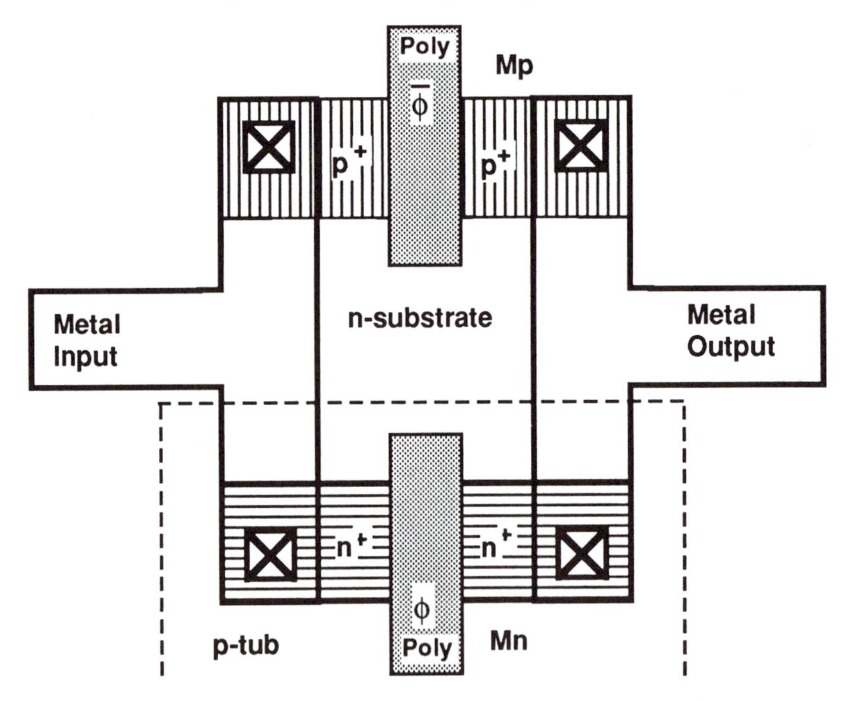

Figure 5.13: Basic TG Layout

then the TG circuit must be large enough to drive the next stage. Although we can equalize β values by adjusting the device sizes, the formula $R_{TG} = (R_n \parallel R_p)$ shows that neither device dominates the circuit. Consequently, there is no reason to use unequal aspect ratios to achieve $\beta_n = \beta_p$ as in the inverter circuit. Instead, equal size transistors are the most common in TG layouts since they are the simplest.

5.4 TG-Based Switch Logic Gates

Transmission gate logic is based on data path control. When the control signals are at $(S, \overline{S}) = (1, 0)$, the TG conducts current in both directions. At the basic level, the logic in Figure 5.14 can be summarized by writing

$$S = 1: \quad B \leftarrow A; \tag{5.23}$$

this notation implies that state A is transferred to B when $S = 1$. When the control states are $(S, \overline{S}) = (0, 1)$, the TG is OFF (a non- conducting state) and the statement must be modified. CMOS nodes tend to be primarily capacitive, indicating the possibility of charge storage and memory. In realistic chip structures, leakage currents exist which drain charge from the storage nodes. Although dynamic logic techniques can work around this limitation, static logic must be configured to avoid ever setting up an isolated node. Violating this rule can lead to significant circuit-level logic errors.

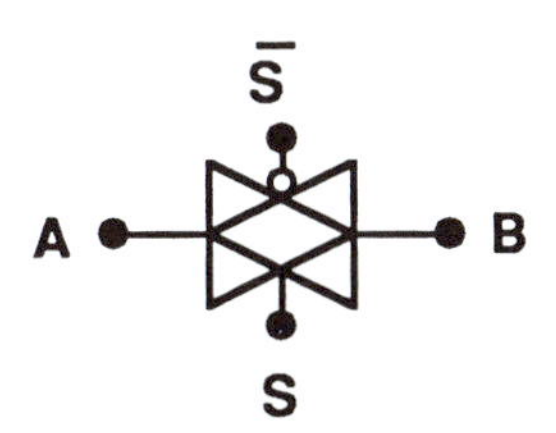

Figure 5.14: Transmission-Gate Logic

Static TG-based logic is achieved by using switched arrays to implement logical functions. Gate control is obtained from either an external source or the input variables themselves. We will present a few of the more common TG-based logic gates here.

5.4.1 Path Selector

A 2-input path selector can be created using the circuit shown in Figure 5.15. In this scheme, the input data lines A and B are controlled by the select variable S such that

$$F = A\overline{S} + BS. \tag{5.24}$$

This can be used as a data path selector as seen by writing

$$\begin{aligned} S = 1 : F &= A, \\ S = 0 : F &= B. \end{aligned} \tag{5.25}$$

This is a special case of a multiplexer network.

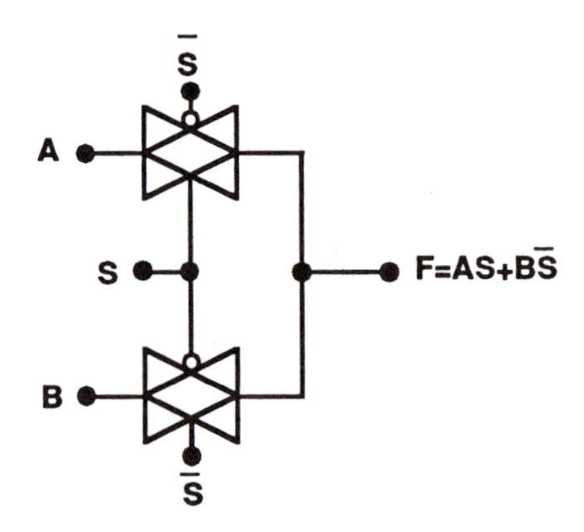

Figure 5.15: 2-Input TG Path Selector

5.4.2 OR Gate

Transmission gate logic directly yields the logical OR function using the circuit shown in Figure 5.16. Input variable A and its complement $\overline{A}$ are used to control both the pMOS pass transistor Mp and the TG. The upper branch transmits when $A = 1$, while the lower TG circuit propagates B to the output when $A = 0$. Since the pMOS transistor only passes a high voltage corresponding a logic value of $A = 1$, the weak logic 0 characteristics are not important.

The OR function itself results from the absorption theorem. Logically, the upper branch gives an output of A, while the lower branch gives an

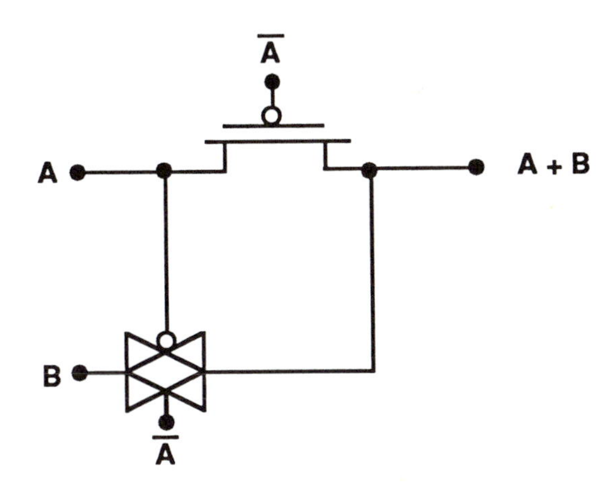

Figure 5.16: TG-Logic OR Gate

output of $\overline{A}B$. The effect of combining the outputs is thus

$$\begin{aligned} F &= A+\overline{A}B \\ &= A+B \end{aligned} \tag{5.26}$$

providing the OR function as advertised.

5.4.3 XOR and Equivalence

The Exclusive-OR (XOR) and Equivalence functions can be implemented by using an input variable to control the transmission gates as shown in Figure 5.17. Recall from Section 4.8 that the XOR function is described by

$$\begin{aligned} F1 &= A \oplus B \\ &= A\overline{B}+\overline{A}B \end{aligned} \tag{5.27}$$

while the Equivalence function (also called the XNOR) is defined by

$$\begin{aligned} F2 &= A \odot B \\ &= AB+\overline{A}\,\overline{B}. \end{aligned} \tag{5.28}$$

The XOR function produces a logic 1 output if either $A=1$ or $B=1$ (but not both); the Equivalence function is 1 if $A=B$ is true. Both the XOR

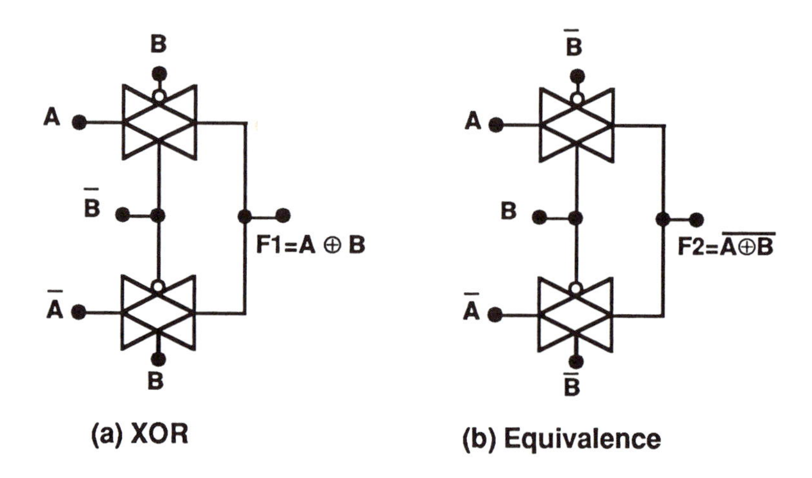

Figure 5.17: TG-Based XOR and Equivalence Functions

and equivalence function are described by sum of product terms which can be directly implemented by TG logic.

An alternate Equivalence function circuit is shown in Figure 5.18. Output control is accomplished by the complementary MOSFET pair (Mp,Mn) which acts like an analog push-pull circuit. Note, however, that the output node does not receive voltage support from the power supply. Rather, the input variable B (and its complement) must supply all necessary current to drive the output capacitance. Reversing B and $\overline{B}$ gives the XOR function directly.

5.4.4 Adders

Consider two logic variables A_0 and B_0. A half-adder produces the sum S_0 and carry C_0 by means of

$$S_0 = A_0 \oplus B_0, \tag{5.29}$$

$$C_0 = A_0 B_0. \tag{5.30}$$

The logic symbol and truth table for a half-adder are shown in Figure 5.19 for future reference. Transmission gate logic allows a direct implementation of the sum calculation using the XOR functions discussed above. The carry bit is most easily generated using a static gate, giving the circuit shown in Figure 5.20.

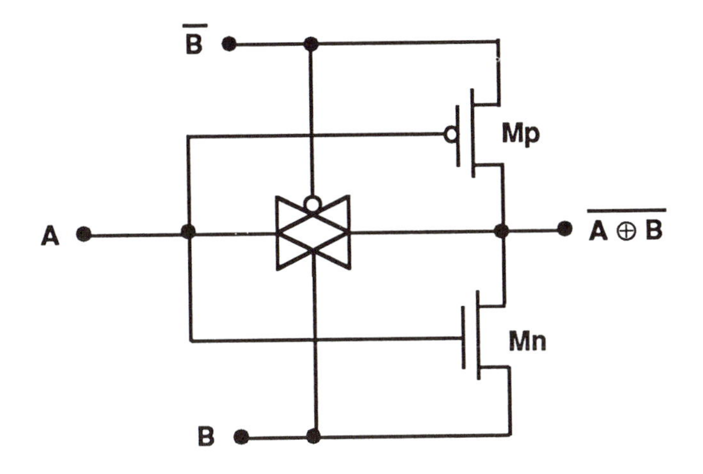

Figure 5.18: Alternate Equivalence Function Circuit

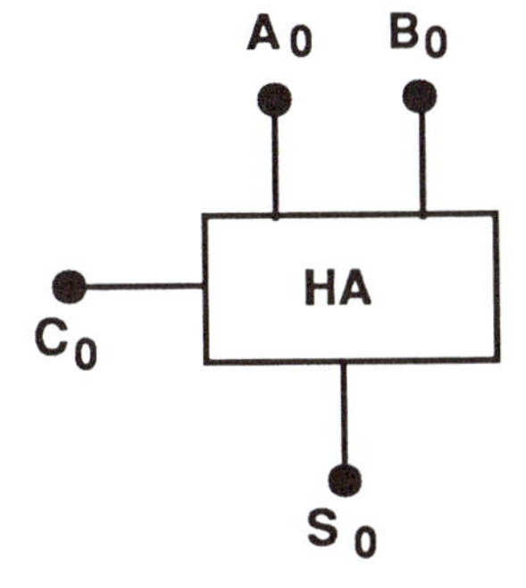

A_0	B_0	S_0	C_0
0	0	0	0
0	1	1	0
1	0	1	0
1	1	0	1

Figure 5.19: Half-Adder Logic Symbol

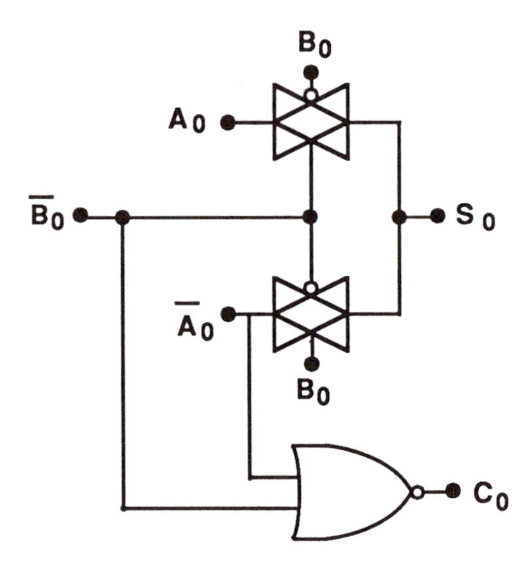

Figure 5.20: TG Half-Adder

A full-adder takes two bits (A_n, B_n) with an input carry C_{n-1} and generates

$$\begin{aligned} S_n &= (A_n \oplus B_n)\overline{C}_{n-1} + \overline{(A_n \oplus B_n)}C_{n-1} \\ C_n &= (A_n \oplus B_n)C_{n-1} + \overline{(A_n \oplus B_n)}A_n \end{aligned} \quad (5.31)$$

as an output. The symmetry of the logic expressions can be directly implemented into a balanced TG circuit as shown in Figure 5.21. Symmetric structuring has the advantage of equalizing the propagation delay for the outputs.

5.5 Latches and Flip-Flops

Transmission gates can be combined with static logic circuits to create latches and flip-flops. These can be clocked or event-driven, and many variations are possible. TG-based flip-flops are very common in CMOS SSI and MSI circuits, and are also popular in ASIC and full-custom designs.

5.5.1 Basic Latch

Data latches can be built using cross-coupled inverters with TG path controllers as shown in Figure 5.22. In this circuit, TG1 allows access to the

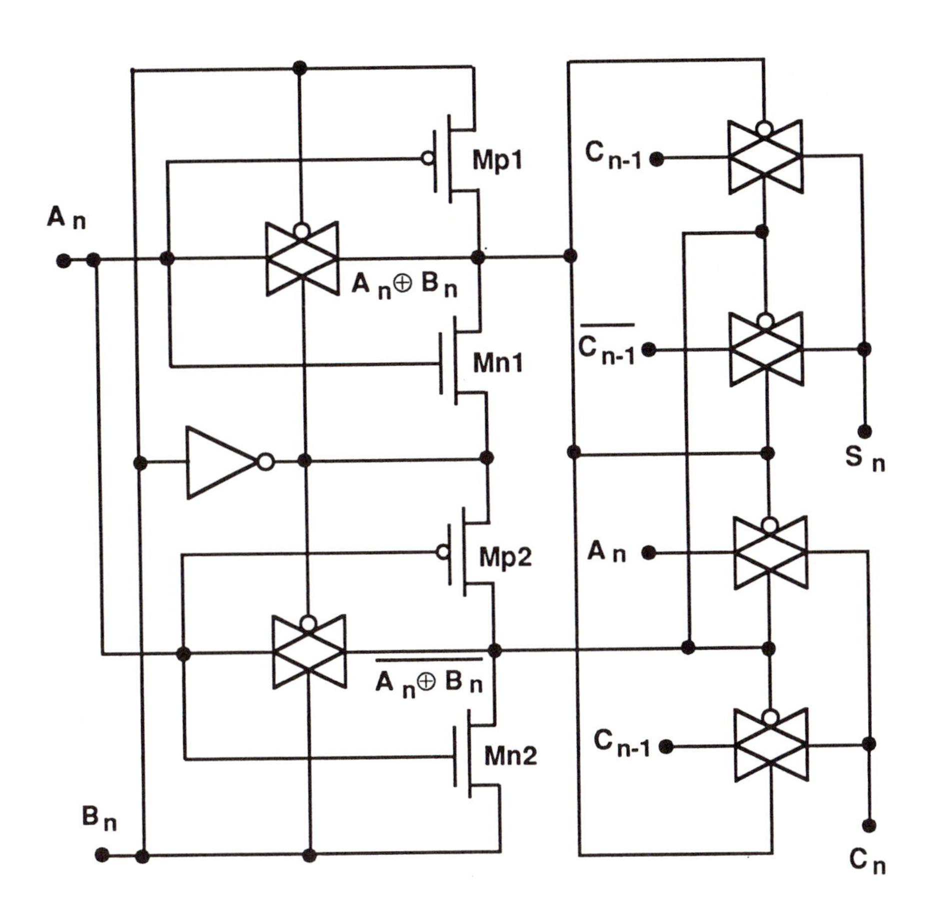

Figure 5.21: Full-Adder Using TGs

latch while TG2 is in the feedback loop. The load signal LD (and its complement) control the gates such that only one TG is conducting at a time. When $LD = 1$, TG1 allows data to enter the latch to the input of inverter 1; during this time, the feedback path through TG2 is open. When the load signal is switched to $LD = 0$, the latch is in a hold state where TG1 is open while TG2 provides feedback. Synchronized loading can be accomplished by

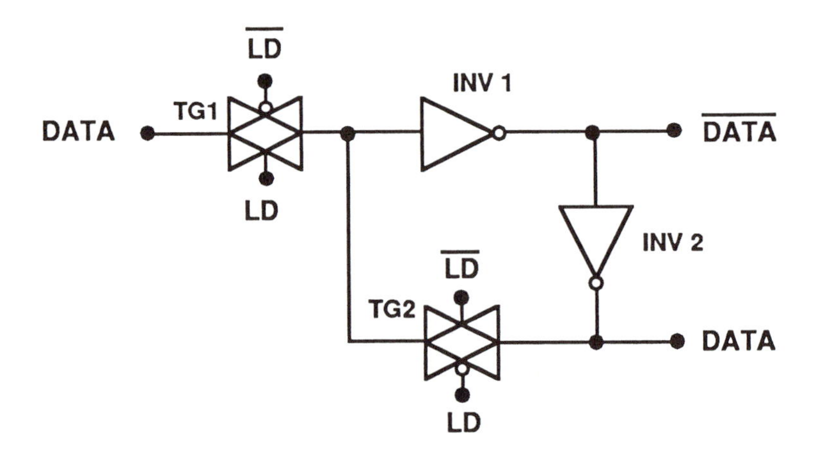

Figure 5.22: Basic TG Latch

ANDing LD with a clock pulse signal CLK so that $(LD \cdot CLK, \overline{LD \cdot CLK})$ are used to control the transmission gates.

5.5.2 D Flip-Flop

A TG-based clocked D flip-flop (DFF) is shown in Figure 5.23. The circuit is designed to load when the clock pulse is at a value of $CLK = 0$. Data input D is valid during this time and provides the input to INV1. When $CLK = 1$, the latch made up of INV1 and INV2 holds the value of D, while $\overline{D}$ is transferred to the second inverter pair consisting of INV3 and INV4. Both complemented and uncomplemented values are available. Since the circuit is made up of cascaded latches, the latency time depends on the frequency of the clocking pulse.

A variation of the DFF is shown in Figure 5.24. This provides a **clear** input CLR which resets the circuit when $CLR = 1$. Although not shown in the circuit diagrams, clock buffers may be added to either circuit if needed. Also, a SET input can be included with minor additions to the basic logic.

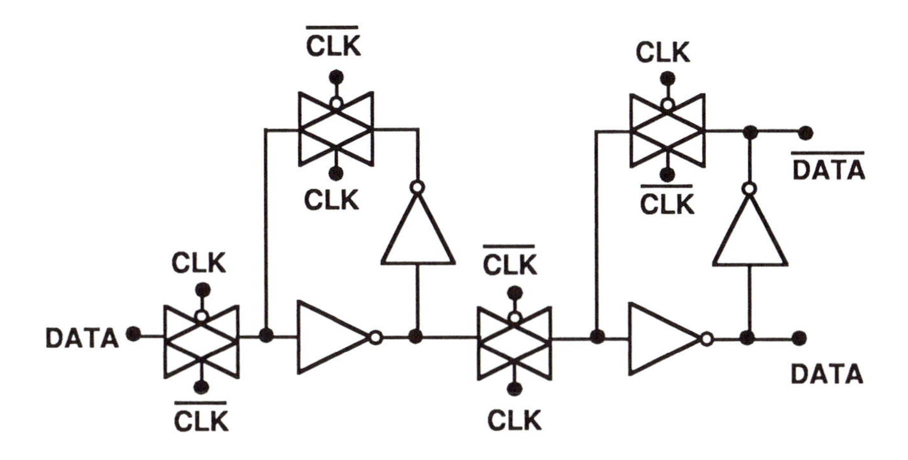

Figure 5.23: D Flip-Flop

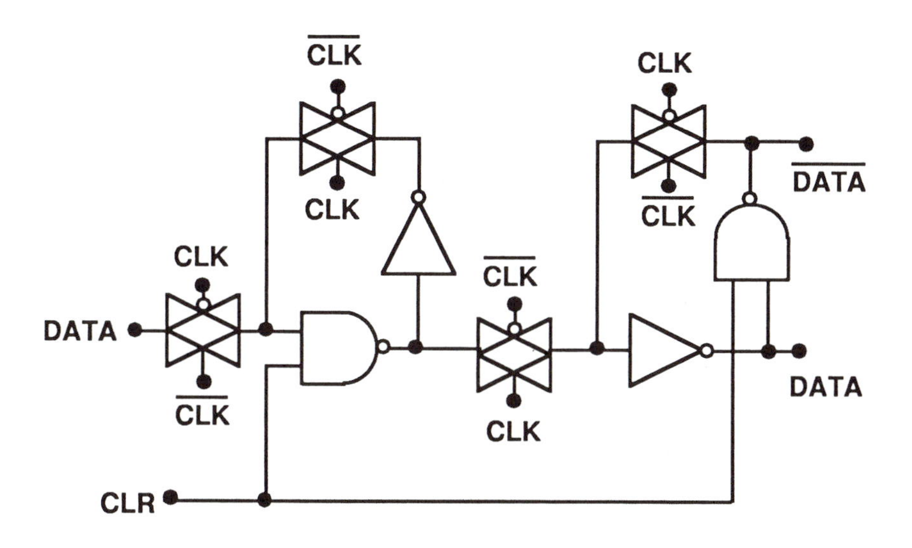

Figure 5.24: DFF with Clear

5.5.3 Toggle Flip-Flop

The toggle flip-flop (TFF) circuit in Figure 5.25 is constructed by using feedback on the cascaded latch. Operation is controlled by the clock signals CLK and $\overline{CLK}$; SET allows an initial state set.

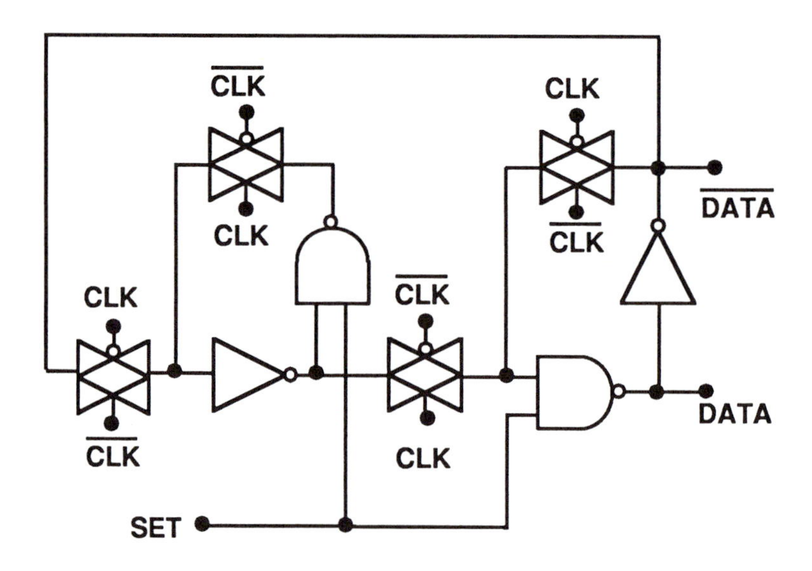

Figure 5.25: Toggle Flip-Flop

5.5.4 JK Flip-Flop

JK flip-flops are obtained by combining the inputs with feedback. The general master-slave structure used in the logic of Figure 5.26 is identical to that found in many textbooks. The only difference is that the implementation uses CMOS TGs for data path control.

5.6 Array Logic

CMOS transmission gates may be used construct switch logic arrays. Ideally, TGs act as voltage-controlled switches; improved transient modeling is obtained by including R_{TG} to account for signal delay.

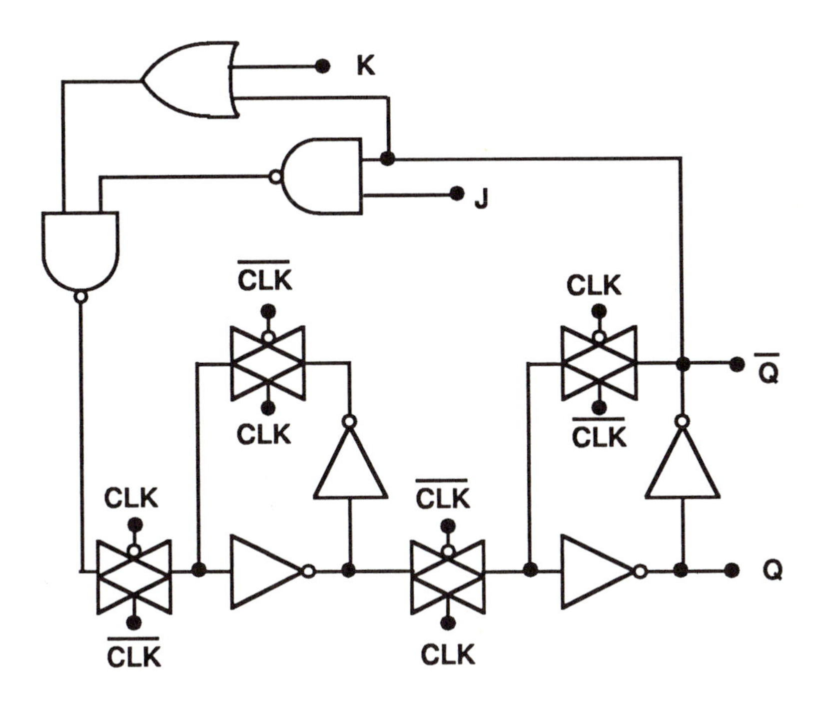

Figure 5.26: JK Master-Slave FF

5.6.1 Multiplexers/Demultiplexers

Multiplexers and demultiplexers can be directly implemented using structured TG arrays. In general, a multiplexer takes multiple inputs and directs a selected data line to a single output; a demultiplexer reverses this function. Figure 5.27 illustrates the block-level view where n data lines are switched using m control signals. In general,

$$2^m \geq n \tag{5.32}$$

must be satisfied; the best design is where $2^m = n$.

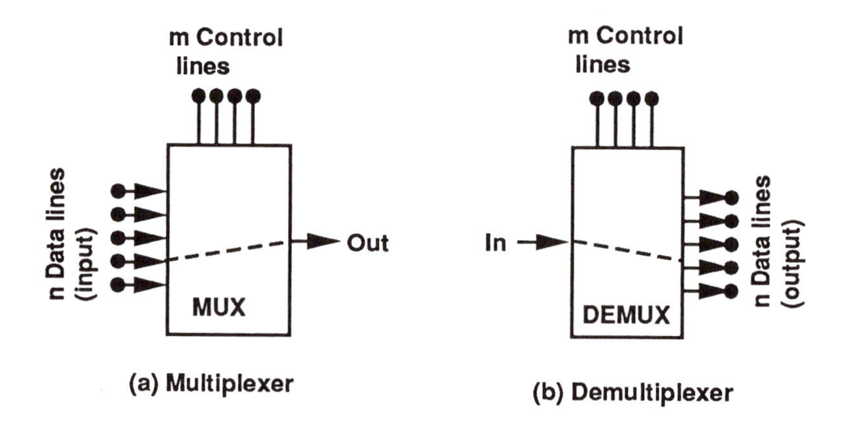

Figure 5.27: Multiplex/Demultiplex Operations

Consider as an example a 4-to-1 multiplexer with input data lines A, B, C, and D. To determine the requirements on the select signals, recall that m variables gives 2^m combinations. Two select signals ($m = 2$) S_0 and S_1 will then be sufficient to select one of the four data inputs. A suitable logic function is given by

$$F = A(S_0 S_1) + B(\overline{S}_0 S_1) + C(S_0 \overline{S}_1) + D(\overline{S}_0 \overline{S}_1). \tag{5.33}$$

The TG implementation of this circuit is shown in Figure 5.28. This type of multiplexing network can be designed for and n inputs, but care must be taken to avoid the possibility of having an isolated output node. This problem is automatically eliminated if the number of data lines is given by $n = 2^m$, with m the number of control signals.

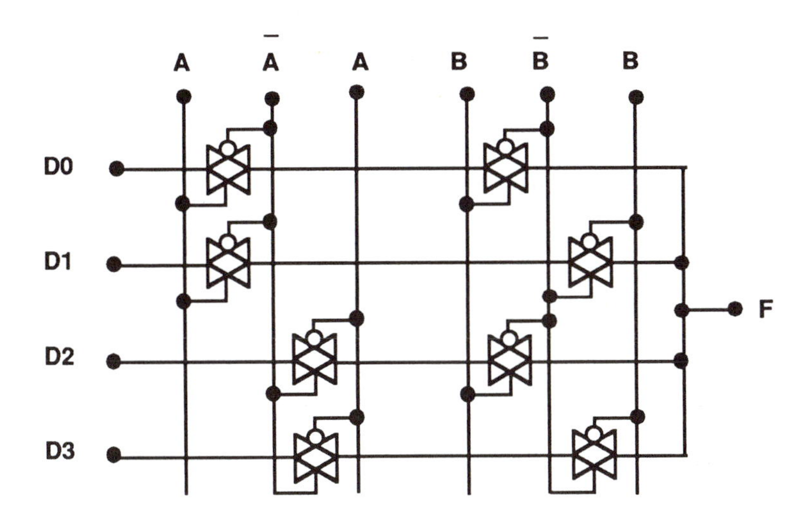

Figure 5.28: 4-to-1 TG-Based Multiplexer

An arbitrary n:1 multiplexer can be constructed in the same manner. For example, and 8:1 MUX will use three select signals $(A, \overline{A})$, $(B, \overline{B})$, and $(C, \overline{C})$ to chose one of eight data lines $(D0, D1, D2, D3, D4, D5, D6, D7)$ for routing to the output. Each data path is specified by a unique (A, B, C) combination by writing the output function in the form

$$\begin{aligned} F &= (\overline{A}\,\overline{B}\,\overline{C}) \cdot D0 + (\overline{A}\,\overline{B}C) \cdot D1 + (\overline{A}B\overline{C}) \cdot D2 + (\overline{A}BC) \cdot D3 \\ &\quad (A\,\overline{B}\,\overline{C}) \cdot D4 + (A\overline{B}C) \cdot D5 + (AB\overline{C}) \cdot D6 + (ABC) \cdot D7; \end{aligned} \tag{5.34}$$

the decimal value of (ABC) determines the selected line number. Since the use of three variables (A, B, C) gives $2^3 = 8$, all control possibilities are defined. The 8:1 MUX circuit has the same structure as that shown for the 4:1 system, with each data line being controlled by three TGs.

When designing a TG MUX, the output must always be connected to a single input line. If two or more input lines are switched to the output, the voltage levels may compete to give an undefined state. In static logic, the output line should never be completely isolated from the input lines. This is due to the problems of **charge leakage** and **charge sharing** at isolated MOS nodes, which are discussed in detail in Chapter 7 in conjunction with dynamic circuit properties.

5.6.2 Split Arrays

Layout problems may arise since transmission gates require both nMOS and pMOS transistors which reside in opposite polarity background regions. Providing individual wells for each TG decreases the logic density, so that alternate structuring may be attractive. One approach is to separate each line into two parallel paths; a chain of n-channel MOSFETs is used to carry logic 0 voltages, while logic 1 levels are propagated along a p-channel transistor path. Figure 5.29 shows the 4:1 MUX with this design. Setting the control variables (A, B) selects both an nMOS and a pMOS path. Duplication is needed to compensate for normal threshold voltage limitations on single polarity MOSFETs. Layout is simplified because the pMOS array can be physically separated from the nMOS array.

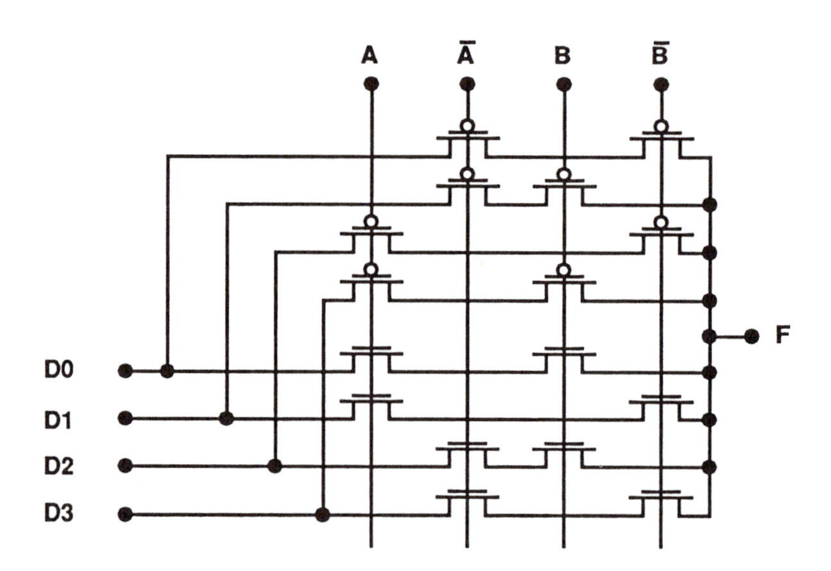

Figure 5.29: Split-Array MUX

5.7 Differential CVS Logic

Differential Cascode Voltage Switch Logic (DCVS logic or CVSL) combines complementary logic arrays with a differential latching circuit; the logic arrays are cascoded to achieve system-level designs. A generic gate is illustrated in Figure 5.30. Two cross-coupled pMOS transistors (Mp1 and Mp2) form the latch, while the logic is performed by complementary nMOS arrays. The logic circuit on the right side uses input variables (A, B, C) to form a logic block $\overline{F}$, while the left hand side has complementary inputs $(\overline{A}, \overline{B}, \overline{C})$ and is denoted by F. A condition of $\overline{F} = 1$ implies that conduction is established through the $\overline{F}$ logic block. Gate level outputs are denoted by $(Q, \overline{Q})$. This illustrates one implementation of **dual-rail** logic

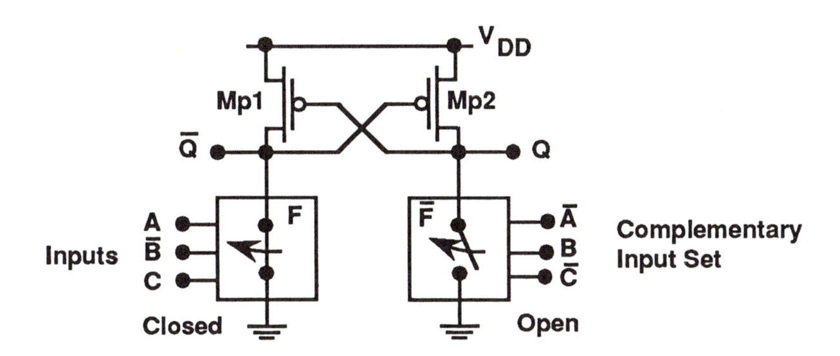

Figure 5.30: General CVSL Gate

where both a function and its complement are generated.

An example of a CVSL logic gate is shown in Figure 5.31 for the function

$$F = AB + C. \tag{5.35}$$

This can be extended to arbitrary logic arrays using the normal nMOS rules. The unique aspect of a CSVL logic gate is the differential arrangement for driving the cross-coupled pMOS latch circuit. This is used to provide gain and increase the switching speed of the circuit, although the performance is not necessarily superior to that possible using standard static CMOS [1].

5.7.1 Basic Operation

The single-variable circuit of Figure 5.32 provides a basis for understanding the operation. The complementary inputs $(D, \overline{D})$ are used to control the

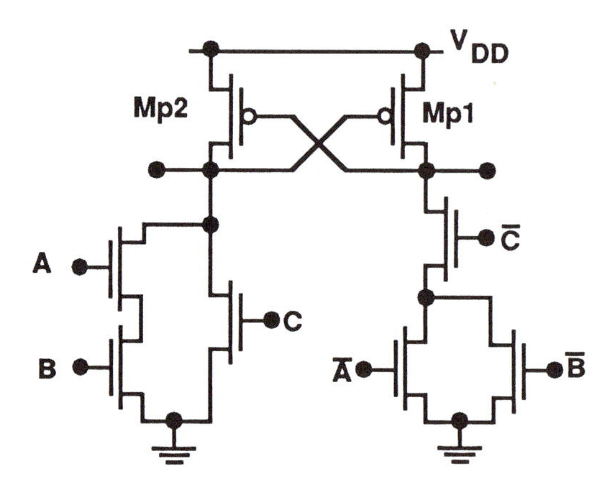

Figure 5.31: CVSL Gate for $F = AB + C$

switching transistors Mn1 and Mn2. Let us assume that initially the inputs are given by $(D, \overline{D}) = (0, V_{DD})$ so that Mn1 is in cutoff while Mn2 is biased active and provides a conducting path to ground. This pulls the voltage across C_2 to $V_2 = 0$ [V], which in turn biases Mp1 into conduction. Charge flows to C_1 so that $V_1 \rightarrow V_{DD}$ while the cross-coupling forces Mp2 into cutoff. When the inputs are switched to $(D, \overline{D}) \rightarrow (V_{DD}, 0)$, the

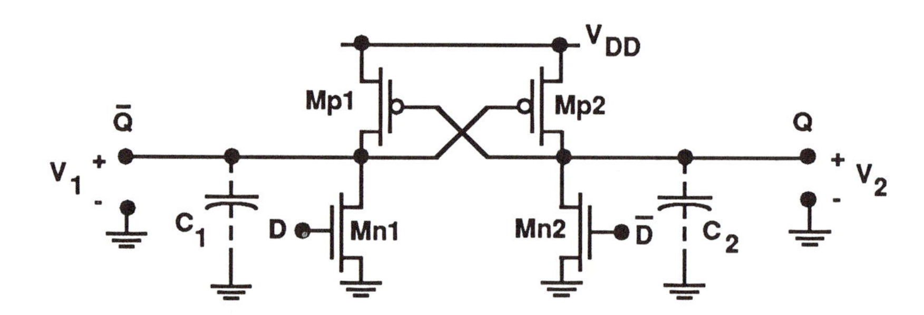

Figure 5.32: CVSL Latch

latch is forced to reverse states. C_1 discharges through Mn1 while C_2 charges through Mp2. After the transients decay away, the voltages are at $V_1 = V_{DD}$ and $V_2 = 0$.

Switching times depend on the node capacitance and device aspect ratios. The pMOS transistors Mp1 and Mp2 are usually designed with equal

β_p values. However, care must be taken when choosing specific aspect ratios. Large values of $(W/L)_p$ allow rapid charging of the output capacitors, but makes it more difficult to "flip" the latching circuit due to the cross-coupled gate-source connections.

While this example illustrates the operation, its simplicity masks an important property. In more complex logic trees, the $\overline{F}$ and F arrays may have different transient characteristics due to the requirement of complementary logic. This can be seen in the circuit for $F = AB + C$ introduced earlier in Figure 5.31. The $\overline{F}$ array has a discharge path through the series transistors AB or through C by itself. The complementary F structure on the other side requires current flow through a series path regardless of the input combination. Asymmetric paths make the switching more difficult, implying that the transient response and power dissipation will be worse than anticipated at first sight.

5.7.2 Logic Design

The examples above illustrate that simple functions such as $F = AB+C$ can be implemented in CVSL using the logic formation rules presented in the last chapter[2] to build independent F and $\overline{F}$ branches. As the complexity of the function increases, it is often more efficient to couple the two logic branches to avoid duplication of the switching.

Consider the XOR/XNOR functions. Writing the expanded forms

$$\begin{aligned} F &= A\overline{B} + \overline{A}B, \\ \overline{F} &= AB + \overline{A}\,\overline{B}, \end{aligned} \tag{5.36}$$

reveals common factors which allow for a reduction in the MOSFET count. Choosing B and $\overline{B}$ as the common transistors gives the logic tree shown in Figure 5.33; this reduces the MOSFET count by 2 over the case where the branches are not coupled. Furthermore, this can be expanded to a 3-input XOR function by exploiting the symmetry again via expressions such as

$$\begin{aligned} F1 &= A \oplus B \oplus C \\ &= (A \oplus B)C + (A \oplus B)\overline{C} \end{aligned} \tag{5.37}$$

to build the XOR3 gate shown in Figure 5.34. This can be extended to an N-input XOR/XNOR array. The XOR/XNOR functions can be used to create the full-adder logic arrays shown in Figure 5.35 [2]; the pMOS latches have been omitted for simplicity. With inputs (A, B, C_i) the sum circuit produces $(\overline{S}, S)$ and the output carry circuit gives $(\overline{C_o}, C_o)$.

[2]See Section 4.7

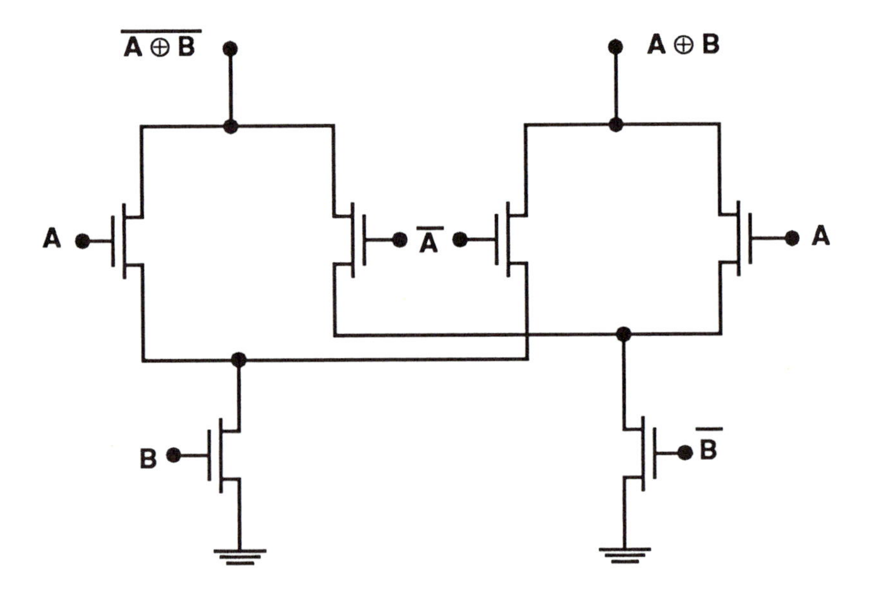

Figure 5.33: XOR/XNOR Logic Tree

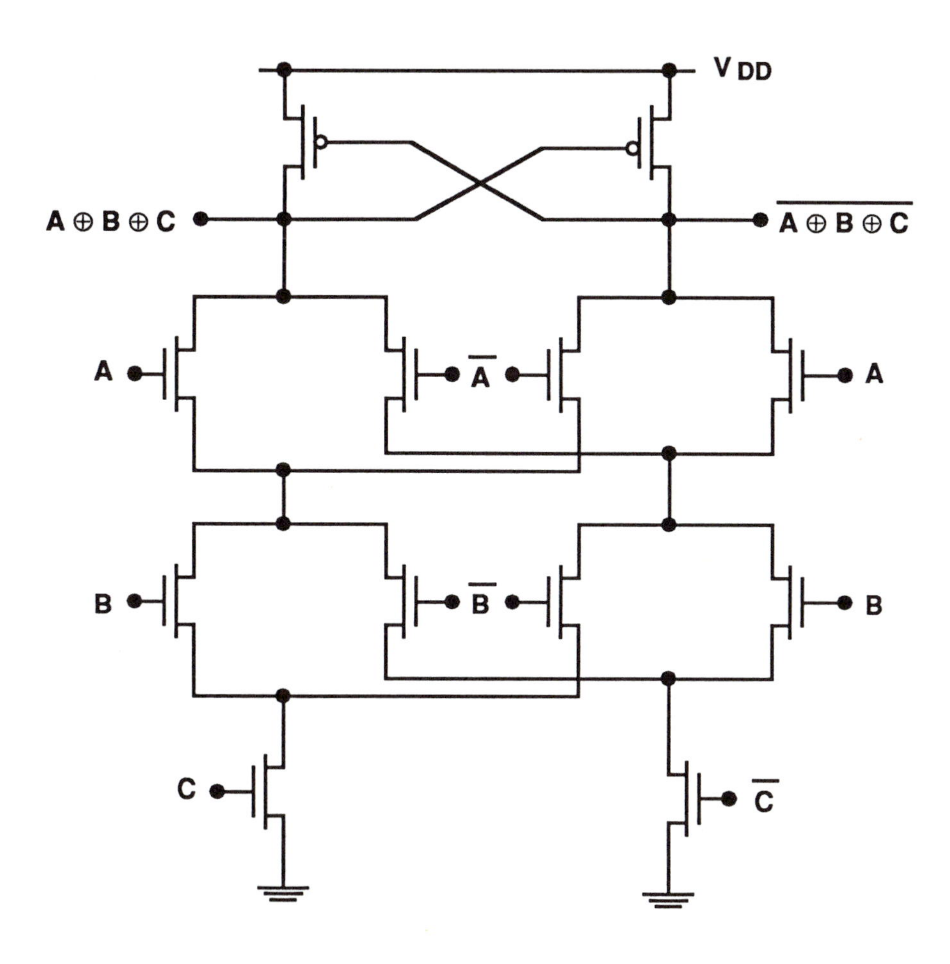

Figure 5.34: 3-Input XOR/XNOR Logic

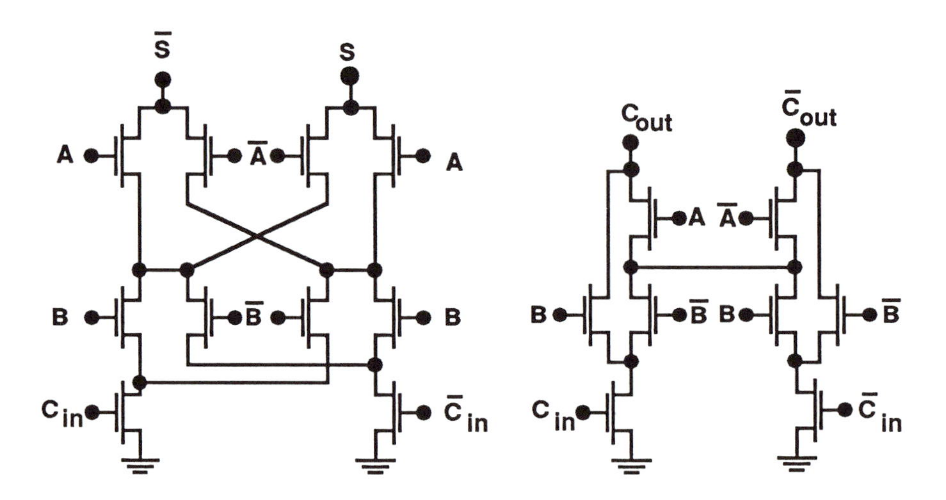

Figure 5.35: CVSL Full Adder Logic

5.8 Complementary Pass-Transistor Logic

Recent work has led to a structured logic approach known as **complementary pass-transistor** logic (CPL) [12]. The logic design style is similar to that used in cascode-voltage switch logic (CSVL). However, CPL eliminates the pMOS latch to increase the switching speed.

The main architectural features of CPL are shown in the block diagram of Figure 5.36. Two complementary sets of inputs are used to drive an nMOS logic array. One set is connected to the transistor gates, while the other group is used for inputs into a multiplexer-type array. The outputs are obtained through buffering inverters. pMOS latches can also be used at the output to reduce the DC power dissipation.

Figure 5.37 illustrates the formation of an AND/NAND logic array in CPL. Basic switch logic can be applied to describe the array function. For example, the left two MOSFETs implement the AND function by means of

$$F1 = AB + B\overline{B} = AB; \tag{5.38}$$

passing this through an inverter yields the NAND. Similarly, a NAND gate is formed by the right MOSFET pair as seen in

$$F2 = \overline{A}B + \overline{B} = \overline{AB}, \tag{5.39}$$

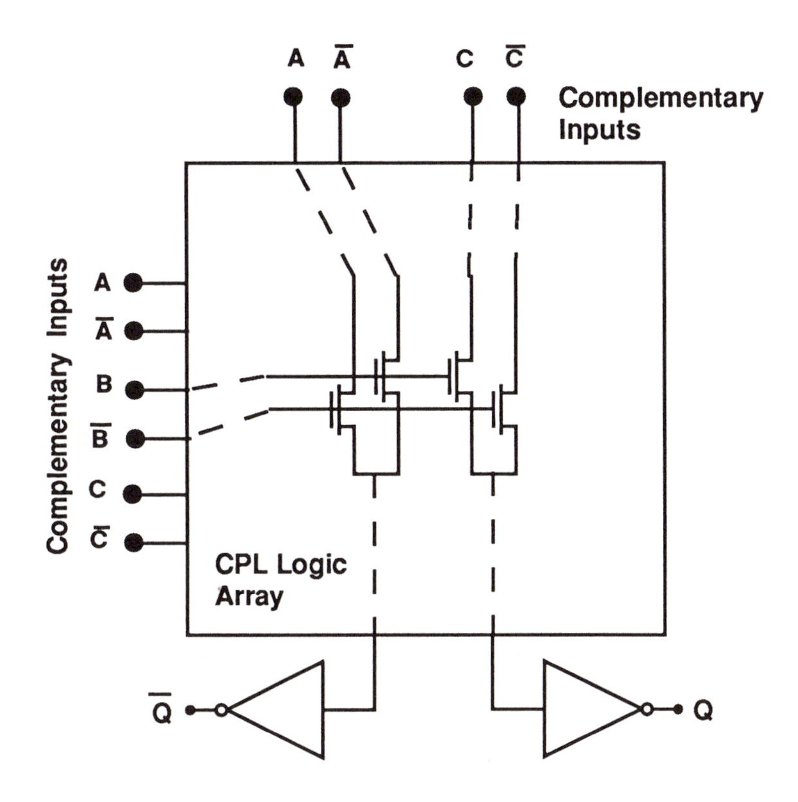

Figure 5.36: Basic CPL Logic Network

so that inverting $F2$ gives the AND operation.

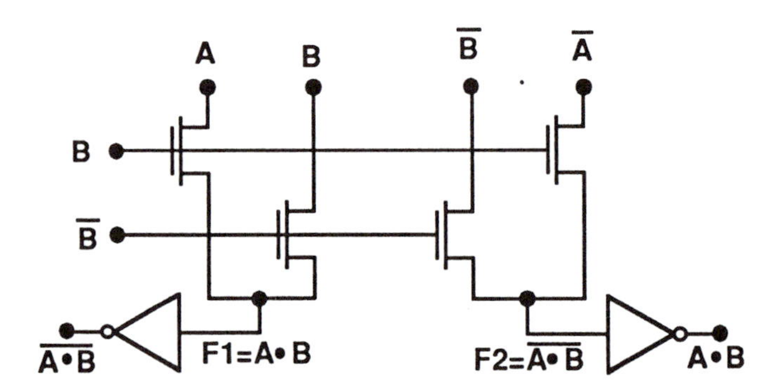

Figure 5.37: CPL AND/NAND Gate

Using nMOSFET logic arrays introduces (logic 1) threshold voltage losses as the signals propagate through the transistor chain. The maximum voltage at the end of an array chain is

$$V_{max} = V_{DD} - V_{Tn}(V_{max}), \tag{5.40}$$

where $V_{Tn} > V_{TOn}$ due to body bias effects. CPL outputs are directed through static inverting buffers to provide full-rail logic swings. The inverter must be designed with $V_{max} > V_{IH}$ to insure proper logic translation.

CPL may be used to develop sets of equivalent logic gates using only nMOS arrays. These can be directly connected to other CPL gates without intermediate buffer amplifiers. Output inverters are required if the array is used to drive other types of logic circuits, or severe charge sharing or charge leakage is present. In addition, one must be careful not to induce more than one threshold voltage loss in a chain, since this degrades the noise margins. Figure 5.38 shows the basic arrays for CPL OR/NOR and XOR/XNOR functions. Gates with 3 or more inputs can be design by straightforward extension. Examples are provided in Figure 5.39.

Let us examine the design of a full adder circuit using wired CPL logic gates. The sum function

$$S = A \oplus B \oplus C, \tag{5.41}$$

where C is the input carry, can be implemented by using a set of cascaded XOR/XNOR arrays. Since CPL provides complementary outputs, $\overline{S}$ is also

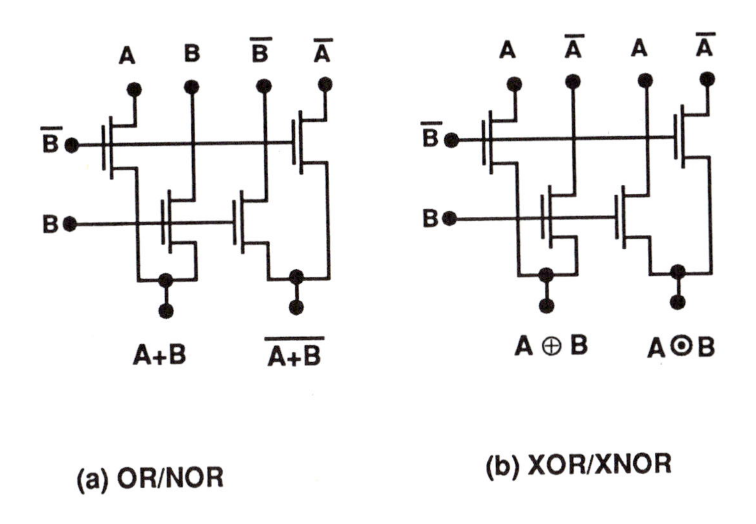

Figure 5.38: 2-Input CPL Arrays

generated. The output carry bit C_{out} can be generated using AND gates via

$$C_{out} = AB + AC + BC; \tag{5.42}$$

one CPL output will be complemented, so that $\overline{C_{out}}$ is obtained without additional effort. Figure 5.40 provides the CPL full adder circuits. Including the output inverters, a total of 28 MOSFETs are required. This is to be compared with the conventional CMOS circuit implementation which uses 40 or more transistors. Moreover, the gates are built in structured nMOS arrays which simplifies layout. CPL is also an excellent candidate for logic design using predefined MOSFET arrays.

One problem with this approach (and, all array logic schemes) is that the MOSFET chains act as RC lines which must be charged and discharged. A long transistor array may introduce excessive transient delays into the network. Adequate buffering becomes an important consideration. Alternately, BiCMOS output drivers[3] can be used as input drivers to speed up the circuit.

[3] BiCMOS is the subject of Chapter 10

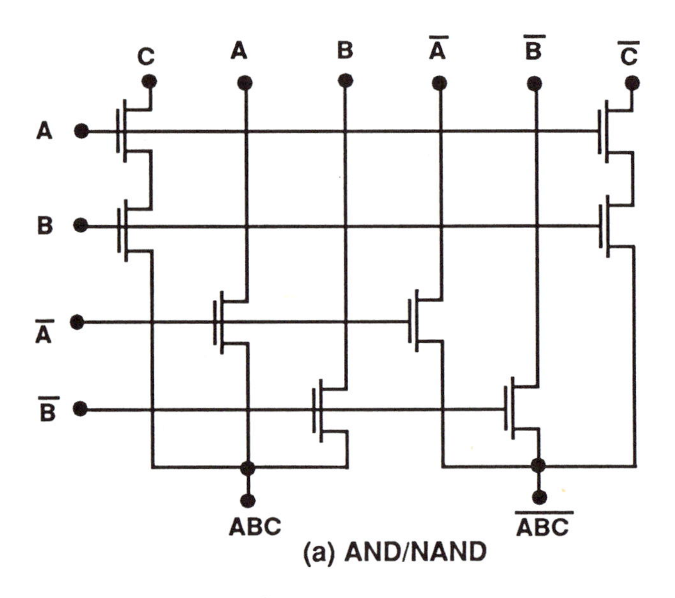

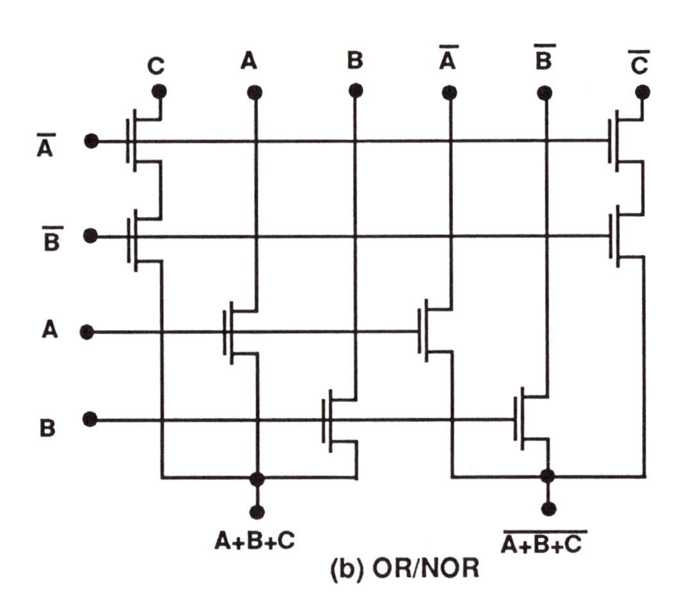

Figure 5.39: 3-Input CPL Arrays

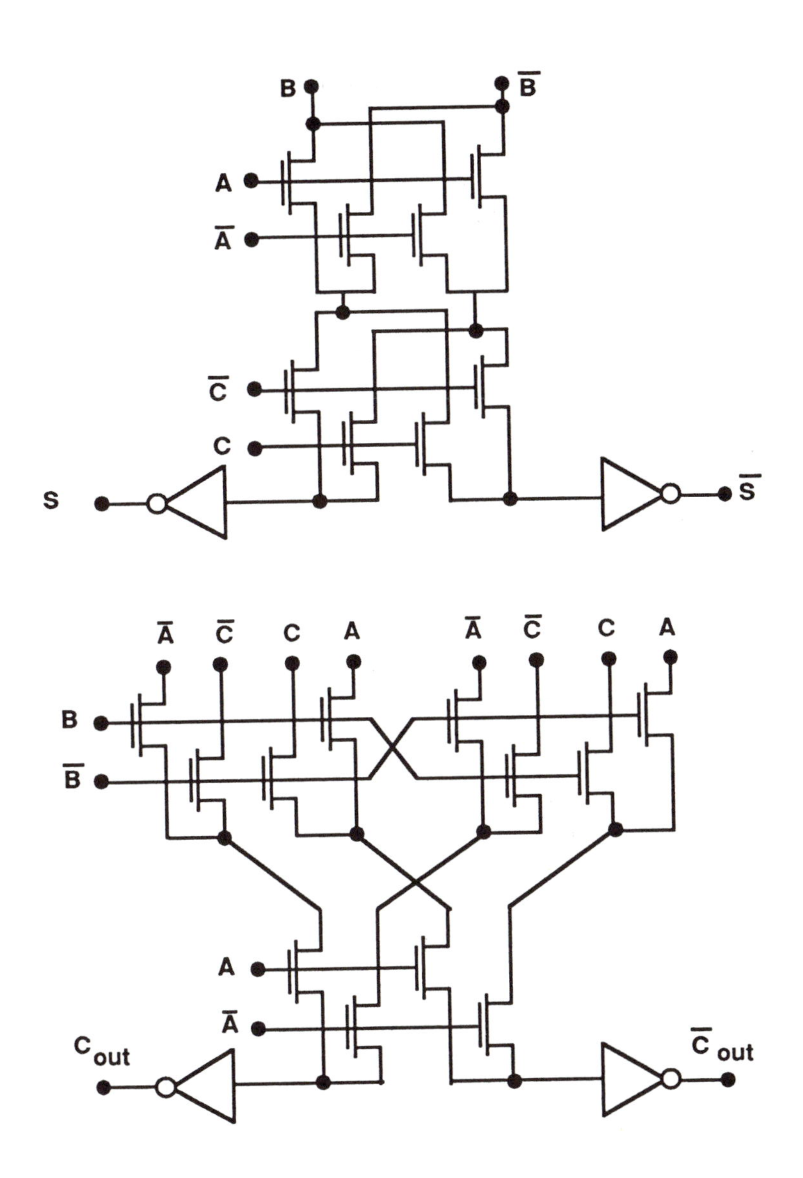

Figure 5.40: CPL Full Adder

5.9 DSL Logic

Differential split-level (DSL) circuits provides another approach to differential logic. The basic DSL switching network in Figure 5.41 can be viewed as a modified DCVS logic circuit. Two MOSFETs MR1 and MR2 have been added between the outputs $(Q, \overline{Q})$ and the internal nodes shown as $(F, \overline{F})$. The gates of MR1 and MR2 are controlled by a reference voltage V_R which is chosen to be

$$V_R = \frac{1}{2}V_{DD} + V_{Tn}; \tag{5.43}$$

this clamps the voltage on nodes F and $\overline{F}$ to $(V_{DD}/2)$. The reduced output voltage swing is introduced in an attempt to improve the switching characteristics. Another reason for using this circuit is that the reduced drain-source voltages allow smaller switching transistors to be used without punchthrough problems. The drawback of the scheme is that V_R provides a DC bias which leads to static power dissipation.

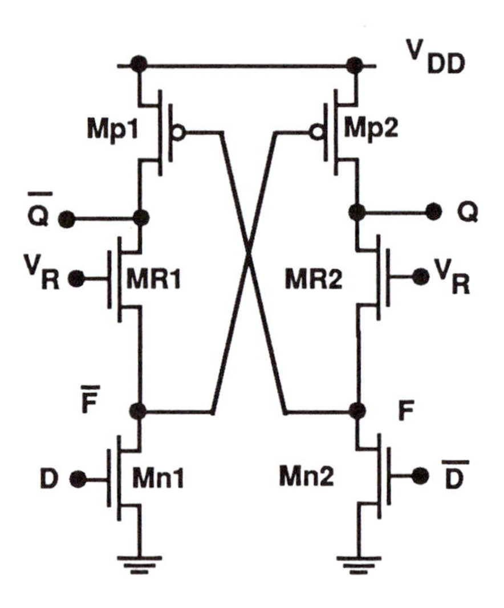

Figure 5.41: DSL Switching Network

Let us analyze the operation of the switching network. We will assume a starting logic condition of $(D, \overline{D}) = (0,1)$ so that initially $(Q, \overline{Q}) = (0,1)$. In terms of voltages, we see that Mn1 is in cutoff so that $\overline{Q} = V_{DD}$.

However, V_R bias the internal node $\overline{F}$ to a voltage of $(V_{DD}/2)$, which allows Mp2 to conduct with

$$V_{SGp2} = V_{DD} - \frac{1}{2}V_{DD} = \frac{1}{2}V_{DD}. \tag{5.44}$$

Power supply current flows to ground through the path created by Mp2, MR2, and Mn2, giving rise to DC power dissipation. The tradeoff becomes clear: increased switching speed is achieved by reducing the voltage swing, which in turn creates DC power dissipation[4].

A working DSL logic circuit is shown in Figure 5.42. Logic is provided

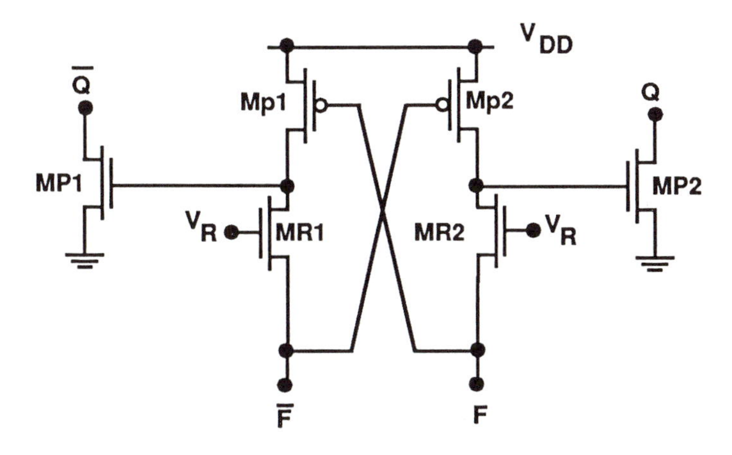

Figure 5.42: DSL Circuit

at the inputs F and $\overline{F}$; these may be direct logic inputs or transistor arrays. The output circuit has been modified such that the basic DSL switching network (denoted by S and $\overline{S}$) now controls the gates of nMOS pass transistors MP1 and MP2. Gate outputs $(Q, \overline{Q})$ are taken at the drains of these devices. In a cascaded DSL network, MP1 and MP2 will be inputs to the next logic tree.

Several types of DSL circuits have been published in the literature [8], and DSL represents an interesting example of modifying an existing logic gate in an attempt to improve performance.

[4]This is similar to the analog situation where a Class C amplifier is biased into Class A operation; the former does not dissipate DC power while the latter uses normally-on transistors.

5.10 References

The literature on switched CMOS logic circuits is quite extensive. The list below provides some of the basic papers and texts for futher reading.

[1] K.M. Chu and D.L. Pulfrey, " A Comparison of CMOS Circuit Techniques: Differential Cascode Voltage Switch Logic Versus Conventional Logic," IEEE J. Solid-State Circuits, vol. SC-22, no. 4, pp. 528-532, Aug., 1987.

[2] K.M. Chu and D.L. Pulfrey, " Design Procedures for Differential Cascode Voltage Switch Logic Circuits," IEEE J. Solid-State Circuits, vol. SC-21, no. 4, pp. 1082-1087, Dec., 1986.

[3] E. D. Fabricius, **Introduction to VLSI Design**, McGraw-Hill, New York: 1990.

[4] H. Haznedar, **Digital Microelectonics**, Benjamin Cummings, Redwood City, CA: 1991.

[5] S-L. Lu, " Implementation of Iterative Networks with CMOS Differential Logic," IEEE J. Solid-State Circuits, vol. 23, no. 4, pp. 1013-1017, Aug., 1988.

[6] N. Kanopoulos and N. Vasanthavada, " Testing of Differential Cascode Voltage Switch (DCVS) Circuits," IEEE J. Solid-State Circuits, vol. SC-25, no. 3, pp. 806-812, June, 1990.

[7] A. Mukherjee, **Introduction to nMOS and CMOS VLSI Systems Design**, Prentice-Hall, Englewood Cliffs, NJ: 1986.

[8] L.C.M.G. Pfennings, W.G.J. Mol, J.J.J. Bastianes, and J.M.F. Van Dijk, " Differential Split-Level CMOS Logic for Subnanosecond Speeds," IEEE J. Solid-State Circuits, vol. SC-20, no. 5, pp. 1050-1055, Oct., 1985.

[9] D. Radhakrishan, S.W. Whitaker, and G.S. Maki, "Formal Design Procedures for Pass Transistor Switching Circuits," IEEE J. Solid-State Circuits, vol SC-20, no. 2, pp. 531-536, April , 1985.

[10] J. P. Uyemura, **Fundamentals of MOS Digital Integrated Circuits**, Addison-Wesley, Reading, MA, 1988.

[11] N. Weste and K. Eshraghian, **CMOS VLSI Design**, Addison-Wesley, Reading, MA: 1985.

[12] K. Yano, *et al.*, " A 3.8 ns CMOS 16 ×16-b Multiplier Using Complementary Pass-Transistor Logic," IEEE J. Solid-State Circuits, vol. SC-25, no. 2, pp. 388-394, April, 1990.

Chapter 6

Chip Design

Integrated circuit performance depends on both the circuit configuration and the chip implementation. Logic, circuit parameters, and fabrication become united when solving the design puzzle.

Chip design and layout is governed by the fabrication process flow. Design rules originate from physical limits imposed by the processing, and also from problems in noise and electromagnetic coupling. In this chapter we will examine chip design in an overall sense, emphasizing the interplay among the circuit parameters. Transmission line properties will be introduced to model high-speed data flow.

Fabricating CMOS integrated circuits requires that we accommodate both n-channel and p-channel MOSFETs on the same wafer. Many CMOS process flow examples can be found in the current literature. Rather than detailing every possibility, we will examine the dominant aspects of CMOS fabrication. Other issues which should be addressed are items such as

- Device isolation
- Interconnect layers
- Design rules
- Latchup prevention

which are all important to the design process. Once the ground rules are established, we can proceed with our discussion of circuit design and layout.

6.1 Isolation

Device isolation deals with electrically decoupling neighboring transistors on a densely-packed integrated circuit. Unwanted conduction channels must be eliminated by preventing both direct and indirect current flow paths. The most common isolation techniques used in bulk CMOS are LOCOS and trench isolation.

6.1.1 LOCOS

The Local Oxidation of Silicon **(LOCOS)** achieves device isolation by selective oxide growth. A typical LOCOS process starts by growing a thin **stress relief** thermal oxide (SiO_2) layer on the silicon surface. Next, silicon nitride (Si_3N_4) is deposited and patterned, keeping nitride in the areas where transistors will be built. The entire surface is then exposed to an oxidizing ambient. Nitride does not oxidize, but any exposed silicon will react to form SiO_2. The resulting LOCOS structure is illustrated in Figure 6.1. Thermal oxidation uses surface silicon in the reactions

$$\begin{aligned} \mathrm{Si} + \mathrm{O_2} &\rightarrow \mathrm{SiO_2} \\ \mathrm{Si} + 2\mathrm{H_2O} &\rightarrow \mathrm{SiO_2} + 2\mathrm{H_2}. \end{aligned}$$

This recesses the oxide into the substrate surface, aiding isolation. Simple analysis shows that

$$X_R = 0.46 X_{FOX} \tag{6.1}$$

where X_R is the depth of recession and X_{FOX} is the thickness of the grown **field oxide** (FOX) which separates device locations. In general, the patterned nitride regions are called **active areas**, while the oxide growth defines the **field regions** between active transistor sections[1].

Inversion of the surface in the field regions is controlled by the **field threshold voltage** $V_{T,F}$. This should be greater than the highest interconnect voltage on the chip to avoid unwanted conduction paths. A **field ion implant** D_F of acceptors is used to increase the field threshold voltage to a valued described by

$$V_{T0,F} = V_{FB,F} + \phi_S + \gamma_F \sqrt{\phi_S} + \frac{qD_F}{C_{FOX}} \tag{6.2}$$

where

$$C_{FOX} = (\epsilon_{ox}/X_{FOX}) \tag{6.3}$$

[1] The two regions are also called the *anti-moat* and *moat*, respectively

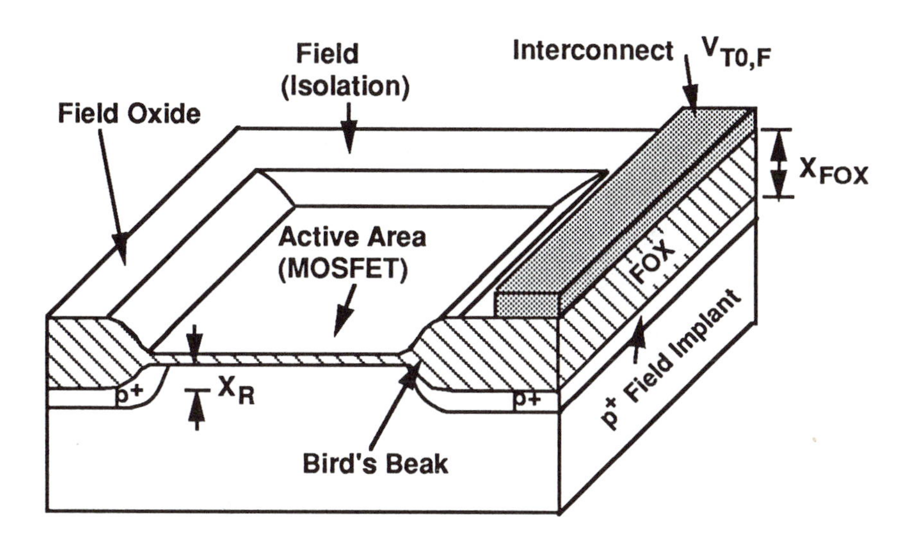

Figure 6.1: LOCOS Isolation

is the field oxide capacitance per unit area,

$$\gamma_F = \frac{\sqrt{2q\epsilon_{Si}N_a}}{C_{FOX}} \tag{6.4}$$

is the field body-bias factor, and $V_{FB,F}$ is the flatband voltage in the field regions. The increased acceptor doping in the field leads to higher drain and source depletion capacitance values for the sidewall contributions. In fact, the sidewall capacitance can dominate the parasitic depletion contributions in a MOSFET.

LOCOS is a widely used isolation technique in many processing lines. However, a major limitation is the problem of **active area encroachment** which occurs during the FOX growth process and reduces the usable (flat) size of the region. The problem is illustrated in Figure 6.2. Even though the nitride protects the silicon surface, oxygen diffuses through the sides of the stress-relief oxide layer during the FOX growth. SiO_2 is thus formed around the edges, lifting the nitride upwards and forming a characteristic **bird's beak** transition region[2] between the active area and the field oxide. Encroachment cannot be avoided and affects the integration density.

[2] So named because of the shape.

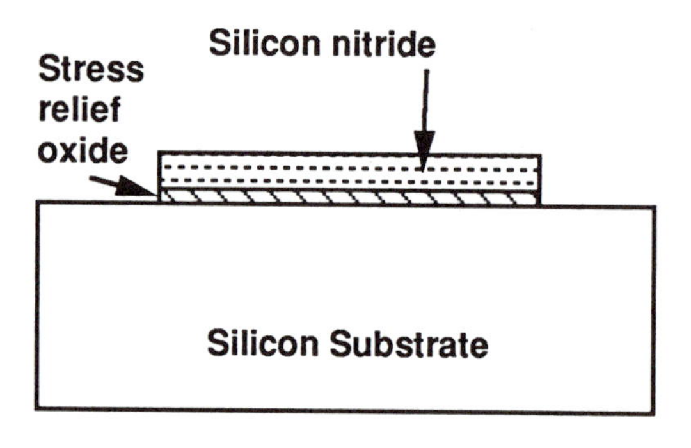

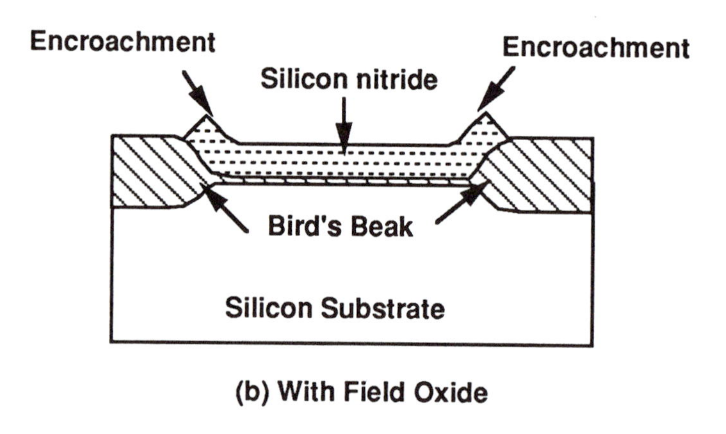

Figure 6.2: Encroachment in LOCOS

6.1.2 Trench Isolation

Trench isolation uses **reactive ion etching** (RIE) to form small trenches in the silicon. The trenches are then filled with oxide and polysilicon to electrically isolate neighboring device regions from one another. High integration levels are possible since the trench widths can be reduced to the order of a few microns. Trench isolation is illustrated in Figure 6.3 A

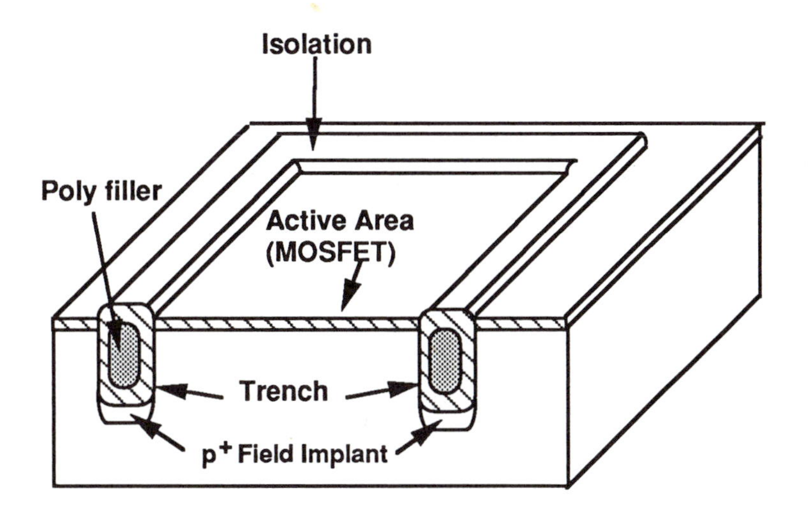

Figure 6.3: Trench Isolation

field implant may be used to increase the trench threshold voltage $V_{T,Tr}$. Small trench dimensions makes this approach particularly important for high-density integration.

The vertical trench regions may also be used to create large-value capacitors without consuming valuable surface real estate. An example geometry which uses doped poly and n^+ as capacitor plates is shown in Figure 6.4. Denoting the trench oxide thickness by X_{TOX}, the capacitance per unit area is given by $C_{TOX} = (\epsilon_{ox}/X_{TOX})$ [F/cm^2], so that the total trench capacitance is

$$C_T = C_{TOX} A_P \tag{6.5}$$

where the plate area follows the contour of the trench. The actual real estate savings is given by (A_P/A_S) where A_S is the value of the surface area. Trench capacitors are commonly used in advanced dynamic RAM (DRAM) cell designs since they conserve surface real estate.

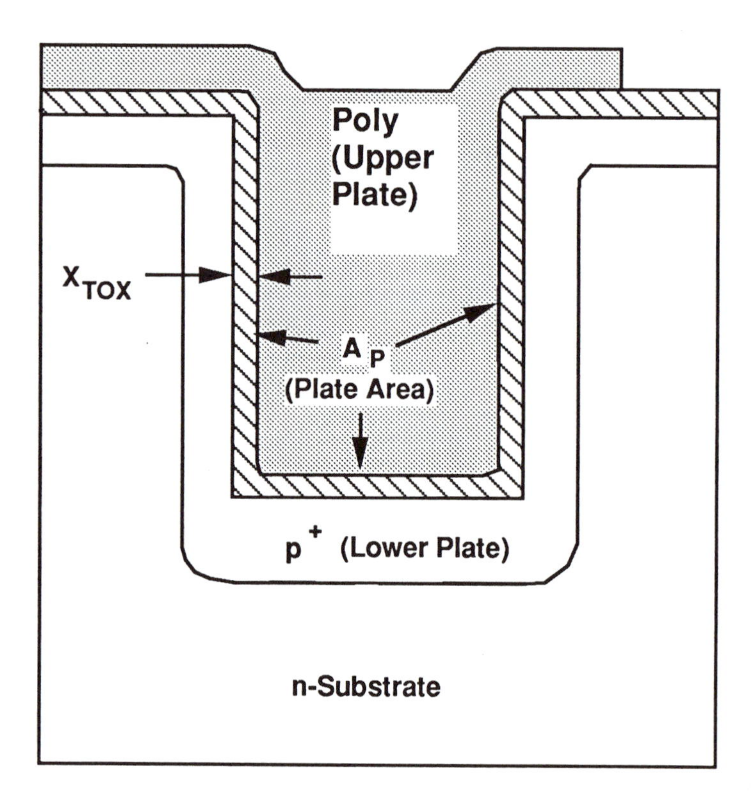

Figure 6.4: Trench Capacitor

Trench isolation has been developed to the point where it is a viable production line technique. It (almost) eliminates the problem of active area encroachment found in LOCOS and is useful when increasing the logic integration density.

6.2 CMOS Process Examples

CMOS requires both nMOS and pMOS transistors. One complication that arises immediately is that an n-channel MOSFET requires a p-type background, while a p-channel MOSFET must be built in an n-type region. Most state-of-the-art processes fabricate the circuit in bulk silicon; a superior but more costly approach is to use an insulating substrate. In this section we will briefly examine the common approaches to creating CMOS integrated circuits. More detailed descriptiosn of the process flows can be found in references [6], [12], and [21].

6.2.1 Bulk CMOS

Bulk CMOS requires that we provide "tubs" or "wells" which have opposite polarity of the substrate to accommodate both n-channel and p-channel transistors. Three approaches are possible; these are compared in Figure 6.5.

n-Well CMOS

Consider a p-type substrate. n-channel MOSFETs can be built directly in the substrate, but we must provide a diffused n-well region to locate pMOSFETs. Figure 6.5(a) shows the cross-section for an n-well CMOS inverter. This type of process allows modification of older nMOS fabrication lines to accommodate CMOS.

p-well CMOS

A p-well process uses an n-type wafer with diffused p-type regions for n-channel FETs. As shown in Figure 6.5(b), this is the electrical complement of n-well CMOS.

Twin-tub CMOS

Twin-tub bulk CMOS provides distinct n-well and p-well regions in a lightly doped epitaxial layer; see Figure 6.5(c) for a cross-sectional view. This is the most complicated approach, but provides self-isolation using reverse-biased

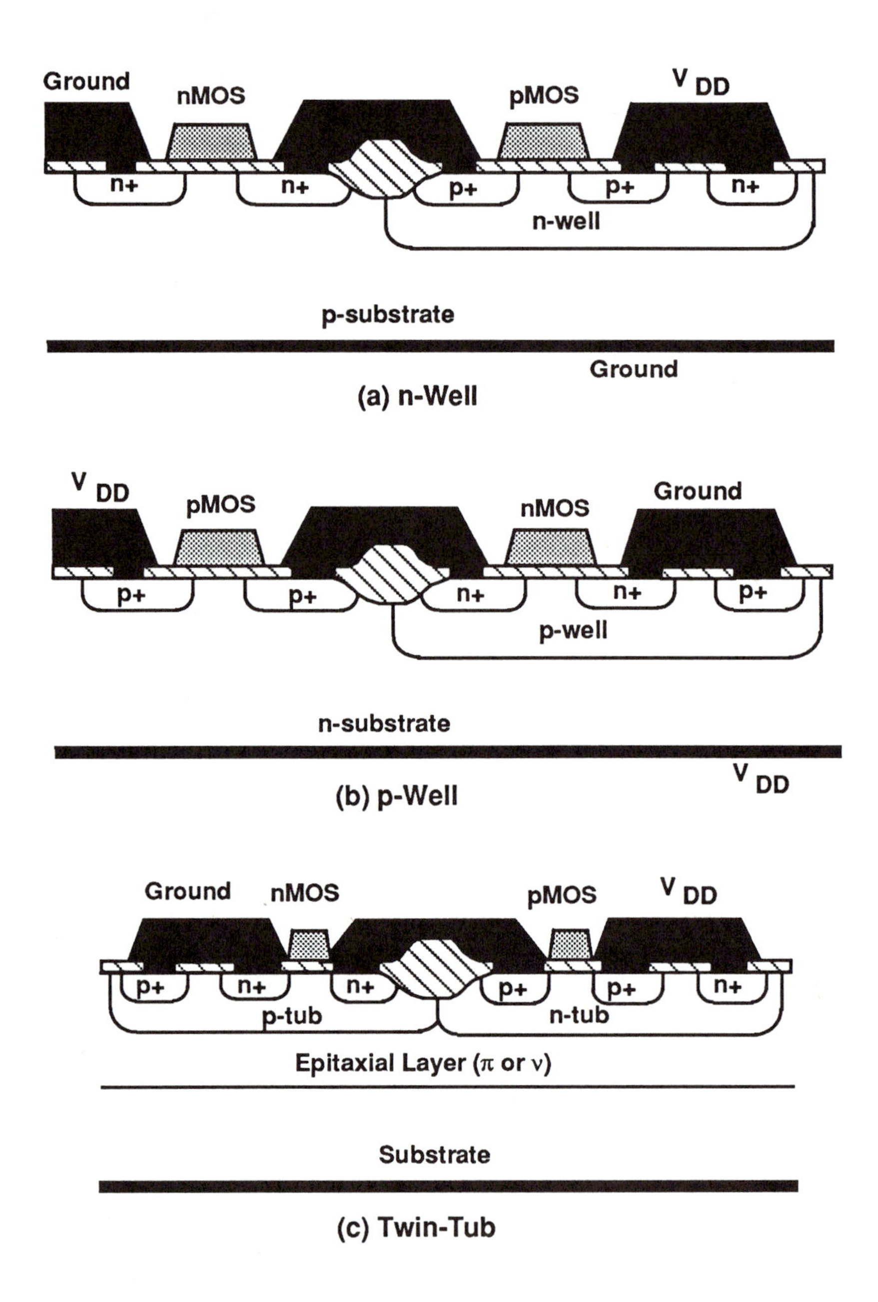

Figure 6.5: Bulk CMOS Processes

pn junctions. In addition, it helps solve the latchup problems discussed below.

6.2.2 Latchup

Bulk CMOS technologies are susceptible to **latchup**. This condition occurs when a parasitic conducting path is established between V_{DD} and ground, directing current away from the circuit. Once latchup occurs, it can only be stopped by removing the power supply and restarting the circuit. In addition to halting the circuit operation, latchup may induce catastrophic failure from heating. We will examine the basic problem in this section. A detailed treatment can be found in Troutman [19].

Figure 6.6 shows the cross-section of a n-well CMOS substrate region where the latchup problem originates. Attention is directed towards the

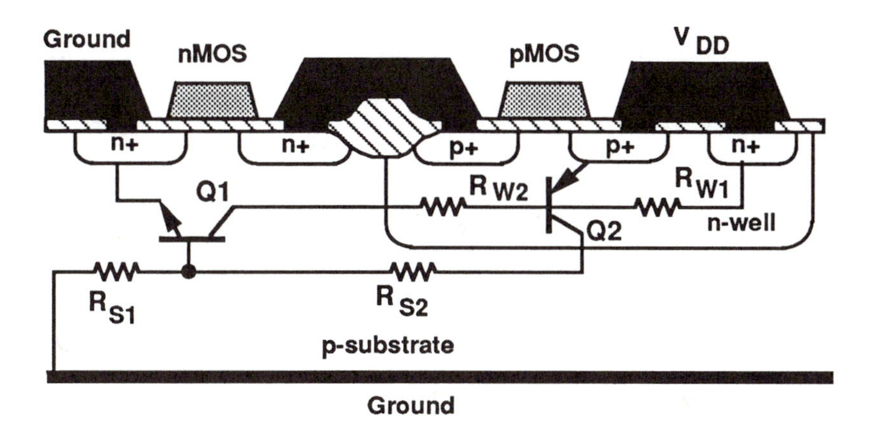

Figure 6.6: Origin of CMOS Latchup

pnp and npn layering. In particular, note the following patterns:

- p^+: pMOS Drain/Source
- n-well: (V_{DD})
- p-sub: Grounded substrate

and

- n^+: nMOS Drain/Source
- p-sub: Grounded substrate
- n-well: (V_{DD})

Grouping the first three and last three layers shows that the layered regions can be modeled using interconnected pnp and npn bipolar junction transistors (BJTs) as shown in the drawing. Parasitic substrate resistances R_{s1} and R_{s2}, and well resistances R_{w1} and R_{w2} are also important in analyzing the behavior. Under normal operation, both bipolar transistors are off and the reverse-biased junctions block current flow from V_{DD} to ground. In this case, the structure is said to be in a **blocking state**.

Latchup can be understood by referencing the circuit in Figure 6.7. We have replaced the power supply with a general voltage V. This device is the basis of several bipolar switching devices[3] and has the $I - V$ characteristic shown. The **breakover voltage** V_{BO} represents a critical breakdown/switching value. When $V < V_{BO}$, the current flow is restricted to leakage levels. However, when V reaches V_{BO}, the device goes into the **breakover** condition and significant current flows through the circuit. This corresponds to a latched condition in the CMOS structure. Once break over occurs, current is diverted away from the CMOS circuitry to the parasitic bipolar transistor subcircuit.

To understand the origin of the latchup problem, note that the voltage across parasitic resistor R_{w1} acts to forward bias the emitter-base junction of Q2. If V_{EB2} reaches the turn-on voltage of about 0.7 volts, I_{C2} flows. This current flowing through R_{s1} develops a forward bias V_{BE1} across the base-emitter junction of Q1, causing I_{C1} to increase. The transistor pair Q1-Q2 are connected to form a positive feedback loop, so that the buildup continues. Recall that the forward current gain in a BJT may be expressed using either the common-base or common-emitter gains:

$$\begin{aligned} \alpha_F &= \frac{I_C}{I_E} < 1 \\ \beta_F &= \frac{\alpha_F}{(1 - \alpha_F)} > 1. \end{aligned} \tag{6.6}$$

Both are current-dependent parameters. Figure 6.8 shows the well-known behavior of β_F as a function of I_C [23]. When the circuit is in a blocking state, the current levels are small; this implies small gain values. If the base-emitter junctions become slightly forward biased, the collector currents increase, giving larger gain. Let us denote the gains of Q1 and Q2 by

[3] Such as SCRs, Diacs, and Triacs.

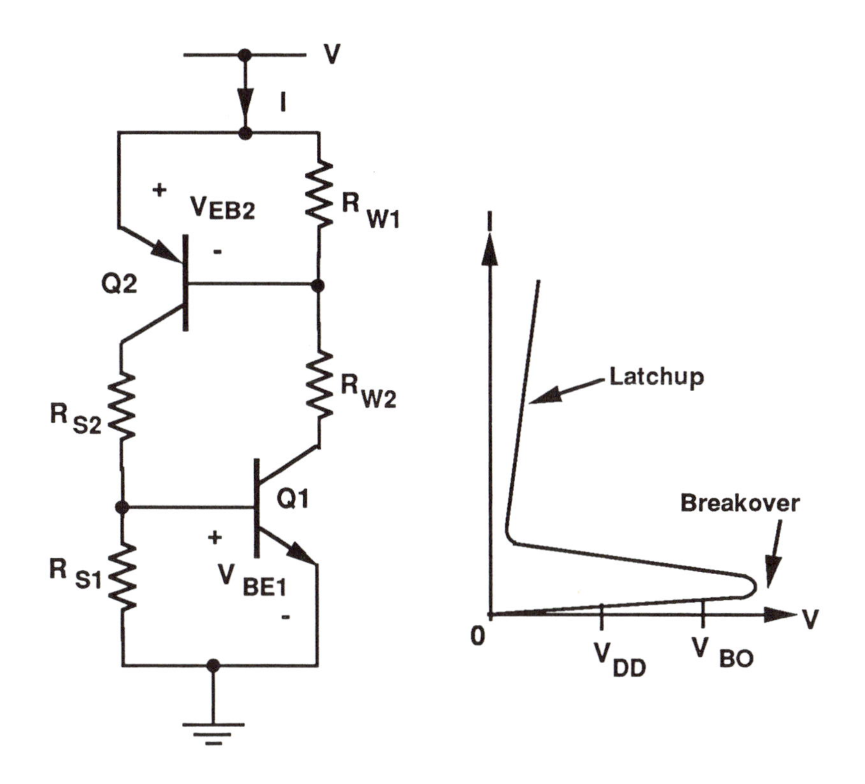

Figure 6.7: Latchup Circuit Model

α_{F1} and α_{F2}, respectively. An analysis of the circuit shows that breakover corresponds to the point where the gains increase such that

$$\alpha_{F1} + \alpha_{F2} = 1, \tag{6.7}$$

or, equivalently, when

$$\beta_{F1}\beta_{F2} > 1. \tag{6.8}$$

Both equations say that once the current reaches a critical value, the feedback loop latches the circuit into a conducting state.

Latchup triggering may occur anytime the circuit voltages exceed normal levels. Causes include

- Voltage overshoot/undershoot

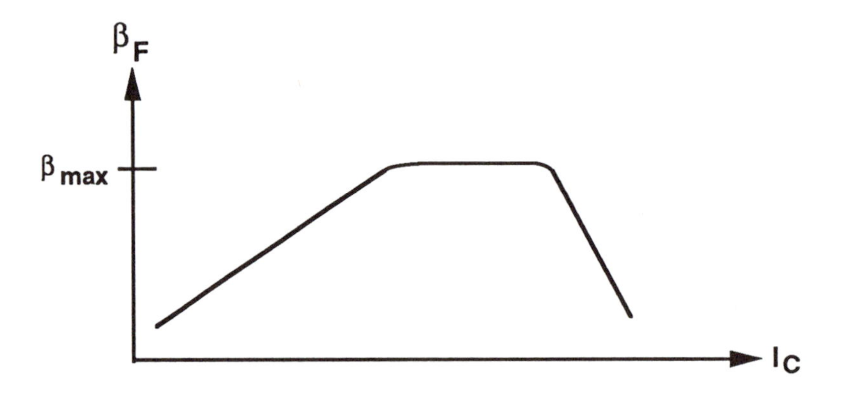

Figure 6.8: Current Gain Variations

- Avalanche breakdown
- Punchthrough
- Parasitic MOSFETs
- Photocurrent

and others. Although careful circuit design may reduce the possibility of inducing latchup, it is generally worthwhile to take extra precautions.

There are two main approaches to dealing with the latchup problem: (a) reduce the transistor current gains, or (b) decouple the transistor feedback loop; it is common to use both in practice. Gain reduction is accomplished in the fabrication and the formulation of design rules. For example, bipolar carrier injection can be reduced by gold doping or neutron irradiation, which in turn decreases the emitter injection efficiency. Insuring a long base region decreases the base transport factor, and this can be accomplished by adequate spacing. Reducing the feedback is most easily achieved through the layout. **Guard rings** are n^+ or p^+ regions which encircle transistor regions to offset latchup. Tying a guard ring to V_{DD} or V_{SS} alters the voltage distributions on the chip so that the parasitic transistors do not turn on; guard rings are discussed in Section 6.4.

Deep trench isolation can also be used to reduce the possibility of latch-up Figure 6.9 illustrates adjacent nMOS and pMOS transistors separated by deep trenches. Parasitic bipolar transistors are not found in the structure since the isolating pn-junctions have been replaced by an oxide barrier.

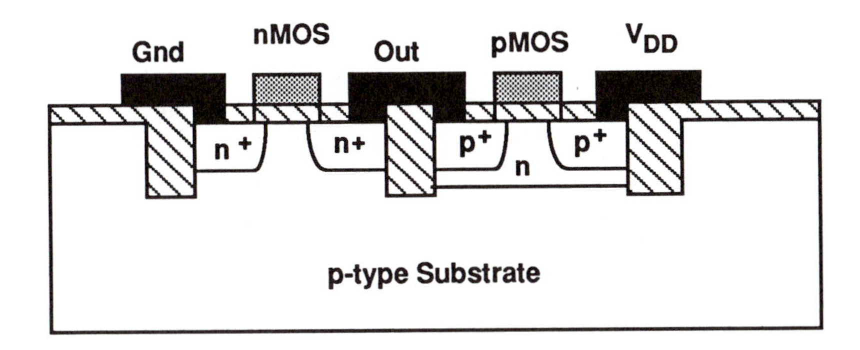

Figure 6.9: Trench-Isolated CMOS

Latchup prevention is an important aspect of CMOS chip layout and design. One should always check to insure that all suggested rules have been followed to guard against the problem.

6.2.3 Silicon-On-Insulator (SOI) Techniques

Insulating substrates such as sapphire provide an alternate approach to CMOS. An example of SOS (silicon-on-sapphire) CMOS is shown in Figure 6.10. Sapphire is chosen as the substrate because it is an excellent insulator and can be lattice-matched to silicon. Integrated circuits are created by first growing an epitaxial layer of silicon on the sapphire. Devices are formed on the epi layer and most interconnects are obtained from subsequent poly or metal patterning. Parasitic conduction between transistors is eliminated by physically separating transistor regions, so that latchup is not a problem.

Using an insulator as a substrate (instead of a semiconductor) reduces the parasitic capacitances in the circuit layout, leading to higher switching speeds. This can be seen by noting that the interconnect capacitance per unit area is approximated be

$$C_{int} = \frac{\epsilon_{int}}{x_{int}} \quad [\mathrm{F/cm^2}] \tag{6.9}$$

as discussed later in Section 6.5. In SOS, x_{int} is the thickness of the sapphire insulator, typically on the order of a millimeter. Using $\epsilon_{int} \simeq (10.5)\epsilon_o$ for sapphire gives an order of magnitude of $C_{int} \sim 100$ [pF/cm^2]. This is much smaller than that computed for standard bulk silicon where $x_{int} \simeq x_{FOX} \sim 0.1[\mu\mathrm{m}]$.

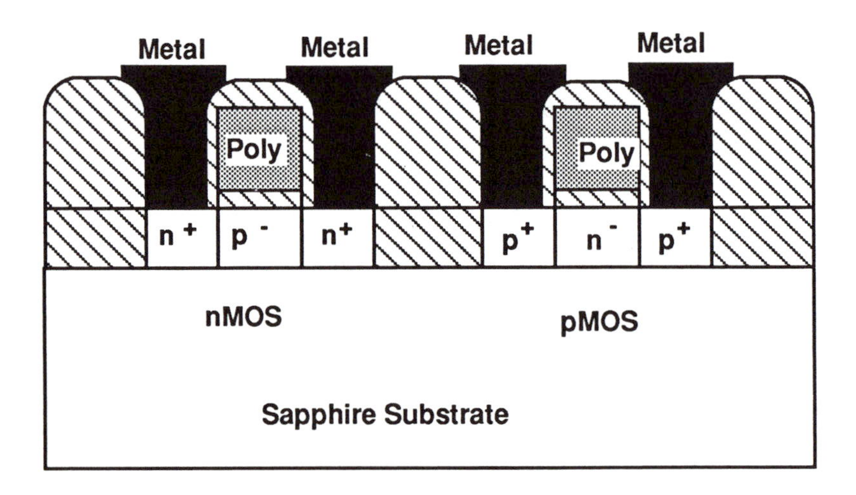

Figure 6.10: SOS CMOS

The main drawbacks of SOS CMOS are (a) the need for a high quality sapphire substrate, and, (b) the requirement for the silicon epitaxial layer. Both increase the cost and complexity of the circuit. SOI techniques are generally reserved for those cases where the higher performance is mandatory.

Recent work has demonstrated the feasibility of an SOI technology which uses an oxide as the insulator. A schematic illustration is shown in Figure 6.11. Silicon dioxide is thermally grown on a silicon substrate. Trenches are etched in the oxide layer and new silicon crystal is grown using selective epitaxy.

6.3 Design Rules

Design rules are sets of geometrical specifications which govern chip design for a given fabrication process. The layout rules are statements of the geometrical limits placed on the mask patterns and includes items such as minimum widths, dimensions, and spacings. Violating the design rules can lead to a geometry which cannot be replicated in the fabrication line, yielding a non-functional circuit. Designers are often saved from simple mistakes by the omnipotent **design rule checker** (DRC) used to find lay-

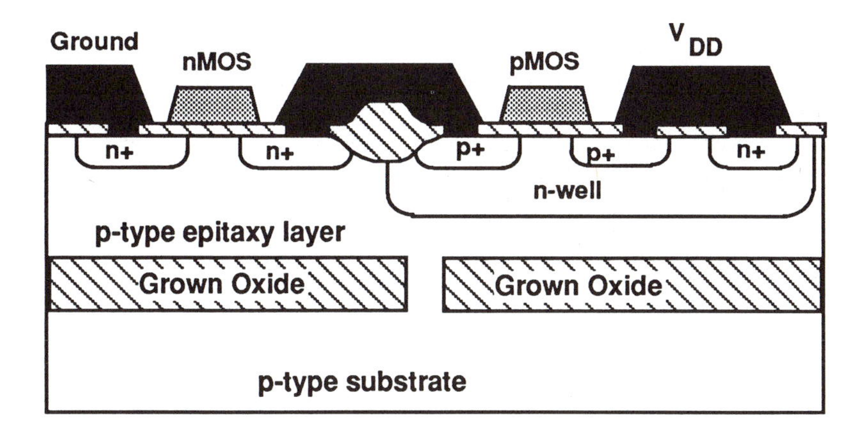

Figure 6.11: Silicon-on-Oxide SOI

out violations. Another important fact is that parasitic circuit component values are a direct consequence of the layout geometry. Since the layout is an integral part of the circuit design, it is important to examine how a design rule set affects the overall performance.

6.3.1 Lithography and Fabrication

Microelectronic lithography is the science (or, art, depending on your viewpoint) of transferring a pattern to each layer of material in an integrated circuit. The **resolution** of the lithography limits the smallest line dimension and constitutes a metric for the surface dimensions. The most common approach is **optical lithography** which uses an ultraviolet light source through a patterned mask to selectively expose a light-sensitive photoresist layer. Alternate approaches include **electron-beam** and **X-ray** sources; these offer finer resolution but introduce other problems. X-ray lithography currently appears to be the likely winner in the next generation, but recent advances in e-beam systems still look promising.

Regardless of the approach, the resolution is limited by *diffraction effects* which occur whenever a wave passes by an opaque edge. This results in the **minimum linewidth** specification in the design rule set and may be viewed as the smallest mask dimension which can be reliably transferred to the chip surface. UV optical lithography has a minimum linewidth on the order of about 0.5 micron; e-beam systems can pattern down to one-tenth of a micron or less.

Diffraction also limits how small we can make the spacing between two lines; this consideration gives a set of minimum spacing allowances in the design rule set. Minimum spacings also are needed to account for misaligned masking steps, lateral spreading, and other problems which occur during the many weeks it takes to fabricate a wafer. Yield enhancement plays an important role in setting the final numbers.

Design rule sets may be **process specific** or **scalable**. A process- specific set gives dimensions in microns and applies to a particular fabrication line. A scalable design rule set, on the other hand, can be adjusted to work on several different process lines. The most common approach is to introduce a generic metric λ [μm], and then specify all dimensions in multiples of λ. Scalable design rule sets are quite useful when employing different silicon foundries.

6.3.2 Basic Design Rule Set

Design rules are best illustrated by example. We have chosen a 2-micron LOCOS p-tub, single-poly, double-metal process which uses 10 masks. The process flow description in Table 6.1 lists the major steps in the fabrication and indicates each mask in proper sequence. Intermediate steps such as the field implant, threshold implants, lithographic sequences, washing, etc., are not listed so as to concentrate on the main issue of design rules.

Geometrical layout rules specify minimum mask feature sizes. Rules are provided for each masking layer, and also for spacings between different layers. The former originates from lithographic constraints or physical considerations (such as the widths of depletion regions), while the latter is needed to account for registration errors in the alignment. Bloats and shrinks may be applied to selected layers during the fabrication process, but the resulting physical overlay of the structure is still represented by the layout drawing.

Table 6.2 provides a listing of design rules for a 2-micron CMOS process. These consist of minimum widths or dimensions, minimum spacings between features on the same or other layers, overlap distances, and other items of importance to the chip layout. Some examples of the design rules are shown in Figure 6.12. Ground rules are usually accompanied by a complete set of drawings to illustrate each specification. Since our interest is directed towards circuit design, we have chosen not to reproduce an entire set here. Rather, the interested reader is directed to the literature[4].

---

[4] See, for example, references [6], [12], and [15]

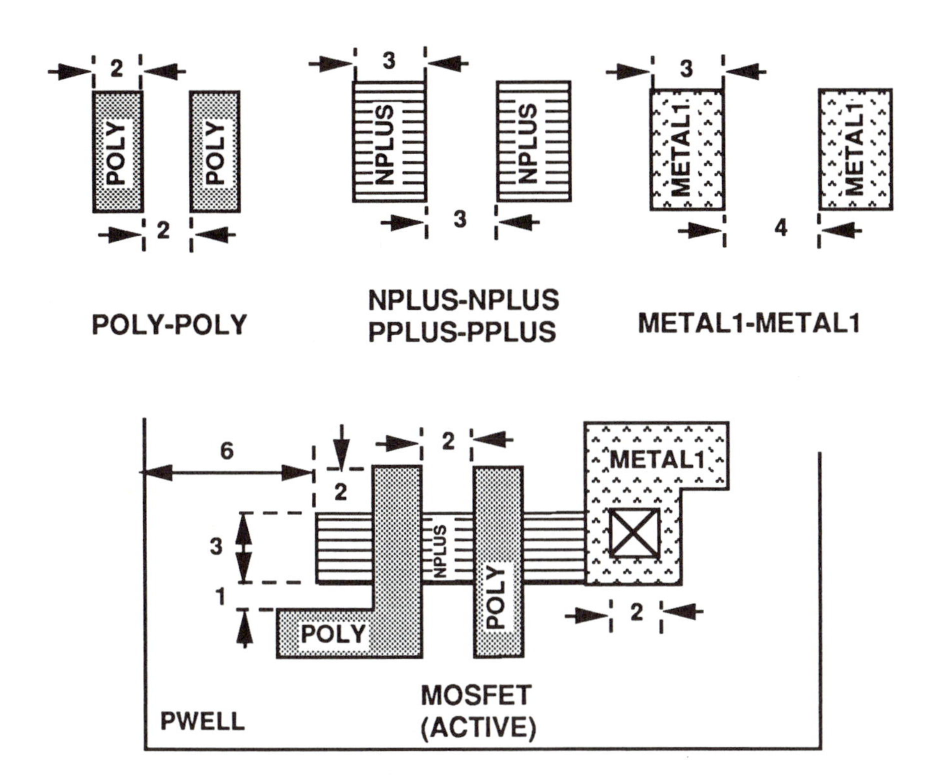

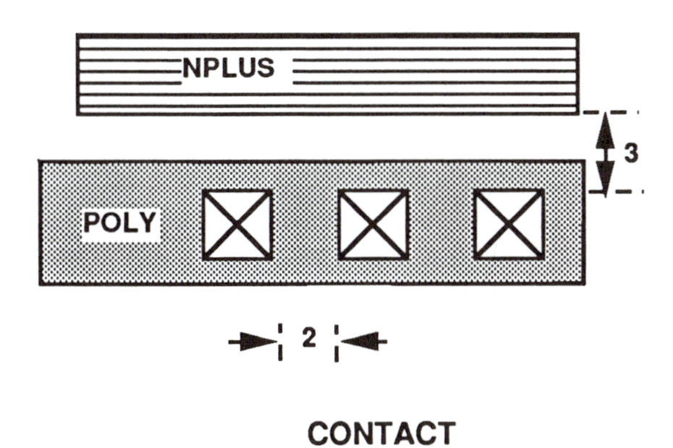

Figure 6.12: Design Rule Examples

STEP NO.	MASK NO.	LAYER NAME	Process Step
0			Start with n-type wafer
1	01	PWELL	p-tub diffusion
2	02	ACTIVE	Active area definition
3		THINOX	Grow gate oxide
4		POLY	Deposit polysilicon
5	03	POLY	Pattern polysilicon
6	04	PPLUS	p^+ implant (pMOS)
7	05	NPLUS	n^+ implant (nMOS)
8			Deposit oxide
9	06	CONTACT	Pattern poly contacts
10		METAL1	Deposit metal 1
11	07	METAL1	Pattern metal 1
12			Deposit CVD oxide
13	08	VIA	Pattern metal 2 contacts
13		METAL2	Deposit metal 2
14	09	METAL2	Pattern metal 2
15		GLASS	Nitride passivation
16	10	PAD	Pattern pad openings

Table 6.1: Basic p-Well CMOS Process Flow

An integrated circuit may be viewed as a set of overlaid geometric patterns. Each layer is shaped to provide the proper characteristics when referenced to every other layer. High-density circuit design requires compacting the geometrical patterns into a small area without violating the design rules. Circuit design is directly affected by items such as

- MOSFET (W/L) values
- Values of parasitics
- Interconnect routings
- Cell placement;

the interplay between layout rules and circuit performance is a critical connection which cannot be underemphasized. The remaining sections of the chapter are directed towards analyzing some of the more important aspects of this relationship.

MASK	VALUE [μm]	Description
01 PWELL	6	Minimum width
	4	Minimum spacing (same V)
	8	Minimum spacing (different V)
02 ACTIVE	2	Minimum width
	3	Minimum spacing
	6	Source-Drain ACTIVE to PWELL
03 POLY	2	Minimum width
	2	Minimum spacing
	2	Gate Overlap with ACTIVE
	2	POLY to NPLUS, PPLUS
	1	Field POLY to ACTIVE
04 PPLUS	3	Minimum width
	3	Minimum spacing
	3	Minimum extent past POLY
05 NPLUS	3	Minimum width
	3	Minimum space
	3	Minimum extent past POLY
06 CONTACT	2×2	Size
	2	Minimum spacing
	5	Spacing POLY to POLY
	3	Spacing POLY to ACTIVE
07 METAL1	3	Minimum width
	4	Minimum spacing
08 VIA	2×2	Size
	2	Minimum spacing
	1	Overlap with METAL 1
	2	Spacing to POLY or ACTIVE
	2	Spacing to CONTACT
09 METAL2	3	Minimum width
	4	Minimum spacing
10 PAD	100×100	Dimensions
	5	Spacing to glass edge

Table 6.2: CMOS 2-Micron Design Rule Example

An exhaustive discussion of the origin and meaning of each design rule is outside of the scope of this book. However, some important points are listed below to illustrate the physical considerations which lead to the numbers.

Active Areas

Dimensional specifications for active device areas are larger than that permitted by the lithography to account for **encroachment** from the isolation. In LOCOS, the active area defines the size of a nitride-on-oxide region which protects the silicon from oxidizing gases; the stress-relief oxide layer is required to alleviate stress which might otherwise crack the oxide. As shown in the sequence of Figure 6.13, growth of the field oxide creates the bird's beak region which must be avoided when patterning the device. The encroachment is typically about 70 % of X_{FOX}[5].

Gate Dimensions

Basic self-aligned MOSFETs are fabricated using the polysilicon gate as a mask for a n^+ or p^+ drain/source ion implant. Lateral doping affects give **effective** channel lengths which are smaller than the **drawn** values shown on the poly mask. The origin of lateral doping effects is shown in Figure 6.14. Ion implantation generally results in a fairly well defined doping edge. However, subsequent heating steps induces diffusion so that the final lateral doping is the distance L_o. Typically, L_o is on the order of $0.75x_j$, where x_j the vertical junction depth. (This, of course, is the origin of the gate overlap capacitance C_o.) This also applies to the case where a lightly-doped drain (LDD) structure is used to combat hot-electron effects.

Gate Overhang

Self-aligned MOSFETs use the gate polysilicon as a mask to the drain and source implants. To insure a functional MOSFET we require that the masks are drawn so that the poly gate extends farther than required in the W direction; Figure 6.15 shows the geometry. Providing for a gate-overhang allowance compensates for mask misalignment between the poly and n^+ or p^+ regions. If the gate overhang is reduced to zero, then even a minor registration error would result in a shorted transistor. Since even a single fault causes a die failure, the gate overhang design rule is well worth the small amount of extra area and capacitance.

---

[5] See reference [20].

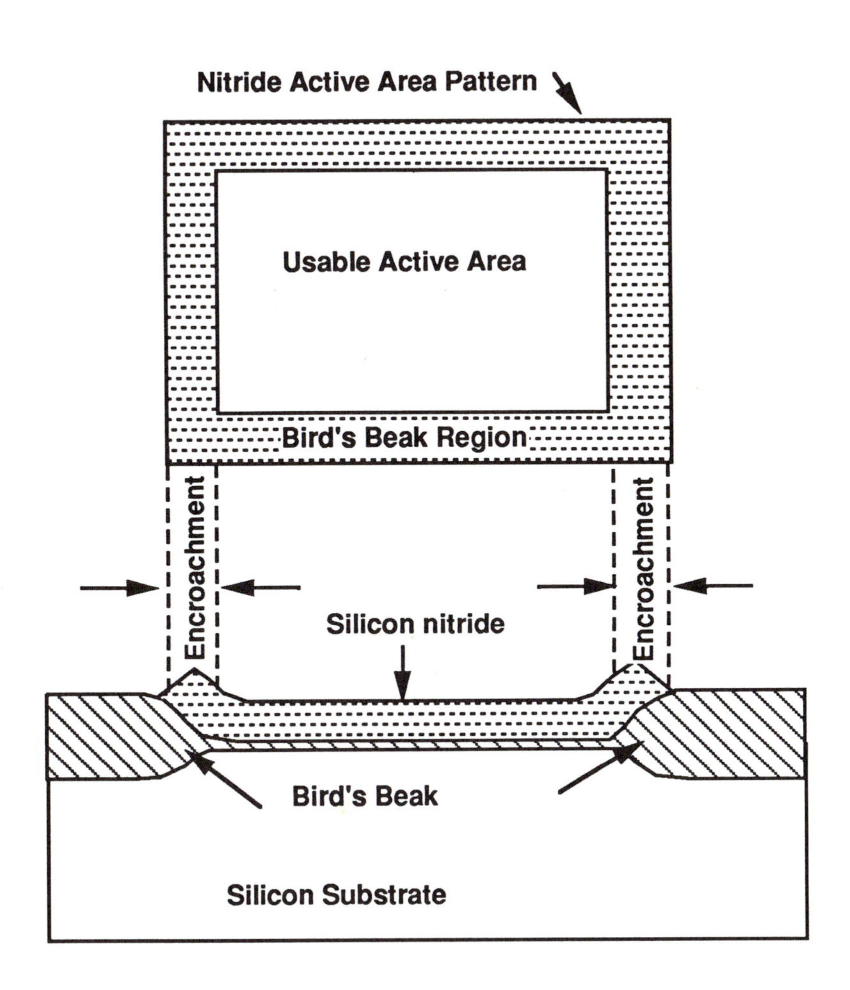

Figure 6.13: Active Area Encroachment in LOCOS

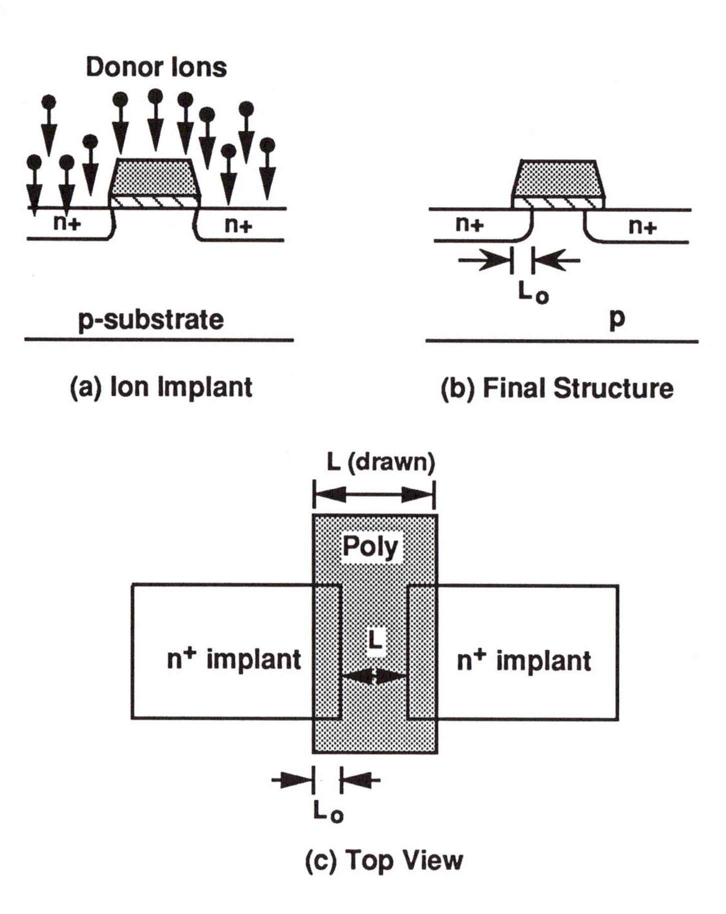

Figure 6.14: Effective Channel Length

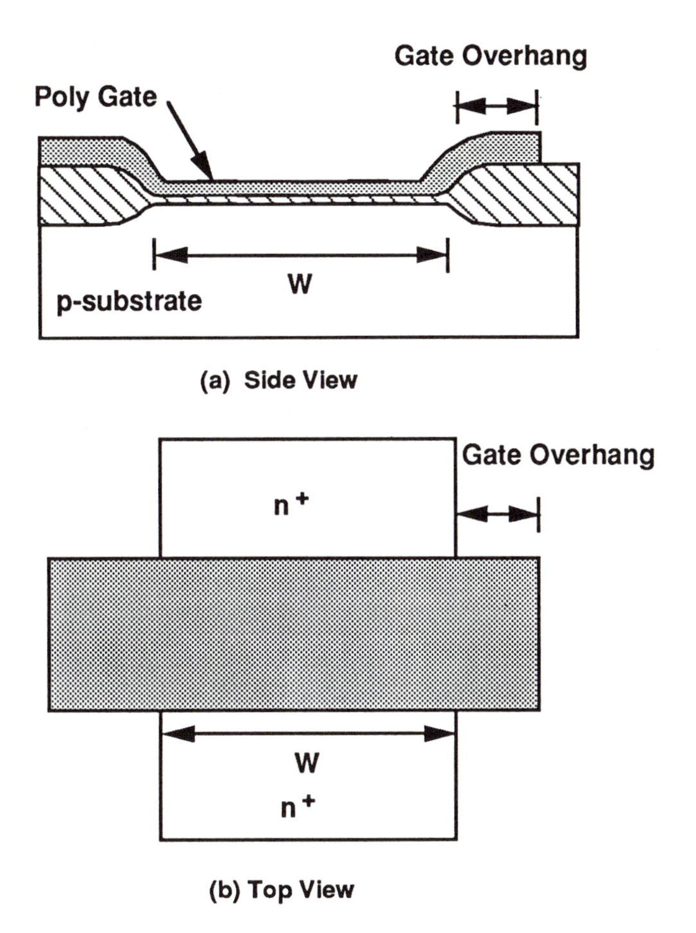

Figure 6.15: Gate Overhang

Contacts and Vias

Contact and via etches in the oxide can be troublesome failure points in a high-density layout. If the contact windows are too large, nonuniform coverage may result in void formation and other problems. The same comment also applies to oxide cuts which are too small. To avoid inducing contact-related failure modes, it is common practice to allow only one size for contact windows; large areas are connected by multiple contacts. This is illustrated in Figure 6.16.

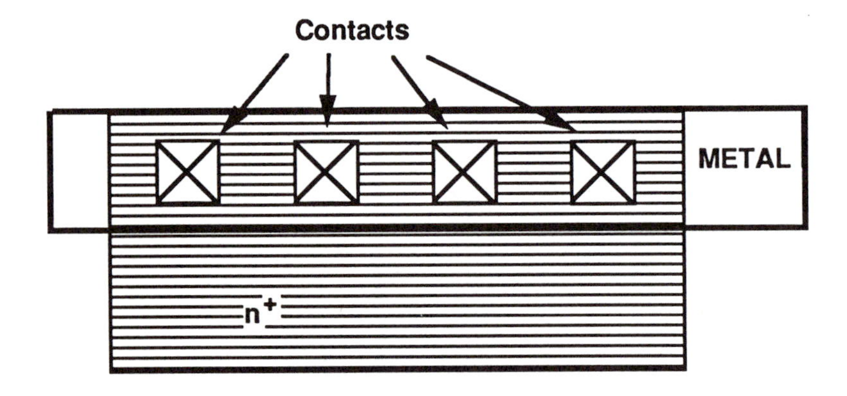

Figure 6.16: Contact Cuts

Metal Dimensions

Metal layers are deposited at the end of the fabrication sequence. They generally encounter a very rugged terrain due to patterning of the previous layers. Owing to this fact, the design rule widths and spacing must be large to insure electrical current flow. Another reason for increased widths is to allow larger current flow levels for power and ground connections. This is particularly important when aluminum (Al) is used, since high current densities create **hillocks** (where the atoms "pile up") and **voids** (where the atoms originate from) which lead to failures.

6.4 Basic Layout

Transforming schematics into physical circuits occurs during the layout process. All aspects of the circuit performance are structured by the patterning. Parasitics, interconnect coupling, and logic integration density are also determined by the geometries used in the layout artwork. Although layout is easy to learn, the interplay between the geometrical shapes and the resulting electrical behavior makes it difficult to master.

In this section we will examine some basic CMOS layout styles. We will maintain the scope of the book by restricting our discussion to simple circuits. Achieving "black-belt" status will be left as an exercise for the reader.

6.4.1 General Layout Strategies

Structured layout is based on the idea of grids and cells. The simplest approaches start with the power distribution lines V_{DD} and V_{SS} and structure the circuits as needed. Each gate is placed in a semi-rectangular cell, and cascaded logic is achieved using adjacent cells. Figure 6.17 illustrates the general idea. Both signal and power lines run horizontally in the network. Logical gates are built between metal V_{DD} and V_{SS} lines, while the signals may move between poly and metal layers when necessary. Minimization of the area is achieved by creative placement and shaping of the MOSFETs, interconnects, and cells in the overall grid structure. It is important to remember that the dimensions

- Set the electrical characteristics
- Must adhere to the design rule set

while being fit together into an interlocking puzzle.

CMOS has the added complications of

- Complementary nMOS/pMOS logic blocks
- Physical separation of nMOS and pMOS transistors

which affect the layout. Complementary structuring is illustrated in Figure 6.18. Each input is connected to both nMOS and pMOS transistors which are physically separated from one another due to the opposite background polarity requirements. For example, if an n-type substrate is used, the pMOSFETs can be placed anywhere but the nMOSFETs must be located inside of p-well regions. Since nMOS and pMOS transistors within a static gate occur in complementary series-parallel arrangements, even simple logic can introduce challenging routing problems.

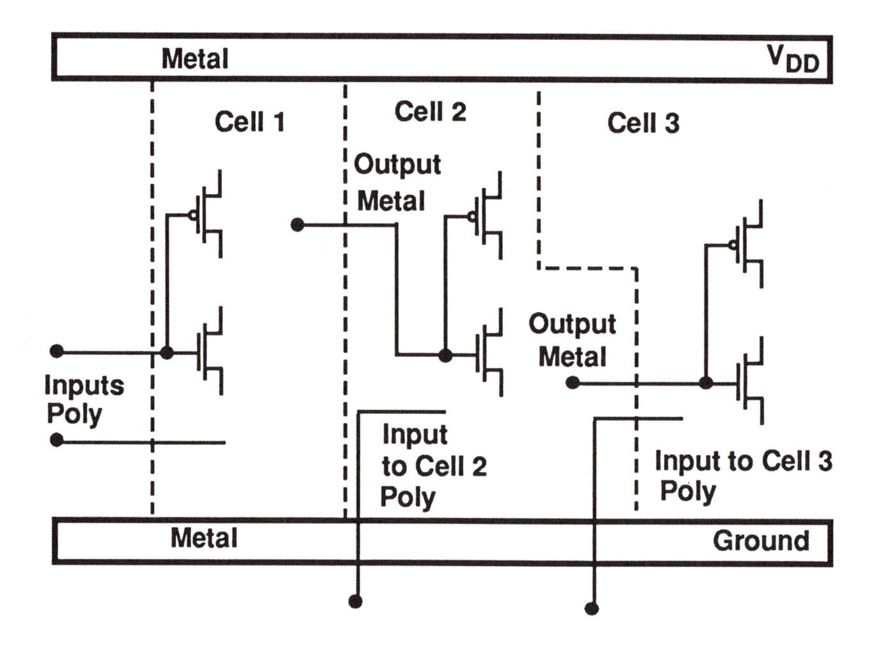

Figure 6.17: General Layout Grid

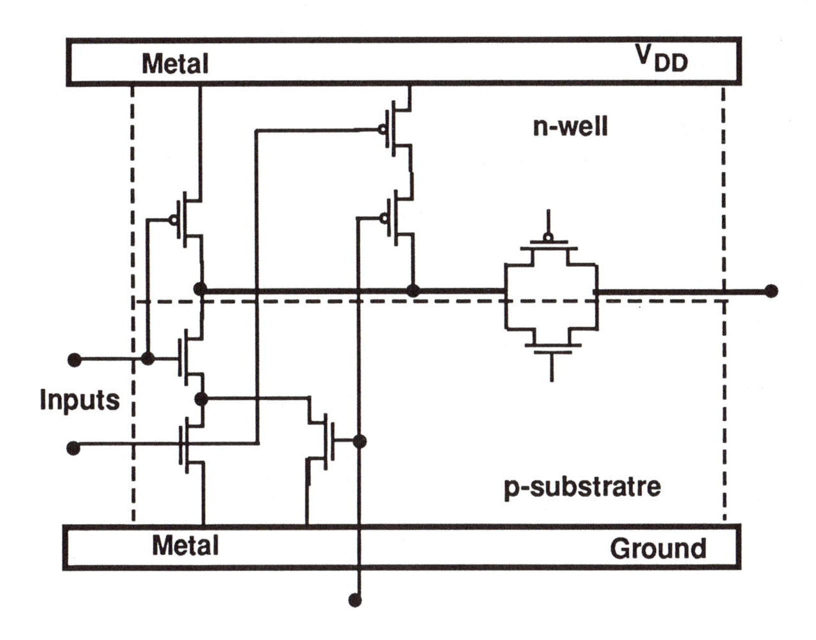

Figure 6.18: Complementary Transistor/Logic Blocks

6.4.2 Equivalent Load Concept

High-speed switching requires

- Large currents
- Small C_{out}

to insure small charging and discharging time constants. It is evident that this leads to a design problem: to increase current flow, we must use large (W/L) values for the MOSFETs[6], which in turn increases the transistor capacitances. Increasing the aspect ratios in a CMOS circuit gives larger values for both C_{in} and C_{out}, affecting the performance of the entire logic chain. In **bottom-up** design, we attempt to optimize each gate, both intrinsically and with respect to its nearest neighbors.

The concept of the **equivalent load** helps the initial layout problem by defining "standard" transistor or logic gate capacitances which are used as a reference. All loads are then specified by the number of equivalent loads. A common choice is a minimum-area transistor as shown in Figure 6.19. Assuming drawn gate dimensions of $(W \times L_d)$, where $L_d = L + 2L_o$ includes the overlap, the gate input capacitance is approximated by

$$C_G \simeq C_{ox} W L_d. \tag{6.10}$$

An inverter made using minimum area nMOS and pMOS transistors has an input capacitance of approximately

$$C_{in} \simeq 2C_G \tag{6.11}$$

which becomes our reference value.

To use the equivalent load concept, we assume that the circuit we are designing must drive a load of value

$$C_L = nC_{in}, \tag{6.12}$$

where n is a scaling factor indicating the size of the transistors used in the next gate. For example, $n = 2$ may imply a single gate with MOSFETs which are twice as large as the reference, or a fan-out $FO = 2$ into two minimum size gates. The circuit is designed according to the assumed load value. After the design of the logic chain is completed, we recheck the circuit to insure that the actual switching performance is acceptable.

---

[6] The aspect ratio becomes large because lithographic and fabrication constraints give a minimum channel length L_{min}.

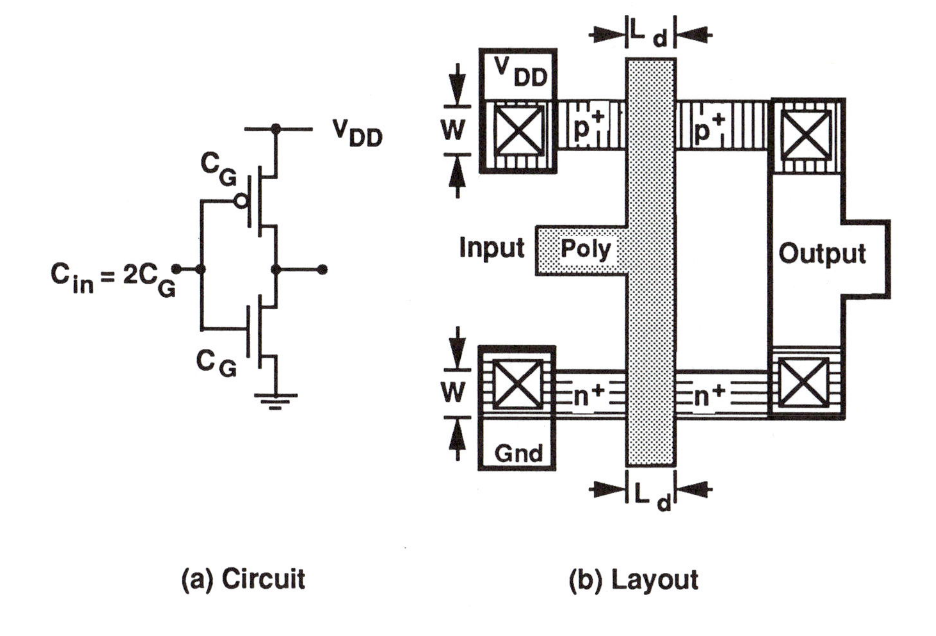

Figure 6.19: Equivalent Load

Other circuit capacitances can be treated in the same manner. Depletion capacitances are estimated by writing

$$C_D = C_{j0}WY, \tag{6.13}$$

where we ignore sidewall contributions for simplicity[7]. An arbitrary transistor is then characterized by

$$\begin{aligned} C_{gate} &= nC_G \\ C_{drain/source} &= mC_D \end{aligned} \tag{6.14}$$

where n and m are scaled according to size. A minimum-area inverter can be used to compute the time constants τ_n and τ_p. Since both are proportional to the output capacitance, the performance of an arbitrary circuit can be estimated using the values of n and m. Interconnect contributions can be included in a similar manner.

Optimization of the circuit performance can also be specified at the system level and then applied to each gate. This type of **top-down** approach

[7] Ignoring the sidewall capacitance may yield significant errors, so care must be used when interpreting the results of this analysis.

has been used to estimate gate sizing rules to speed up the response of a static logic chain; the details are presented in Section 8.3 of Chapter 8. While other problems of this type have been investigated in the literature, the complexity of the parasitic elements which contribute to C_{out} limits the analyses to coarse approximations. The usefulness of the results depends on the desired level of accuracy[8]. Often the overall system complexity dictates that we sacrifice details to get to first silicon; "tweaking" can be implemented in the next mask set if needed.

In general, combining the two views offered by bottom-up (circuit level) and top-down (system level) design provides the most powerful approach to high-performance design. Large digital networks contain both critical and non-critical logic paths so that intermixing design philosophies is often required.

6.4.3 Latchup Prevention

Circuits which are fabricated in bulk CMOS (n-well or p-well) require additional safeguards to avoid latchup. A common approach is to use **guard rings** which are heavily doped n^+ or p^+ regions around MOSFETs as shown in Figure 6.20. Guard rings reduce the transistor current gain and offset the potential and are effective in preventing latchup [19]. Another common preventative measure is providing substrate bias contacts next to every MOSFET which is connected to the power supply or ground.

6.4.4 Static Gate Layout

Static CMOS gates are based on complementary nMOS/pMOS logic blocks. Cell design can be split into two tasks: transistor placement and interconnect routing. Real estate budgets often have priority status, so that some thought may be required to fit the subsystem into the allocated area. The main limitations are usually due to design rule spacings and the complexity of the interconnect topology. Other considerations which may come into play include the shape of the allocated area, location of input and output lines relative to neighboring logic units, and clock distribution[9].

Several standard layouts have been published in the literature and can be used directly or with only minor modifications (see, for example, [8]). Some of the more interesting designs are based on the complementary placement of opposite polarity MOSFETs. Consider a NOR2 gate. This circuit

[8] A related question occurs in transistor modeling: when is the simple resistor-switch model of a MOSFET sufficient?

[9] Clocking problems are discussed in Chapters 7 and 8

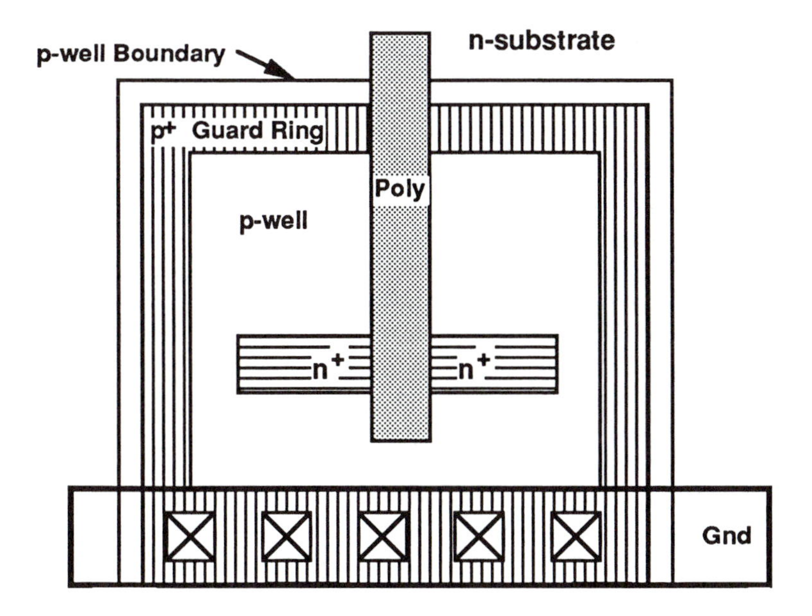

Figure 6.20: Guard Ring Example

uses 2 nMOS transistors in parallel and 2 pMOS transistors in series. Figure 6.21 shows how the complementary arrangement can be implemented by using similar transistor arrays with different interconnect patterning. Reversing the transistors in the NOR2 gate in Fig. 6.21(a) directly yields the NAND2 gate shown in Figure 6.21(b).

AOI, OAI, and non-series/parallel gates can be constructed using similar techniques. Although some layouts are based on the schematic patterning, these do not generally yield minimum-area circuits. Thoughtful use of transistor arrays and interconnect routing is usually required; intelligent CAD/CAE tools may also prove helpful.

6.4.5 TG-Based Logic

The layout of transmission-gate logic circuits is complicated by the TG itself. The switch uses parallel-connected nMOS and a pMOS transistors which reside in opposite-polarity backgrounds. Consider, for example, a p-well process. The p-channel transistor is located on the n-substrate, while the nMOSFET is in a p-well region[10]. Two extreme layout philosophies are (a) use a p-well for every TG, or, (b) use a single p-well for all transmission gates in the circuit. These are illustrated in Figure 6.22. Approach (a) reduces integration density due to the p-well spacing requirement, but is easy to replicate on a CAD systems; (b), on the other hand, may provide higher logic density, but has a larger capacitance from the extra interconnect. Although both are used in practice, minimizing the number of wells is usually the preferred strategy. Since each well requires a connection to either V_{DD} or V_{SS}, this also aids in power distribution.

A critical aspect of high-speed CMOS layout is control of the parasitic capacitance values.

6.5 Interconnects

Interconnect limitations become painfully apparent in high-density design. In "olden days", concern was directed towards the device number and sizes when calculating real estate budgets. In modern designs with micron-sized transistors, device count is no longer a limiting problem. Instead, the interconnect requirements begin to dominate the layout.

Important aspects of interconnects in chip design include

- Width

[10] It is this problem that led us to examine the split-polarity array in Figure 5.29 of the previous chapter.

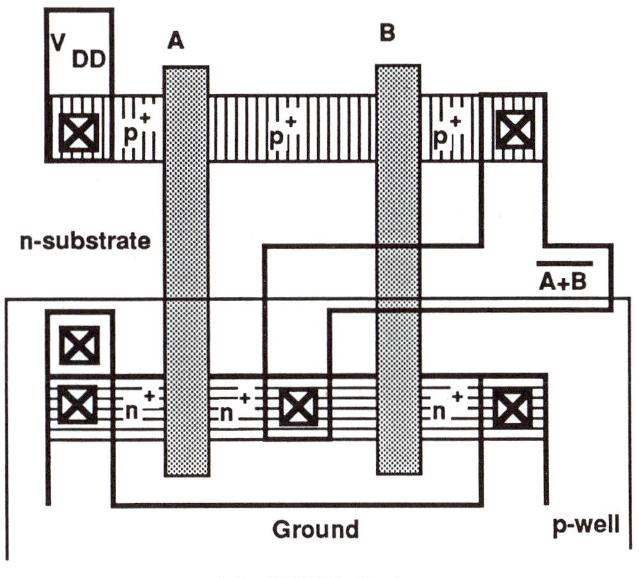

(a) NOR2 Gate

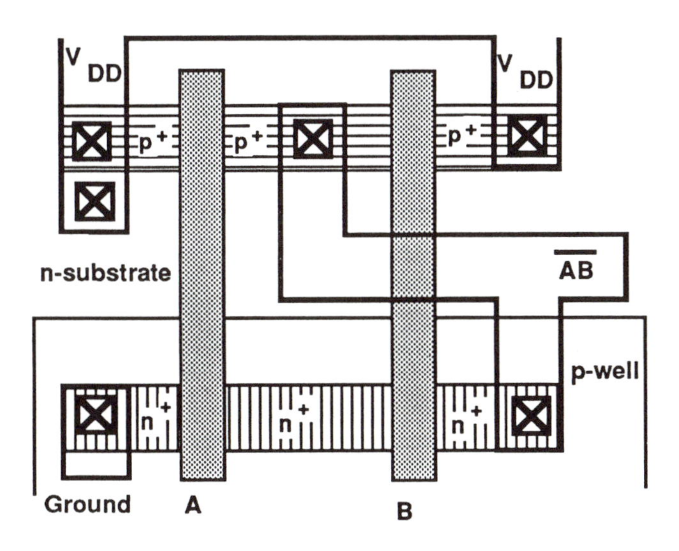

(b) NAND2 Gate

Figure 6.21: Complementary Static Gates

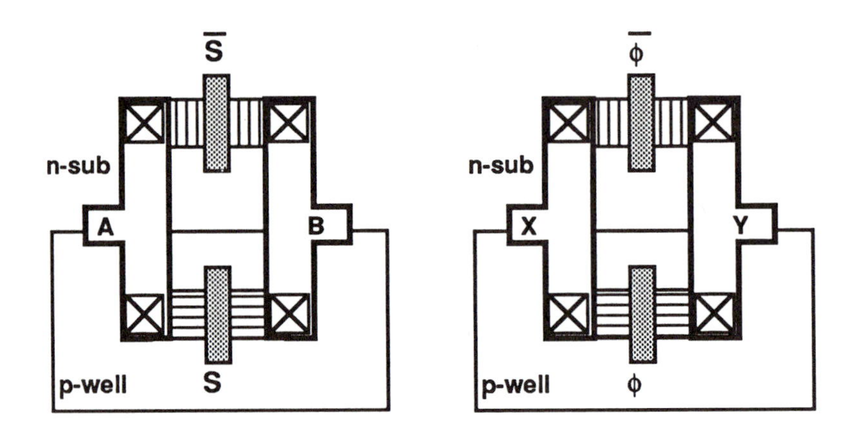

(a) Individual wells

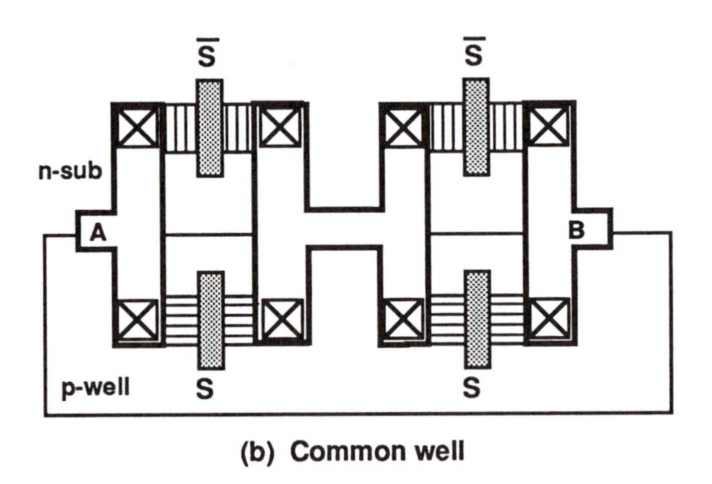

(b) Common well

Figure 6.22: TG Layouts

- Spacing
- Parasitics
- Routing and vias

all of which are important in the layout. Design rules deal with the fabrication technology and address the first three items, while the fourth deals with approaches working within a given process. We will discuss the problem by assuming that minimum width and spacing dimensions are given, and concentrate on parasitics and routing strategies.

6.5.1 Parasitics

Parasitic elements arise in the form of stray resistance, capacitance, and inductance. These are unavoidable in a realistic chip environment and generally constitute a limiting factor in the circuit performance.

Resistance

Every region on an integrated circuit has finite conductivity σ $[\Omega - \text{cm}]^{-1}$ [or, equivalently, a resistivity $\rho = (1/\sigma)$] which gives a parasitic resistance. For the simple rectangular geometry shown in Figure 6.23, we define the **sheet resistance** R_s $[\Omega/\Box]$ by

$$R_s = \frac{1}{\sigma t}, \tag{6.15}$$

where t is the thickness of the layer. R_s is the value of the "end-to- end" resistance of a square with dimensions $w \times w$. In general the line resistance is found using

$$R_{line} = R_s n, \tag{6.16}$$

with $n = (L/w)$ as the **number of squares** seen by the current. Corners and bends can be included by introducing "equivalent squares" which account for current crowding and bending electric fields[11].

Capacitance

The most critical interconnect parasitic is the line capacitance. C_{line} is always important to the previous driving circuit, and may constitute a significant fraction of the total output capacitance of the stage. Capacitance calculations can be quite involved, particularly for inhomogeneous problems

[11] In practice, corners are often counted as 1 square for simplicity.

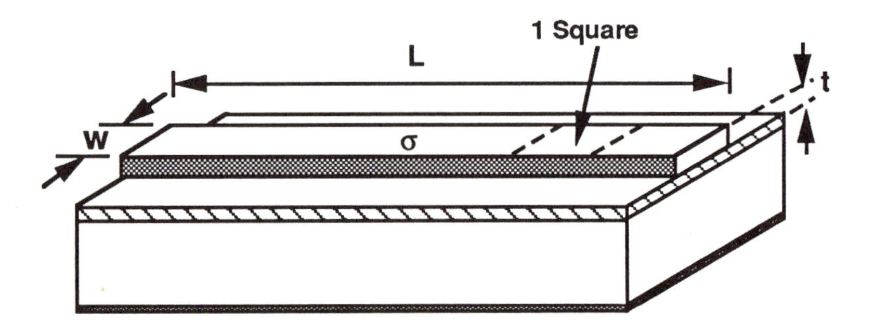

Figure 6.23: Interconnect Resistance

where different dielectrics are present. We will examine three approximations for the line capacitance with increasing precision.

The simplest model is based on the parallel-plate geometry shown in Figure 6.24. Denoting the oxide thickness underneath the interconnect by x_{int}, the interconnect capacitance per unit area is given by

$$C_{int} = \frac{\epsilon_{ox}}{x_{int}} \quad [\mathrm{F/cm^2}]. \tag{6.17}$$

For a line with length L and width w, the line capacitance is

$$C_{line} \simeq C_{int} L w \quad [F], \tag{6.18}$$

which neglects fringing electric fields and differences in materials.

A more accurate equation is obtained by modeling the line as having a microstrip geometry [16]. Figure 6.24 shows the basic geometry; the silicon substrate is taken as the ground plane for the conducting line. Empirically, the capacitance per unit length is approximated by

$$C \simeq \epsilon_{ox}[1.15(\frac{w}{x_{int}}) + 2.8(\frac{t}{x_{int}})^{0.222}] \quad [\mathrm{F/cm}], \tag{6.19}$$

where t is the thickness of the interconnect material. The additional factors account for fringing effects and can be significant.

The most precise model for the line capacitance is obtained by including the presence of the silicon substrate which has a thickness X_{Si} and recognizing that the true "ground plane" is the metal coating at the bottom of the substrate. This is illustrated by the drawing in Figure 6.26. If $w < x_{int}$,

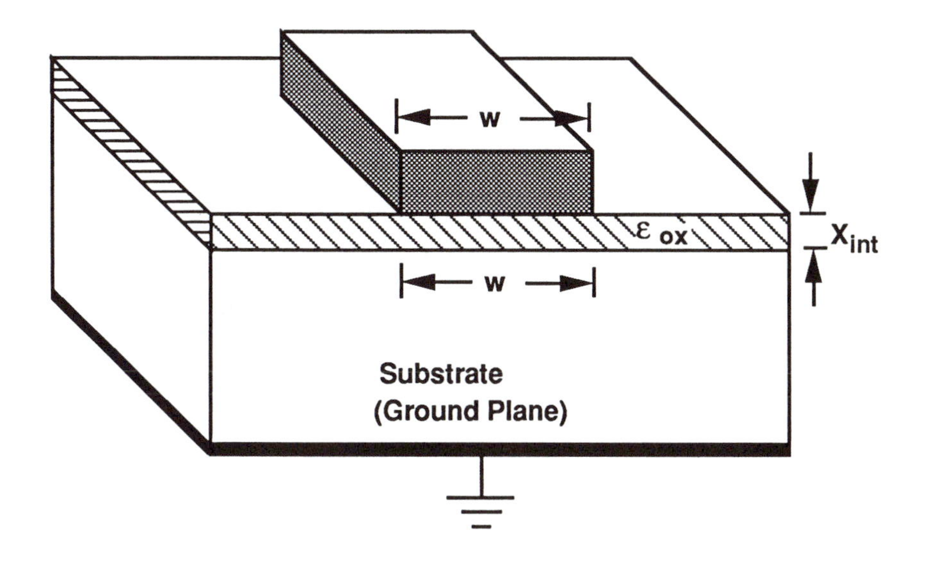

Figure 6.24: Parallel-Plate Capacitor Model

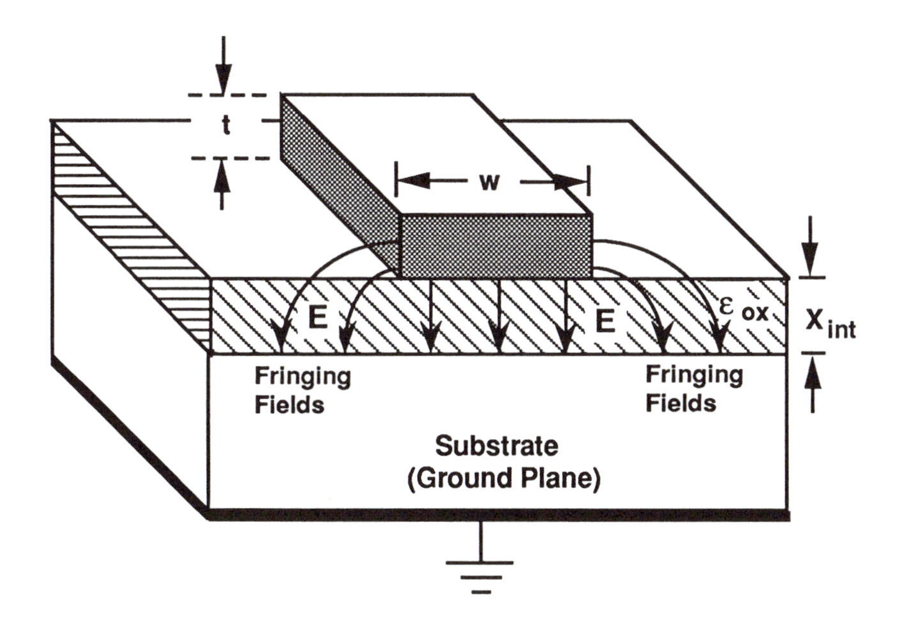

Figure 6.25: Microstrip Model for Interconnect

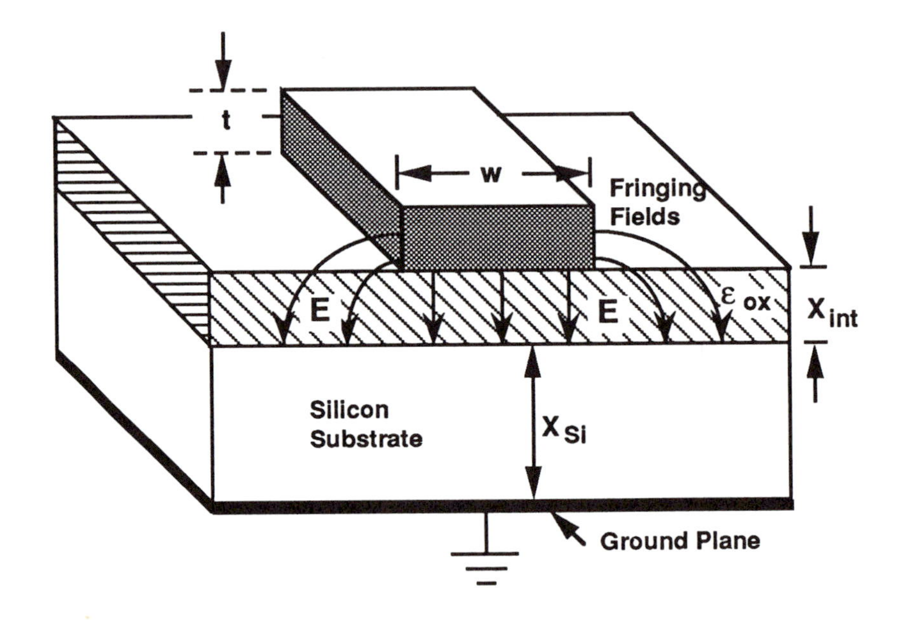

Figure 6.26: Microstrip with Silicon Substrate

then the capacitance per unit length is given by

$$\mathcal{C} \simeq \frac{2\pi\epsilon_{eff}}{\ln\left[(8x_{int}/w) + (w/4x_{int})\right]} \quad [\mathrm{F/cm}], \tag{6.20}$$

where the effective permittivity is

$$\epsilon_{eff} = \left[\frac{1}{2}(\epsilon_r + 1) + \frac{(\epsilon_r - 1)}{\sqrt{1 - (10X/w)}}\right]\epsilon_o \tag{6.21}$$

with $\epsilon_r = 3.9$ for silicon dioxide, and

$$X = x_{int} + X_{si} \tag{6.22}$$

is the total thickness of the material. For $w \geq x_{int}$,

$$\mathcal{C} \simeq \epsilon_{ox}\left[2.42 + \frac{w}{x_{int}} - 0.44\frac{x_{int}}{w} + (1 - \frac{x_{int}}{w})^6\right] \quad [\mathrm{F/cm}]. \tag{6.23}$$

This allows a more exact calculation of C_{line} in a critical circuit.

Inductance

The self-inductance of the line can also be important for high-speed switching networks. For the silicon substrate geometry we find that

$$\mathcal{L} = \frac{\mu_o}{2\pi} \ln\left[\frac{8X}{w} + \frac{w}{4X}\right] \quad [\text{H/cm}] \tag{6.24}$$

is a reasonable approximation to the inductance. Since $\mu_o = 4\pi \times 10^{-9}$ [H/cm], this reduces to

$$\mathcal{L} = 2\ln\left[\frac{8X}{w} + \frac{w}{4X}\right] \quad [\text{nH/cm}] \tag{6.25}$$

for direct application.

Self-inductance is important because of the basic $i - v$ relation

$$v = L\frac{di}{dt}. \tag{6.26}$$

High-performance CMOS circuits necessarily require fast current switching, i.e., (di/dt) can be large. The induced voltage across the parasitic inductance slows down the logic propagation through the system. Although the actual voltage is small for chip-level geometries, the noise margins and the overall performance may be affected.

EXAMPLE 4.3: Parasitic Line Elements

Consider a metal interconnect which has a width of $w = 2$ [μm] over a field oxide of thickness $x_{int} = 0.45$ [μm]. The silicon substrate is 500 [μm] thick. Assume that the interconnect has a length of $L = 50$ [μm] and a sheet resistance of $R_s = .002$ [$\Omega/\Box$]

First note that $X = 500.45$ [μm]. Since $w \geq x_{ox}$, we compute

$$\begin{aligned} C &\simeq (3.45 \times 10^{-13})\left[2.42 + \frac{2}{.45} - 0.44\frac{.45}{2} + (1 - \frac{.45}{2})^6\right] \\ &\simeq 2.379 \quad [\text{pF/cm}]. \end{aligned}$$

The inductance per unit length is

$$\begin{aligned} L &= 2\ln\left[\frac{4003.6}{2} + \frac{2}{2001.8}\right] \\ &\simeq 15.2 \quad [\text{nH/cm}] \end{aligned}$$

which provides the values for the calculations of the parasitic elements.

To apply this to the 50 [μm] interconnect, we simply multiply L and C above by the length $L = 50$ [μm] to obtain the values

$$\begin{aligned} C_{line} &\simeq 3.33 \text{ [fF]}, \\ L_{line} &\simeq 76 \text{ [pH]}. \end{aligned}$$

As a comparison, we note that using the simple parallel plate capacitance formula gives $C_{line} \simeq 7.67$ [fF]. Finally, the line resistance is given by

$$\begin{aligned} R_{line} &= R_s n \\ &= (.002)\frac{50}{2} \\ &= .05 \ [\Omega] \end{aligned}$$

where we calculated the number of squares as $n = 25$.

6.5.2 Interconnect Levels

Most state-of-the-art CMOS processes allow for at least three interconnect levels. For example, doped POLY, METAL1, and METAL2 are common in many process flows to provide for adequate layout freedom. Each level has distinct electrical characteristics so that the routing can be important to the circuit performance.

Each level possesses parasitics which couple the line to the substrate (self or intrinsic values) and to every other line. Interline coupling is primarily capacitive as shown in Figure 6.27 where a METAL line crosses over a POLY line. If we ignore fringing effects, the coupling capacitance is estimated by

$$C_{P-M1} = \frac{\epsilon_{ox} A}{x_{P-M1}}, \tag{6.27}$$

where x_{P-M1} is the thickness of the separating oxide, and A is the common overlap area shown in the drawing. This type of parasitic is more difficult to add to the data transmission analysis since it directly couples two signal lines. It may be important in noise and crosstalk analysis.

6.6 Data Transmission

Line parasitics affect the transmission of data from one gate or logic section to the next. Timing problems can often be traced to parasitic elements

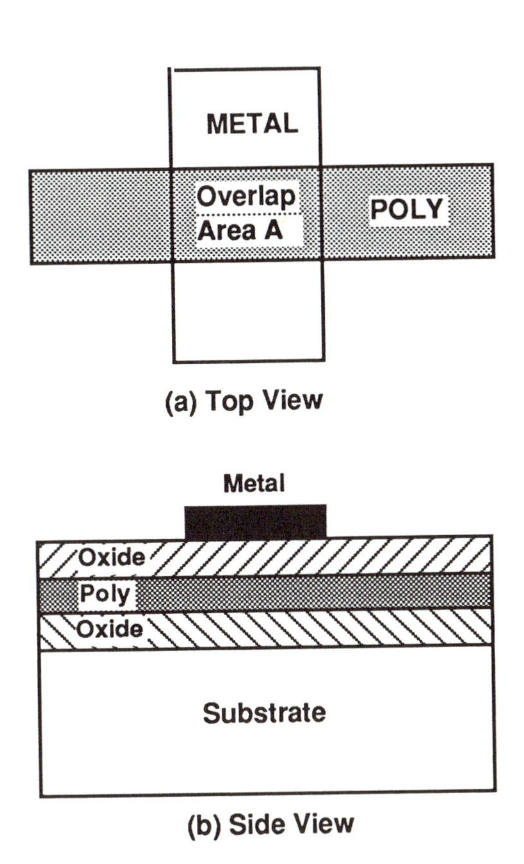

Figure 6.27: Interconnect Overlap Area

which were ignored in the initial design. As data rates increase, the effects of the line parameters become more important.

Modelling is accomplished using either lumped circuit elements or transmission lines. Transmission line theory is the most fundamental, but is much more complicated than using the lumped element approximation. Simplified circuit theory is sufficient to describe on-chip data transfer. However, output driver circuits and chip-to-chip communication paths often experience transmission line effects.

6.6.1 Basic Model

Transmission lines are characterized by the presence of distributed resistance, capacitance, inductance, and (leakage) conductance per unit length. Figure 6.28 shows the most primitive model for a differential segment of a transmission line. The parasitic elements are denoted by $\mathcal{R}$ [Ω/m], $\mathcal{C}$ [F/m], $\mathcal{L}$ [H/m], and $\mathcal{G}$ [℧/m], while dz represents the length of the line. An **ideal lossless line** is defined by having $\mathcal{R} = 0$, $\mathcal{G} = 0$; although all transmission structures exhibit losses, the ideal line is often a good approximation at the chip level, especially if the line resistance is small. We will use the ideal

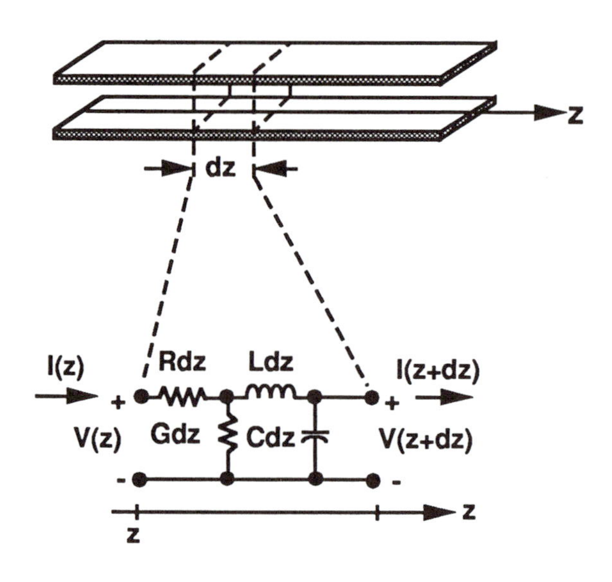

Figure 6.28: Transmission Line Model

transmission line approximation to discuss the basic properties of the line; resistive effects will be added when necessary.

A lossless line is characterized by a **characteristic impedance**

$$Z_0 = \sqrt{\frac{\mathcal{L}}{\mathcal{C}}} \quad [\Omega] \tag{6.28}$$

which gives the ratio between the voltage and current at any point z at any time t on the line:

$$\frac{V(z,t)}{I(z,t)} = Z_0. \tag{6.29}$$

The voltage and current travel as **waves** with a **phase velocity** v_p. On a simple transmission line,

$$v_p = \frac{1}{\sqrt{\mathcal{LC}}} = \frac{c}{\epsilon_r}, \tag{6.30}$$

where $c = 3 \times 10^{10}$ [cm/sec] is the speed of light in free space and ϵ_r the dielectric constant of the material between the conductors. The **delay time** t_d for a wavefront to propagate along a line of length D is given by

$$t_d = \frac{D}{v_p}; \tag{6.31}$$

for silicon dioxide ($\epsilon_r = 3.9$), the delay time is approximated by

$$t_d \simeq 65.83\ D\ [\text{ps}] \tag{6.32}$$

where D is in units of centimeters. This can be used to determine whether the circuit approximations are sufficient, or if a full transmission line approach is required.

Consider a circuit with an output low-to-high time of t_{LH}. It can be shown that transmission line effects are important if

$$t_{LH} \leq 2.5 t_d. \tag{6.33}$$

On the other hand, if

$$t_{LH} \geq 5 t_d, \tag{6.34}$$

then a lumped element approach is sufficient [4]. This rule of thumb indicates that for CMOS switching times on the order of nanoseconds, dimensions less than about 6 [cm] can be treated using lumped element approximations. On-chip interconnects, therefore, do not generally require transmission line descriptions.

The above criteria can be understood using qualitative reasoning. Consider the basic transmission system shown in Figure 6.29 where a source drives a transmission line of length D which is connected to the input of logic gate; the terminating circuit is referred to as the load. The delay time t_d represents the time required for a pulse to move from the source to the load. When the pulse reaches the load, some of the energy is reflected

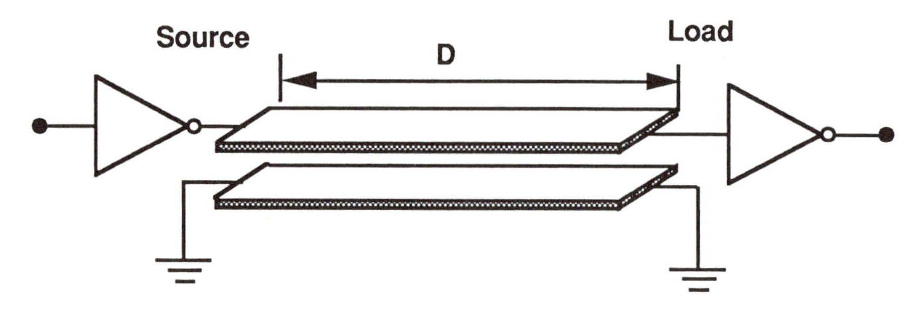

Figure 6.29: Basic Transmission System

back towards the source (as discussed in detail later). The reflected pulse reaches the source at a time $2t_d$, the round-trip delay time. This affects the operation of the source circuit since the reflected pulse changes the voltage and current, but KVL and KCL must still be satisfied.

Short lines give small values of $2t_d$. The condition that $t_{LH} > 5t_d$ for using lump element analysis is equivalent to saying that (dV_{out}/dt) is large compared to 2-plus round trip line delays. Qualitatively, this implies that reflections are occurring, but that (a) the voltage per reflection is very small, and (b) the reflections occur so rapidly that they are averaged into the overall response. The opposite extreme where transmission line effects are observed is when $t_{LH} < 2.5t_d$. In this situation, (dV_{out}/dt) is very large so that the entire pulse has been transmitted before the first reflected pulse arrives back at the source. Since the amplitude of the pulses are large, their effects will be much more noticeable and must be treated with transmission line theory.

6.6.2 Lumped-Element Analysis

Most on-chip circuits can be analyzed using lumped-element interconnect models. This approach accounts for presence of parasitic capacitance, resistance, and inductance on the line using discrete circuit components. Figure 6.30 illustrates the general problem for the basic inverter. Before proceeding to analyze the transient response, let us compare the model to the basic circuit in Figure 3.5 of Chapter 3.

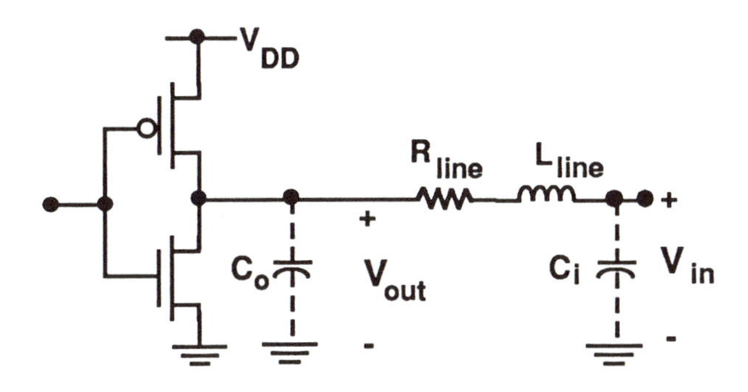

Figure 6.30: Lumped-Element Interconnect Model

In the basic inverter analysis, C_{out} was assumed to be a single element which dominates the output node. This simplifies the transient response calculations. However, if we include the interconnect properties as shown, two modifications must be made. First, C_{out} is now split into at least two distinct contributions by means of

$$C_{out} \rightarrow C_o + C_i \tag{6.35}$$

where C_o is the capacitance seen directly at the gate output node, and $C_i = C_{line} + C_{in}$ with C_{in} the input capacitance of the next stage. In a simple model, C_o and C_i are separated by the interconnect. Second, both R_{line} and L_{line} slow down the signal. The parasitic resistance always induces a voltage drop, while the parasitic inductance blocks rapid changes in the current flow. Implementing these two changes gives the output circuit shown in Figure 6.30

To understand the effects of the line parasitics, we apply KVL and the capacitor $i - v$ relation to obtain

$$V_{out} = V_{in} + R_{line}C_i\frac{dV_{in}}{dt} + L_{line}C_i\frac{d^2V_{in}}{dt^2} \tag{6.36}$$

as the second-order equation describing the interconnect circuit. Note the basic resonant frequency

$$\omega_o = \frac{1}{\sqrt{L_{line}C_i}} \tag{6.37}$$

of the LC combination. During an output transition from 0 to V_{DD}, the pMOS transistor conducts according to

$$I_{Dp} = C_o \frac{dV_{out}}{dt} + C_i \frac{dV_{in}}{dt}. \tag{6.38}$$

Combining the (6.37) and (6.39) gives a differential equation for $V_{in}(t)$. A high-to-low discharge through the inverter nMOS transistor is described by a similar equation.

Several important points can be made using simplified modeling (thus allowing us to avoid a brute-force analysis !). First, define

$$\alpha = \frac{1}{2R_{line}C_i}, \tag{6.39}$$

so that the roots of the homogeneous second-order equation are given by

$$\begin{aligned} s_1 &= -\alpha + \sqrt{\alpha^2 - \omega_o^2}, \\ s_2 &= -\alpha - \sqrt{\alpha^2 - \omega_o^2}. \end{aligned} \tag{6.40}$$

This gives rise to the standard analysis for overdamped, underdamped, and critically damped response [3]. For the limiting case where $R_{line} \to 0$ such as for a metal interconnect, high-frequency LC oscillations may occur.

Next consider the effect of R_{line}. The time constant

$$\tau_{line} = R_{line}C_i \tag{6.41}$$

provides a measure of the RC delay introduced by the parasitics. Poly interconnects tend to have sheet resistances on the order of $R_s \sim 20$ [$\Omega/\Box$], so that τ_{line} may be large. On the other hand, silicides or metal layers exhibit sheet resistances of only a few milliohms/square, so that R_{line} will usually be negligible.

Including multiple discrete elements in the interconnect model increases the accuracy of the simulation. Figure 6.31 shows an equivalent circuit where we have broken down the resistance and capacitance contributions according to $R_{line} = R_1 + R_2 + R_3$ and $C_i = C_1 + C_2 + C_3$. The time constant for this circuit can be approximated using the RC charging model introduced in Chapter 4 for MOSFET chains. Ignoring L_{line} gives a $(1/e)$ delay time of

$$t_D = R_1C_1 + (R_1 + R_2)C_2 + (R_1 + R_2 + R_3)C_3 \tag{6.42}$$

for this arrangement. The last term dominates the expression, and is itself smaller than the value of τ_{line} obtained using the simpler model. Line inductance L_{line} can also be decomposed in a similar manner; however, the performance of on-chip interconnects is not generally limited by inductive effects, so that we will avoid the added complexity.

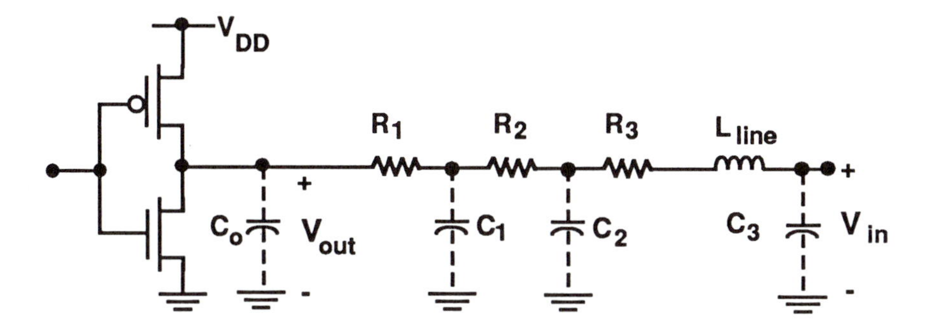

Figure 6.31: Multiple Lumped-Element Model

6.7 Transmission Line Analysis

Transmission line effects are important when the length of the interconnects exceeds a few centimeters. This condition generally holds for chip-to-chip and higher system-level connections. Finite signal propagation speeds and reflections from mismatched connections combine to slow the data rate. These considerations are of particular concern when designing off-chip driver circuits.

6.7.1 Wave Properties

An analysis of the transmission line segment in Figure 6.28 gives that the voltage $V(z,t)$ and current $I(z,t)$ satisfy the coupled equations [14]

$$\begin{aligned} \frac{\partial V(z,t)}{\partial z} &= -\mathcal{R}I(z,t) - \mathcal{L}\frac{\partial I(z,t)}{\partial t} \\ \frac{\partial I(z,t)}{\partial z} &= -\mathcal{G}V(z,t) - \mathcal{C}\frac{\partial V(z,t)}{\partial t}. \end{aligned} \tag{6.43}$$

For most structures of interest to chip designers, $\mathcal{G}$=0 since SiO_2 is an excellent insulator. Eliminating the current yields

$$\frac{\partial^2 V}{\partial z^2} = \mathcal{RC}\frac{\partial V}{\partial t} + \mathcal{LC}\frac{\partial^2 V}{\partial t^2} \tag{6.44}$$

for the voltage $V(z,t)$ on the line. Assuming that line losses are negligible, we approximate $\mathcal{R} \sim 0$ to arrive at the ideal voltage **wave equation**

$$\frac{\partial^2 V}{\partial z^2} = \frac{1}{v_p^2}\frac{\partial^2 V}{\partial t^2} \tag{6.45}$$

where $v_p = (1/\sqrt{\mathcal{LC}})$ is the phase velocity.

The second-order wave equation has two linearly independent solutions. We construct

$$V(z,t) = V^+(z - v_p t) + V^-(z + v_p t) \tag{6.46}$$

as the general wave form. In this expression, V^+ and V^- are arbitrary functions of the composite variables $(z - v_p t)$ and $(z + v_p t)$, respectively. The phase fronts are defined by points where the arguments are constants, i.e.,

$$z \mp v_p t = \text{constant}. \tag{6.47}$$

Using this interpretation, we see that the first term $V^+(z - v_p t)$ represents a wave with constant phase points which travel according to $z = v_p t$; this corresponds to a wavefront moving in the $+z$ direction. The second term $V^-(z + v_p t)$ is a wave in the $-z$ direction since the wavefronts move with $z = -v_p t$. Current and voltage on the line are related by means of the basic equations, which give the general solution

$$\begin{aligned} I(z,t) &= \frac{V^+(z - v_p t)}{Z_0} - \frac{V^-(z + v_p t)}{Z_0} \\ &= I^+(z - v_p t) + I^-(z + v_p t), \end{aligned} \tag{6.48}$$

with Z_0 as the characteristic impedance. The first term is a positively-moving $(+z)$ current wave; the second term represents current flow in the $-z$ direction as indicated by the negative sign.

6.7.2 Basic Properties

The characteristics of transmission lines are best illustrated by example. Consider first the basic problem shown in Figure 6.32 where an ideal voltage source sends a voltage wavefront of height V_0 down a transmission line of length D. The line has a characteristic impedance Z_0 and is terminated with a load resistor R_L. The wavefront travels at a velocity v_p, so that the leading edge of the pulse does not reach the end of the line until a time

$$t_d = \frac{D}{v_p}. \tag{6.49}$$

Introducing the unit step function

$$u(z - v_p t) = \begin{cases} 0 & (z < v_p t) \\ 1 & (z \geq v_p t) \end{cases} \tag{6.50}$$

allows us to write

$$\begin{aligned} V^+(z,t) &= V_0\, u(z - v_p t), \\ I^+(z,t) &= \frac{1}{Z_0} V_0\, u(z - v_p t), \end{aligned} \tag{6.51}$$

to describe the motion of the leading pulse edges. This illustrates the basic delay in data transmission.

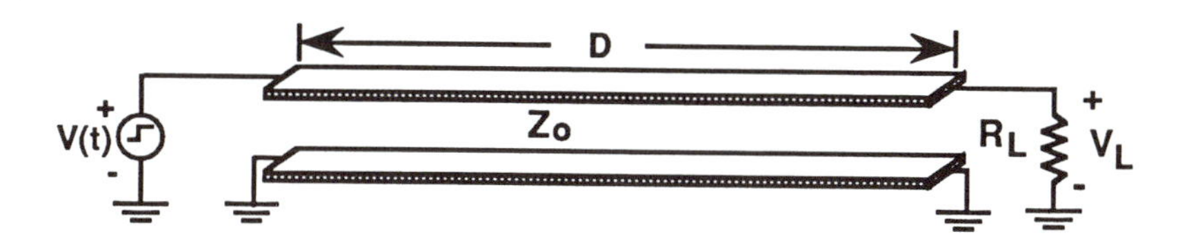

Figure 6.32: Transmission Line Example

When the wavefront reaches the load, Ohm's Law must be satisfied:

$$V_L = I_L R_L. \tag{6.52}$$

To meet this constraint and also satisfy KCL, we allow for the possibility of both positive (incident) and negative (reflected) voltages and currents in the form

$$\begin{aligned} V(z = L) &= V_0 + V_r, \\ I(z = L) &= \frac{V_0}{Z_0} - \frac{V_r}{Z_0}, \end{aligned} \tag{6.53}$$

where V_r is the amplitude of the reflected voltage wave. Substituting into Ohm's Law gives the load **reflection coefficient**

$$\Gamma_L = \frac{V_r}{V_0} = \frac{R_L - Z_0}{R_L + Z_0}. \tag{6.54}$$

If $R_L = Z_0$, then we have a **matched load** and there is no reflected wave: $\Gamma = 0$ and $V_r = 0$. However, if the load is not matched to the line ($R_L \neq Z_0$), then a reflected wavefront is established with an amplitude

$$V_r = \Gamma_L V_0. \tag{6.55}$$

This adds to the voltage V_0 which was set up during the first positive pass. Since Γ_L can be positive or negative, the reflected wave may increase or decrease the voltage on the line.

The reflected wave reaches the voltage source at a time $2t_d$. Since we have modeled the source as ideal, $R_S = 0$. The source reflection coefficient is then

$$\Gamma_S = \frac{R_S - Z_0}{R_S + Z_0} = -1, \tag{6.56}$$

which says that the reflected wave "bounces" off the source with a change in polarity. This second-pass forward wave has an amplitude $V^{2+} = -V_r$ and moves towards the load. Summing the voltage at the input gives

$$\begin{aligned} V_S &= V_0 + \Gamma_L V_0 + \Gamma_S(\Gamma_L V_0) \\ &= V_0 + V_r + (-V_r), \end{aligned} \tag{6.57}$$

so that the source reflection cancels the effect of the previous negatively travelling wave and reestablishes $V = V_0$ on the line at the source end (as required). When V^{2+} hits the load, it reflects an amount

$$V^{2-} = \Gamma_L V^{2+} = \Gamma_L[\Gamma_S(\Gamma_L V_0)]. \tag{6.58}$$

The bouncing motion continues in time and $V_L(t)$ can be obtained by summing the voltage waves at the load.

Analyzing the problem from a physical viewpoint we see that after the transient effects have decayed away, the load voltage must converge to

$$V_L = V_0 \quad (t \rightarrow \infty) \tag{6.59}$$

since the transmission line just becomes a wire. This can be verified by the reflection coefficient formalism by summing all of the waves across the load. The algebra leads to the infinite series

$$\begin{aligned} V_L &= V_0\Big[1 + (\Gamma_L\Gamma_S) + (\Gamma_L\Gamma_S)^2 + \ldots\Big] \qquad (6.60) \\ &\quad + V_0\Gamma_L\Big[1 + (\Gamma_L\Gamma_S) + (\Gamma_L\Gamma_S)^2 + \ldots\Big], \end{aligned}$$

which sums to a value of

$$V_L = V_0\Big(\frac{1 + \Gamma_L}{1 - \Gamma_L\Gamma_S}\Big). \tag{6.61}$$

Substituting the values of Γ_S and Γ_L shows that the proper asymptotic value is reached.

6.7.3 Capacitive Load

A common MOS transmission problem is a transmission line terminated in a load capacitance C_L as shown in Figure 6.33. This can be used to model the input to a MOSFET gate, so it is quite useful to analyze the problem in some detail. It is assumed that the step voltage source is matched to the load with $R_S = Z_0$. At time $t = 0$, the wavefront "sees" an impedance of Z_0 at the input to the transmission line. Voltage division thus gives

$$V^+ = \frac{1}{2}V_0 \tag{6.62}$$

for the launched wavefront amplitude. Using physical reasoning we see that

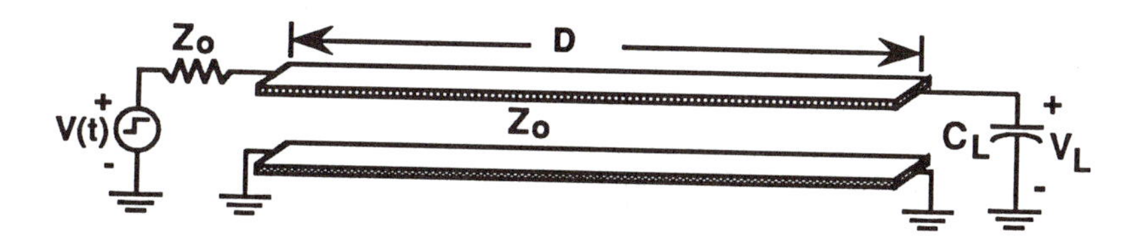

Figure 6.33: Transmission Line With a Capacitive Load

the capacitor initially acts as a short circuit when the wavefront strikes at $t = t_d$. A reflected wave is generated to satisfy the load current condition

$$i_L = C_L \frac{dV_L}{dt} \tag{6.63}$$

where both V_L and i_L are made up of incident and reflected wave contributions. As $t \to \infty$, C_L goes to an open circuit and $V_L \to V_0$.

To analyze the reflection, we Laplace transform and work in s-domain where the load impedance is given by $Z_L = (1/sC_L)$. The reflection coefficient is then

$$\rho(s) = \frac{(1/sC_L) - Z_0}{(1/sC_L) + Z_0} \tag{6.64}$$

such that

$$V^-(s) = \rho(s)V^+(s) \tag{6.65}$$

gives the s-domain reflected wave. The reflected wave is found by first writing the time-domain incident wave as $V^+(t) = (V_0/2)u(t - t_d)$ which Laplace transforms to

$$V^+(s) = \frac{V_0}{2s}e^{-st_d} \tag{6.66}$$

in the s-domain. Substituting and inverse transforming yields

$$V^-(t) = V_0[\frac{1}{2} - e^{-(t-t_d)/\tau}] \quad (t \geq t_d) \tag{6.67}$$

for the reflected wave. In this equation,

$$\tau = Z_o C_L \tag{6.68}$$

is the load time constant. Since we have matched the source resistance to the line, there are no more reflections to contend with and this completes the wave analysis.

The capacitor voltage is easily computed as

$$\begin{aligned} V_L(t) &= V^+ + V^- \\ &= V_0[1 - e^{-(t-t_d)/\tau}]u(t - t_d), \end{aligned} \tag{6.69}$$

which is just a delayed RC charging event. This is illustrated in Figure 6.34; again, the limit where $t \to \infty$ just corresponds to the transmission line being replaced by a simple wire. There are two time intervals of importance in

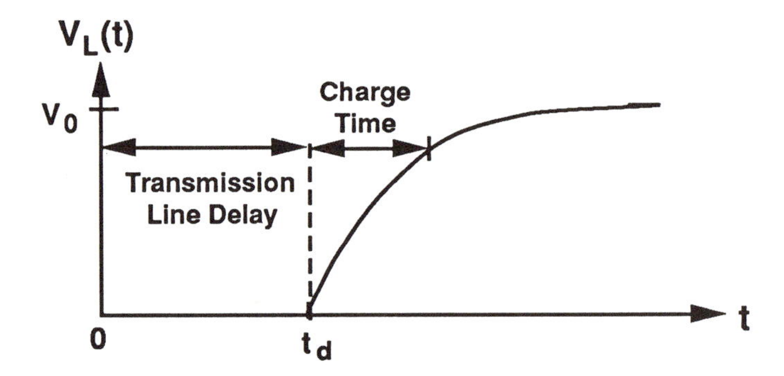

Figure 6.34: Capacitor Voltage

this problem: the transmission delay time t_d and the capacitor charge or discharge time as characterized by τ.

6.7.4 Transmission Line Drivers

CMOS driver circuits can be designed around the properties of transmission lines. Consider the circuit shown in Figure 6.35 where an inverter is connected to a transmission line with characteristic impedance Z_0. In order to minimize the problem of ringing, transistors Mn and Mp can be designed to provide an approximate impedance match. Denoting the output of the inverter as the voltage source, we set

$$\Gamma_S = \frac{R_S - Z_0}{R_S + Z_0} = 0, \tag{6.70}$$

so that mismatches at the load end will not induce long-term ringing. Note, however, that matching the source resistance to Z_0 implies that launched voltage wave only has an amplitude of

$$V^+ = \frac{1}{2} V_{out} \tag{6.71}$$

due to normal voltage division. This can affect the actual transmission rate of high voltage levels.

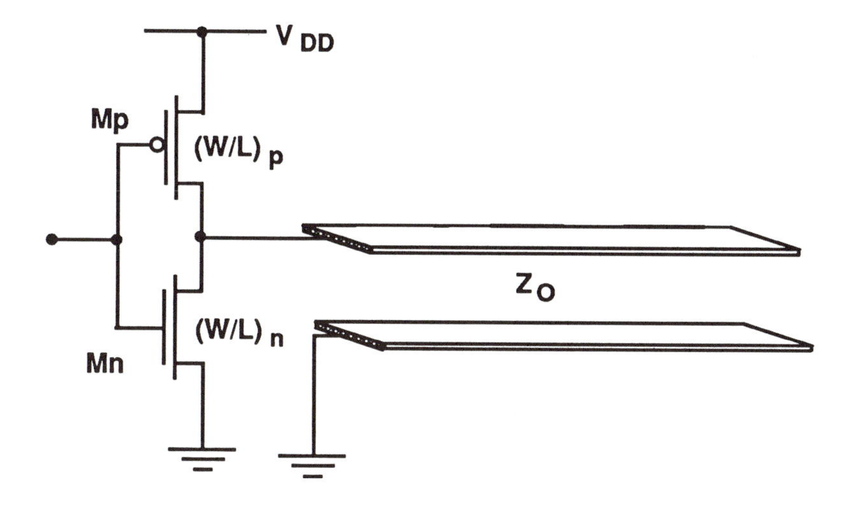

Figure 6.35: Transmission Line Driver

Designing a matched circuit is achieved using the concept of equivalent resistances. Consider a low-to-high transition at the line input. MOSFET Mp conducts so that we will match the equivalent transistor resistance $R_p \simeq [\beta_p(V_{DD} - |V_{Tp}|)]^{-1}$ to the line impedance. This requires an aspect ratio of

$$\left(\frac{W}{L}\right)_p \simeq \frac{1}{k'_p Z_0 (V_{DD} - |V_{Tp}|)} \tag{6.72}$$

which is sufficient for a first approximation. When there is a high-to-low transition, the n-channel transistor Mn conducts. Applying the same approach gives

$$\left(\frac{W}{L}\right)_n \simeq \frac{1}{k'_n Z_0 (V_{DD} - V_{Tn})} \tag{6.73}$$

for choosing the nMOS aspect ratio.

Matching the individual transistors to Z_0 assumes that the gate input voltage has step-like characteristics. In a realistic switching event, simultaneous conduction in Mn and Mp occurs, altering the design formalism. Suppose we choose the midpoint voltage $V_{out} = (1/2)V_{DD}$ as reference. Since R_n and R_p are in parallel[12], our design equation is

$$Y_0 = k'_n \left(\frac{W}{L}\right)_n (V_{DD} - V_{Tn}) + k'_p \left(\frac{W}{L}\right)_p (V_{DD} - |V_{Tp}|), \tag{6.74}$$

where $Y_0 = (1/Z_0)$ is the characteristic admittance. Individual aspect ratios are chosen to satisfy the switching time constraints. The pMOS value $(W/L)_p$ is the most critical in this regard since $k'_p < k'_n$.

The values of $(W/L)_n$ and $(W/L)_p$ calculated using either approach should only be viewed as approximations. Computer-aided design and simulation results allows for more precise matching.

6.7.5 RC Lines

Most on-chip interconnect structures exhibit small values for the line inductance $\mathcal{L}$. If we assume that $\mathcal{R}$ and $\mathcal{C}$ are the dominant parasitics, then the differential line model can be simplified to that shown in Figure 6.36. Analyzing the behavior of voltage on a distributed RC network provides an understanding of an important signal delay contribution. The coupled equations describing voltage and current are now given by

$$\begin{aligned} \frac{\partial V(z,t)}{\partial z} &= -\mathcal{R} I(z,t), \\ \frac{\partial I(z,t)}{\partial z} &= -\mathcal{C}\frac{\partial V(z,t)}{\partial t}, \end{aligned} \tag{6.75}$$

which combine to give the **voltage diffusion equation**

$$\frac{\partial^2 V(z,t)}{\partial z^2} = \mathcal{RC}\frac{\partial V(z,t)}{\partial t}. \tag{6.76}$$

[12] In an incremental sense.

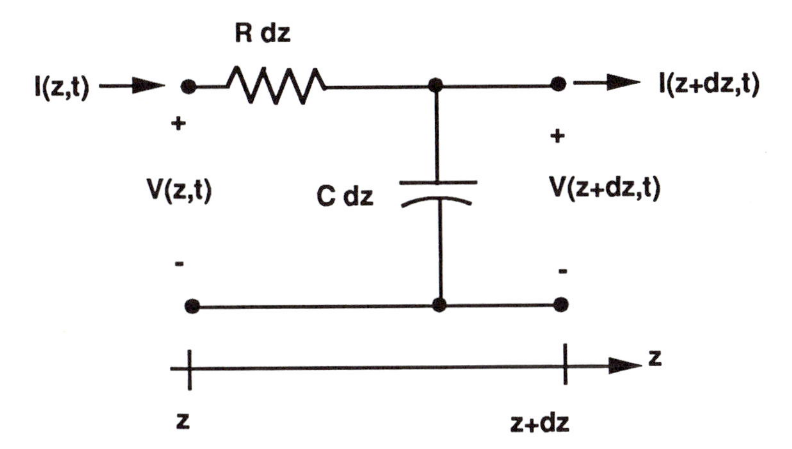

Figure 6.36: RC Line Model

Since the resonant $\mathcal{L}-\mathcal{C}$ circuit has been altered, this does not admit wave solutions. Instead, this equation describes the buildup of voltage along a purely distributed RC interconnect.

Consider the problem illustrated in Figure 6.37 where an input pulse is applied to the distributed RC line. Laplace transforming the diffusion

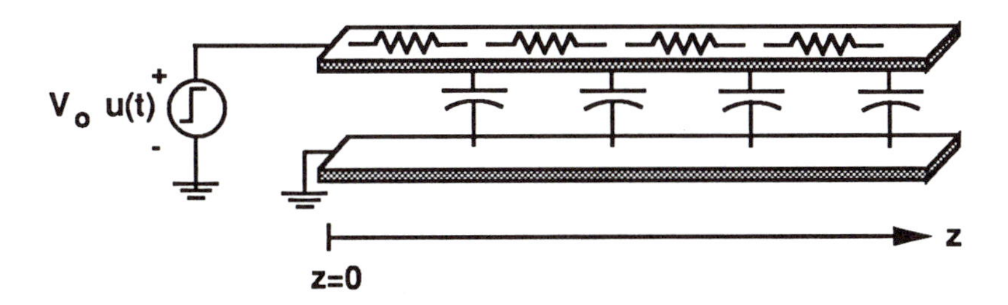

Figure 6.37: Voltage on a Distributed RC Line

equation gives an s-domain expression of

$$\frac{\partial^2 V(z,s)}{\partial z^2} - s\mathcal{RC}V(z,s) = -V(z,0) \tag{6.77}$$

with $V(z, 0)$ the initial condition. The homogeneous equation has general solutions of

$$V(z, s) = V^+ e^{-\alpha z} + V^- e^{+\alpha z} \tag{6.78}$$

with $\alpha = \sqrt{s\mathcal{RC}}$, and V^+ and V^- as constants. Using a time-domain input of

$$V(0, t) = V_0 u(t) \tag{6.79}$$

and assuming an infinitely-long line so that $V(z \to \infty, t) < \infty$ gives us the s-domain expression

$$V(z, s) = \frac{V_0}{s} e^{-\alpha z}. \tag{6.80}$$

Inverse transforming back into time-domain results in

$$V(z, t) = V_0 \operatorname{erfc}\left(\frac{z}{2}\sqrt{\frac{\mathcal{RC}}{t}}\right) u(t), \tag{6.81}$$

where erfc(y) is the **complementary error function**. The current is given by

$$I(z, t) = V_0 \frac{\mathcal{C}}{\pi \mathcal{R} t} \exp\left(\frac{-\mathcal{RC}z^2}{4t}\right) u(t), \tag{6.82}$$

as can be verified from the basic $I - V$ equations.

Let us summarize some of the important properties of the complementary error function erfc(y). First, it is the complement of the **error function** erf(y) such that

$$\operatorname{erfc}(y) = 1 - \operatorname{erf}(y); \tag{6.83}$$

this, of course, is the origin of its name. The error function can be defined by the integral representation

$$\operatorname{erf}(y) = \frac{2}{\sqrt{\pi}} \int_0^y e^{-\alpha^2} d\alpha \tag{6.84}$$

with the limiting values

$$\begin{aligned} \operatorname{erf}(0) &= 0, \\ \operatorname{erf}(\infty) &= 1. \end{aligned} \tag{6.85}$$

The second equation allows us to write the complementary error function as

$$\operatorname{erfc}(y) = \frac{2}{\sqrt{\pi}} \int_y^\infty e^{-\alpha^2} d\alpha. \tag{6.86}$$

Note that

$$\frac{d[\operatorname{erfc}(y)]}{dy} = -\frac{2}{\sqrt{\pi}} e^{-y^2} \tag{6.87}$$

by direct differentiation.

Now consider the voltage expression $V(z,t)$. To understand the diffusion of the voltage we choose the argument to be some value

$$\frac{z}{2}\sqrt{\frac{\mathcal{RC}}{t}} = \xi, \tag{6.88}$$

where ξ is a constant. The voltage point

$$V_\xi = V_0 \text{erfc}(\xi) \tag{6.89}$$

moves in space z as time t increases according to

$$z = \frac{2\xi}{\sqrt{\mathcal{RC}}}\sqrt{t}, \tag{6.90}$$

which allows us to trace arbitrary values. Figure 6.38 shows the progression of the voltage for different values of the transmission line $\mathcal{RC}$ time constant. In general, the larger the value of $\mathcal{RC}$, the longer it takes the voltage to

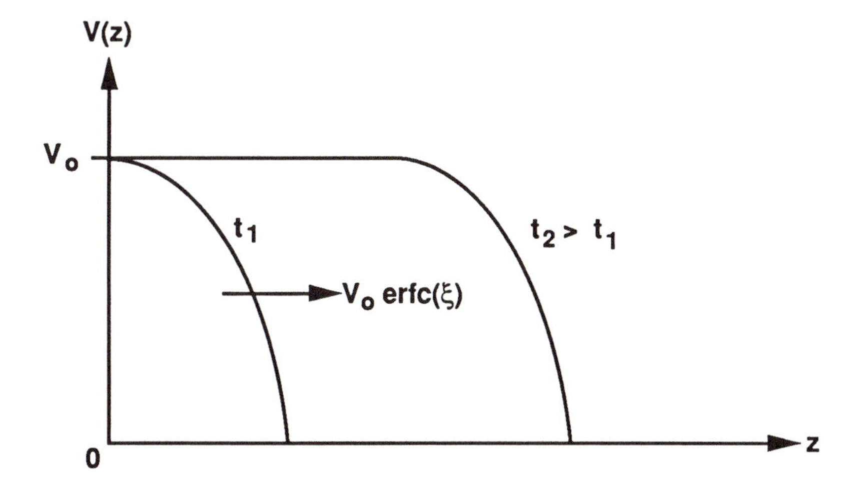

Figure 6.38: Transferral of Voltage on an RC Line

diffuse down the line as expected from physical reasoning. Lines with large $\mathcal{R}$-losses are generally avoided due to this type of behavior.

6.8 Crosstalk

Crosstalk can be a troublesome problem in high-density layouts. Consider the situation shown in Figure 6.39 where two adjacent lines are electromagnetically coupled by a parasitic capacitance C_m and a mutual inductance L_m. A voltage pulse on one line induces transients on the other. In the worst-case scenario, an incorrect or false transition will occur. Minimization of crosstalk problems is important to insure the integrity of the data transmission path. In this section we will examine the physical basis for electromagnetic coupling and analyze some basic models for typical coupling environments.

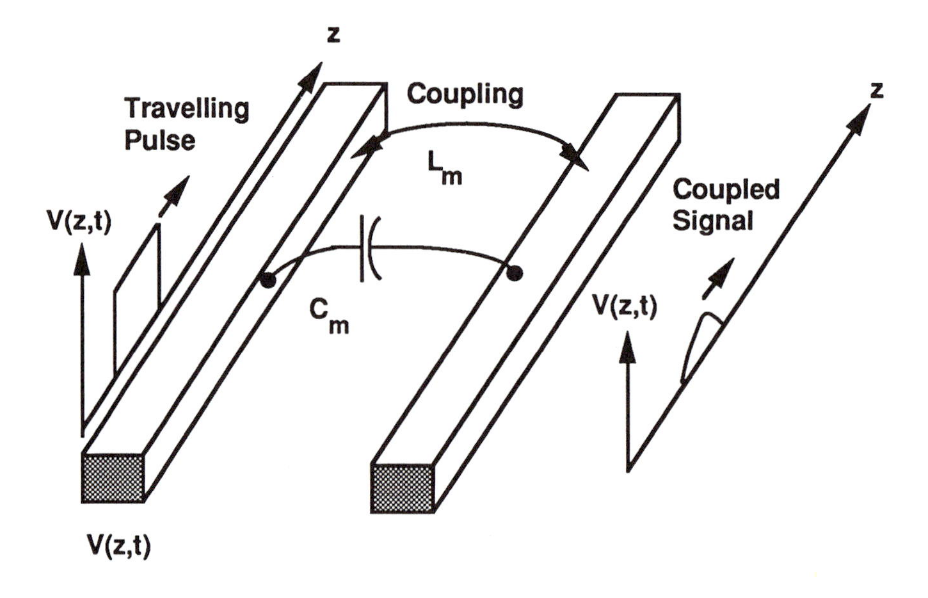

Figure 6.39: Crosstalk Example

6.8.1 Origin of Crosstalk

Crosstalk occurs because of electric and magnetic field coupling among conductors. Consider the situation where we have two interconnect lines separated by a distance d, as illustrated in Figure 6.40(a). We will denote the line voltages by V_1 and V_2, both of which are referenced to ground. In general, electric (capacitive) coupling exists between any two conductors at different voltages. To describe the coupling effect, we write the charges Q_1 and Q_2 on conductors 1 and 2, respectively, as

$$\begin{aligned} Q_1 &= C_{11}V_1 + C_{12}(V_1 - V_2) \\ Q_2 &= C_{21}(V_2 - V_1) + C_{22}V_2, \end{aligned} \tag{6.91}$$

where the parameters $C_{\alpha\beta}$ $(\alpha, \beta = 1, 2)$ are the capacitances as determined by the geometry and material properties[14]. C_{11} and C_{22} are the **intrinsic** (isolated) values, while $C_{12} = C_{21} = C_m$ is the **coupling** capacitance between the two lines. Since $I = (dQ/dt)$,

$$\begin{aligned} I_1 &= C_{11}\frac{dV_1}{dt} + C_m\frac{d(V_1 - V_2)}{dt}, \\ I_2 &= C_m\frac{d(V_2 - V_1)}{dt} + C_{22}\frac{dV_2}{dt}, \end{aligned} \tag{6.92}$$

shows that a pulsed voltage on a line can induce current flow on a neighboring interconnect. The capacitors are shown in Fig. 6.40(b).

Magnetic coupling also exists between the two lines. Let I_1 and I_2 be the currents in lines 1 and 2, respectively. The total magnetic flux Φ_α through each line is expressed by

$$\begin{aligned} \Phi_1 &= L_{11}I_1 + L_{12}I_2, \\ \Phi_2 &= L_{21}I_1 + L_{22}I_2, \end{aligned} \tag{6.93}$$

where L_{11} and L_{22} are the **self inductances**, while $L_{12} = L_{21} = L_m$ gives the **mutual inductance** which describes the magnetic coupling. Differentiating the flux expressions gives

$$\begin{aligned} V_1 &= L_{11}\frac{dI_1}{dt} + L_m\frac{dI_2}{dt}, \\ V_2 &= L_m\frac{dI_1}{dt} + L_{22}\frac{dI_2}{dt}, \end{aligned} \tag{6.94}$$

illustrating the interplay caused by the presence of magnetic coupling. Figure 6.40(c) shows the magnetic flux lines which lead to the inductance values.

6.8.2 Interconnect Coupling

Figure 6.40(a) provides the basic geometry for the coupling problem. Two interconnects (assumed to be on the same layer) are separated by a distance d. To simplify the analysis, we assume that the silicon substrate is a good

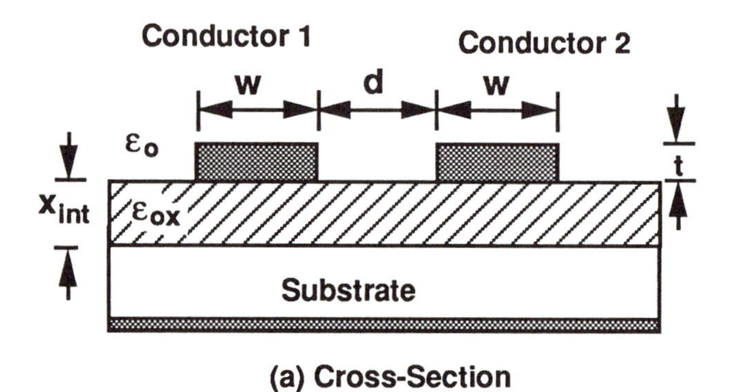

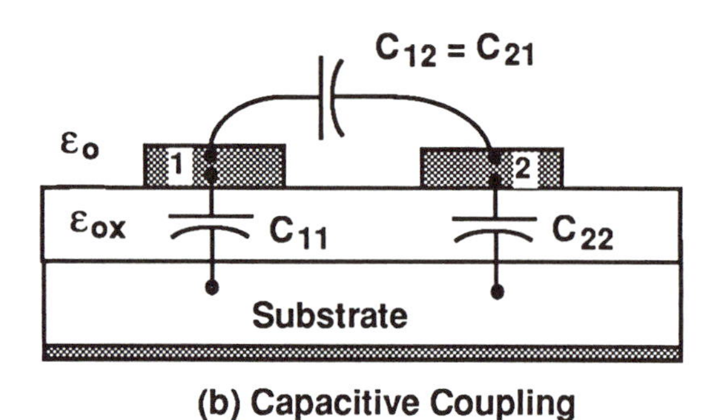

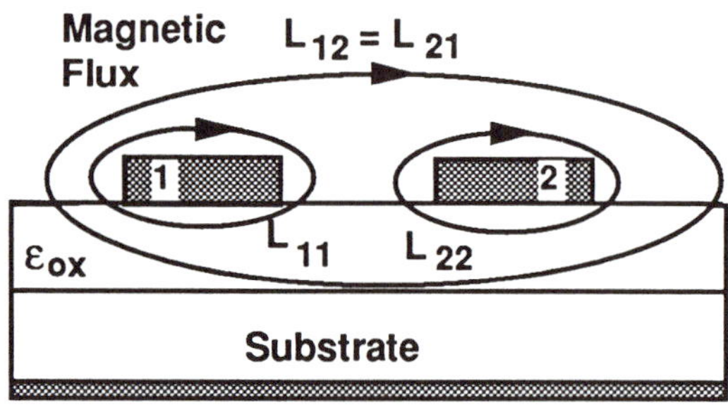

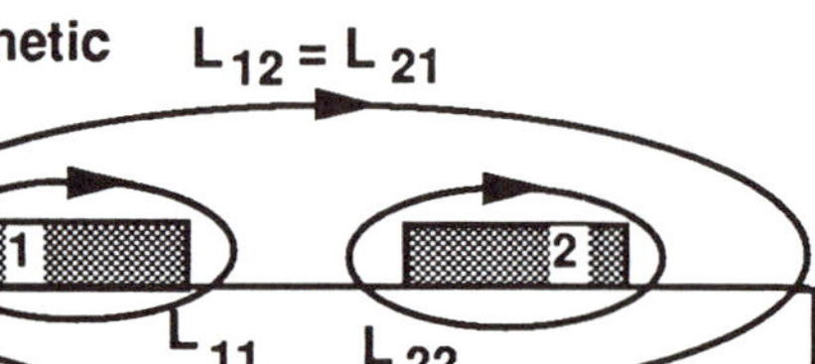

(c) Inductive Coupling

Figure 6.40: Generalized Coupling Problem

conductor and acts as a reasonable ground plane, and that the permittivity above the lines is given by ϵ_o.

Consider first the capacitors shown in Fig. 6.40(b). The intrinsic capacitance can be estimated using the equation

$$\mathcal{C} = \epsilon_{ox}\left[1.15(\frac{w}{x_{int}}) + +2.8(\frac{t}{x_{int}})^{0.222}\right] \text{ [F/cm]} \tag{6.95}$$

as quoted in Section 6.5.1. Within the same order of approximation, the coupling capacitance per unit length is estimated by

$$\mathcal{C}_m = \epsilon_{ox}\left[0.03(\frac{w}{x_{int}}) + 0.83(\frac{t}{x_{int}}) - 0.07(\frac{t}{x_{int}})^{0.222}\right](\frac{d}{x_{int}})^{-4/3} \text{ [F/cm]}. \tag{6.96}$$

This typically gives an error less than about 10 % for parametric ratios in the ranges

$$\begin{aligned} 0.3 &< (\frac{w}{x_{int}}) < 10, \\ 0.3 &< (\frac{t}{x_{int}}) < 10, \\ 0.5 &< (\frac{d}{x_{int}}) < 10. \end{aligned} \tag{6.97}$$

The coupling inductance per unit length can be estimated from

$$\mathcal{L}_m \simeq \frac{\mathcal{L}}{\mathcal{C}}\mathcal{C}_m; \tag{6.98}$$

this expression is, at best, a zero-order approximation for the present geometry due to the varying permittivity.

The limiting factor in using the above expressions arises from the observation that most interconnects in a chip environment will have a dielectric material on top; silicon dioxide or silicon nitride coatings are common. The dielectric layer increases the capacitance to a larger value than predicted by the above equations.

With the formulas above, an error in capacitance leads to an error in inductance. Even with these limitations, the formulas are sufficient for first-case analyses. In problem situations, 2D and 3D computer simulation codes, such as those based on finite-element or finite-difference techniques, are required to obtain more accurate capacitance and inductance values.

6.8.3 Circuit Coupling

The effects of crosstalk depend on the level of coupling and the switching environment. As an example, consider an inverter chain which is coupled to a neighboring line as illustrated in Figure 6.41. The presence of C_m and L_m indicates that energy can be transferred between the two circuits. The

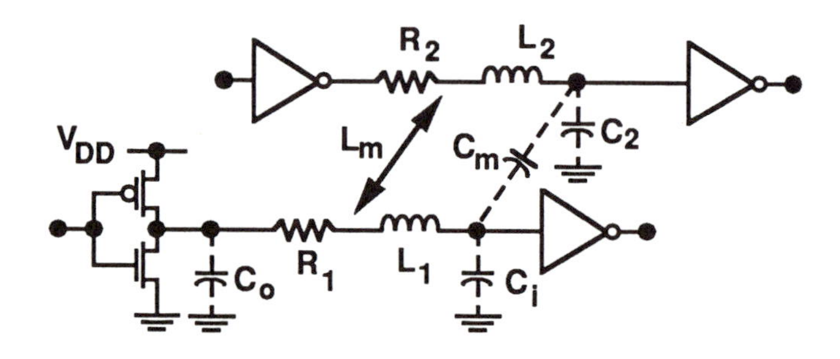

Figure 6.41: Lumped-Element Coupling Model

parasitic capacitance C_m dominates the coupling mechanism in CMOS circuits, so L_m is usually ignored. To approximate the effects using a lumped-element model, we assume that a transient voltage V_i across C_i acts as a source while line 2 is idle. The coupling through C_m induces a voltage V_2 across capacitor C_2. If the amplitude of the induced voltage exceeds V_{IH} of the inverter, a false trigger will occur. Crosstalk is usually bidirectional, so we must also be concerned with the reverse case where a pulse on line 2 upsets the voltage on line 1. Both can be approximated using standard circuit analysis.

Long coupling lengths exhibit transmission line effects, so that a distributed analysis must be used. This is illustrated in Figure 6.42; parallel lines such as these are common in data bus layouts. Although these geometries are purposely used to design directional couplers in rf and microwave circuits, they create problems in chip design where coupling is a problem (not the basis for a device). It is possible to analyze transmission line coupling for idealized geometries [17]. However, an accurate analysis of realistic problems requires computer simulations.

6.9 Gate Arrays in CMOS

Up to this point, we have examined how the layout and interconnect properties affect the performance of a CMOS circuit. In full-custom and semi-custom IC design, the location of each MOSFET is arbitrary. However, there is a large class of integrated circuits which have predefined transistor

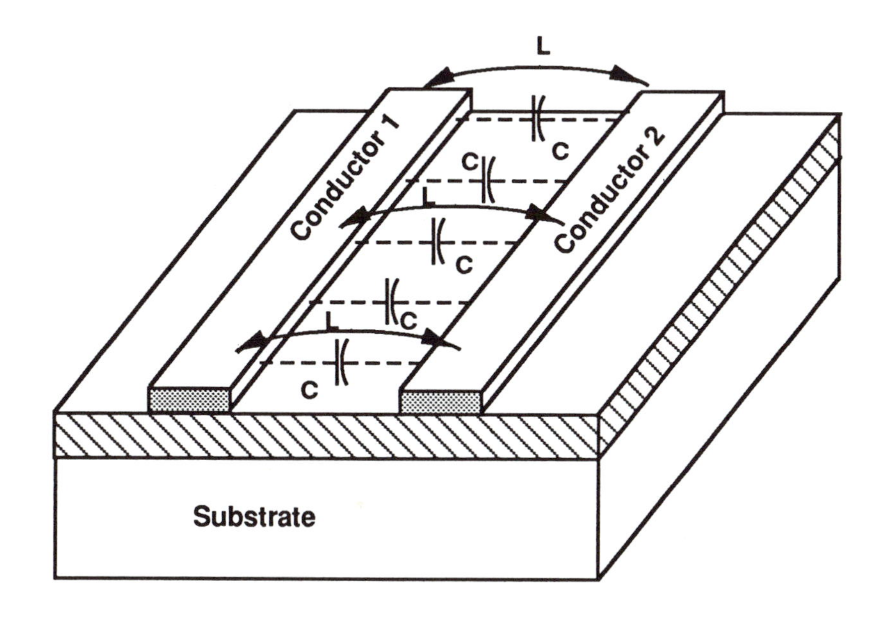

Figure 6.42: Distributed Crosstalk

and interconnect patterns. Logic design is accomplished by "wiring" the devices together. It is interesting to briefly look at one of these circuits known as the **gate array** IC.

Gate arrays are structured grids of transistors; logic gates are wired according to user specifications. The logic designer's job is to determine the connections among the MOSFETs and gates to form the desired functions. Once the interconnection paths are established, a mask set is made and used to pattern a metal layer (or two). Gate arrays are on the borderline between custom-VLSI and programmable logic devices (PLDs). Since our treatment deals with circuit aspects of CMOS, they are included here for completeness.

At the circuit level, a gate array consists of groups of transistors in a pre-defined geometry. Figure 6.43 illustrates the basic idea. This particular segment has 3 gates, each of which is connected to an nMOS and a pMOS transistor. Contact cuts are permitted only at the grid locations shown. Interconnnect lines around the transistor array are provided to aid in wiring the circuit. Power distribution is selected from the vertical V_{DD} and V_{SS} lines by etching the contacts.

Logic circuits are formed by specifying interconnect contacts and routing. Consider a static NAND3 CMOS gate. This can be implemented in the gate array grid using the patterning shown in Figure 6.44. A NOR3 is obtained by reversing the wiring patterns of the nMOS and pMOS arrays. Gate arrays are quite easy to use. As such, they are quite popular in many application environments [9].

An important limiting factor on the performance of gate arrays is switching speed. This an intrinsic problem which arises from the use of set layout geometries. Parasitic interconnect capacitances are determined by the routing paths. Since one cannot move the transistors, it is not possible to change the locations of the logic cells to reduce line capacitance. Switching times and propagation delays are limited by the very structure of the array.

Gate arrays are very useful for quick prototyping of a logic design, or as a lower-cost alternate to full-custom or application-specific integrated circuit (ASIC) cell designs. Even though the gate array design philosophy is greatly simplified, it is based on the same ideas that we have examined here for generalized CMOS.

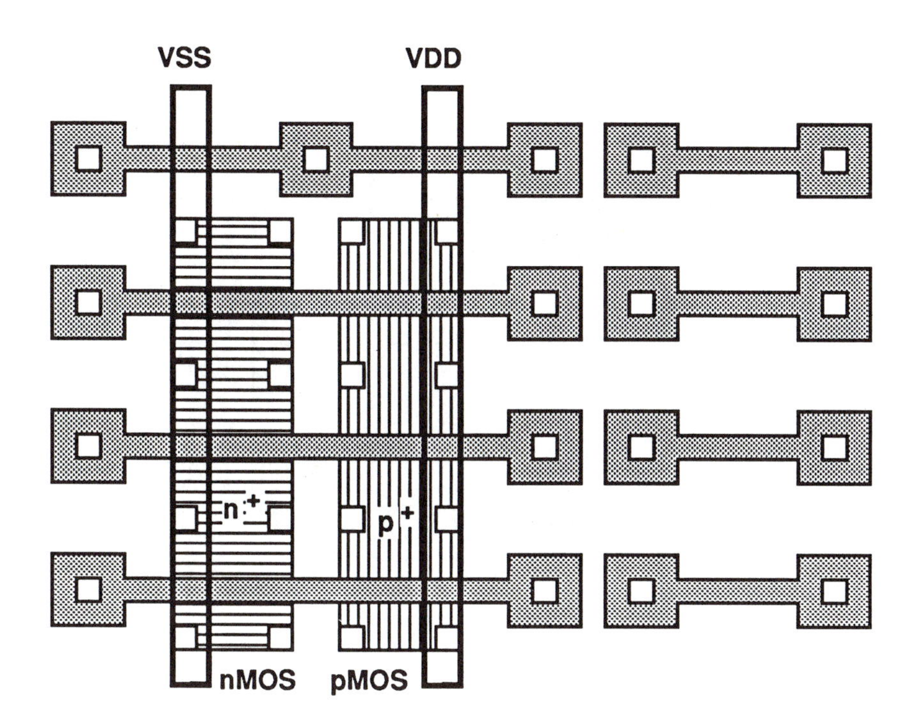

Figure 6.43: Gate Array Layout

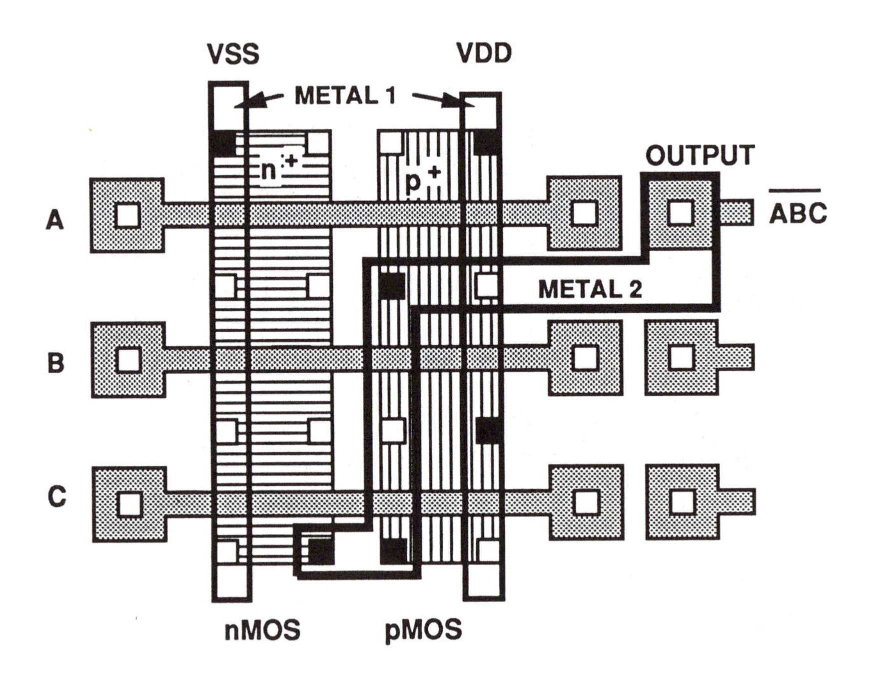

Figure 6.44: NAND3 Gate

6.10 References

The references below include discussions on processing, layout, interconnect modelling and coupling, and chip design in general.

[1] M. Annaratone, **Digital CMOS Circuit Design**, Kluwer Academic Publishers, Boston, 1986.

[2] R. J. Antinone and G.W. Brown, "The modeling of resistive interconnects for integrated circuits," IEEE J. Solid-State Circuits, vol. SC-18, no. 2, pp. 202-210, April, 1983.

[3] Chua, L.O., C.A. Desoer, and E.S. Kuh, **Linear and Nonlinear Circuits**, McGraw-Hill, New York, 1987.

[4] H.B. Bakoglu, **Circuits, Interconnections, and Packaging for VLSI**, Addison-Wesley, Reading, MA, 1990.

[5] E.G. Fabricius, **Introduction to VLSI Design**, McGraw-Hill, New York, 1990.

[6] R.L. Geiger, P.E. Allen, and N.R. Strader, **VLSI Design Techniques for Analog and Digital Circuits**, McGraw-Hill, New York, 1990.

[7] H. Hasegawa, M. Furukawa, and H. Yanai, "Properties of microstrip line on Si-SiO_2 system,", IEEE Trans. Microwave Theory and Tech., vol. MTT-19, pp. 869-881, Nov., 1971.

[8] D.V. Heinbuch, **CMOS3 Cell Library**, Addison-Wesley, Reading, MA, 1988.

[9] E.H. Hollis, **Design of VLSI Gate Array ICs**, Prentice-Hall, Englewood Cliffs, NJ, 1987.

[10] R. L. Liboff and G.C. Dalman, **Transmission Lines, Waveguides, and Smith Charts**, Macmillan, New York, 1985.

[11] H. Guckel, P.A. Brenna, and I. Palocz, "A parallel-plate waveguide approach to microminiaturized, planar transmission lines for integrated circuits," IEEE Trans. Microwave Theory and Tech., vol. MTT-15, pp. 468-476, Aug., 1967.

[12] W. Maly, **Atlas of IC Technologies: An Introduction to VLSI Processes**, Benjamin-Cummings, Menlo Park, CA, 1987.

[13] A. Mukherjee, **Introduction to nMOS and CMOS VLSI Systems Design**, Prentice-Hall, Englewood Cliffs, NJ, 1986.

[14] S. Ramo, J.R. Whinnery, and T. Van Duzer, **Fields and Waves in Communication Electronics**, 2nd ed., Wiley, New York, 1984.

[15] W.R. Runyon and K.E. Bean, **Semiconductor Integrated Circuit Processing Technology**, Addison-Wesley, Reading, MA, 1990.

[16] M.V. Schneider, "Microstrip lines for microwave integrated circuits," Bell Syst. Technical J., vol. 48, pp. 1421-1444, May-June, 1969.

[17] T. Shibata and E. Sano, "Characterization of MIS Structure Coplanar Transmission Lines for Investigation of Signal Propagation in Integrated Circuits, " IEEE Trans. Microwave Theory and Tech., vol. 38, no. 7, pp. 881-890, July, 1990.

[18] S.M. Sze, **VLSI Technology**, 2nd ed., McGraw-Hill, New York, 1988.

[19] R.R. Troutman, **Latchup in CMOS Technology**, Kluwer Academic Publishers, Boston, 1986.

[20] H. Umimoto and S. Odanaka, "Three-Dimensional Numerical Simulation of Local Oxidation of Silicon," IEEE Trans. Electron Dev., vol. 38, pp. 505-511, March, 1991.

[21] J. P. Uyemura, **Fundamentals of MOS Digital Integrated Circuits**, Addison-Wesley, Reading, MA, 1988.

[22] S. Veeraghavan, J.G. Fossum, and W.R. Eisenstadt, "SPICE Simulation of SOI MOSFET Integrated Circuits," IEEE Trans. Computer-Aided Design, vol. CAD-5, no. 4, pp.653-658, October, 1986.

[23] E.S. Yang, **Microelectronic Devices**, McGraw-Hill, New York, 1988.

[24] H-T. Yuan, Y-T. Lin, and S-Y. Chiang, "Properties of Interconnections on Silicon, Sapphire, and Semi-Insulating Gallium Arsenide Substrates," IEEE Trans. Electron Devices, vol. ED-29, no. 4, pp. 639-644, April, 1982.

Chapter 7

Synchronous Logic

Clocking is used to synchronize logic flow though a digital network. CMOS admits to several different clocked design styles; some approaches work with individual gates, while the more advanced implementations are directed toward overall systems design.

7.1 Clock Signals

The most basic CMOS timing scheme is a single-clock **2-phase** system which uses non-overlapping clock signals $\phi(t)$ and $\overline{\phi}(t)$ to control data flow. Figure 7.1 illustrates the clocking patterns. The **clock frequency** is given by $f = (1/T)$ [Hz] where T [sec] is the **period**.

An alternate dual-clock 2-phase approach uses two pairs of clock signals $(\phi_1, \overline{\phi_1})$ and $(\phi_2, \overline{\phi_2})$ such that

$$\phi_1(t) \cdot \phi_2(2) = 0 \tag{7.1}$$

as are shown in Figure 7.2. **Pseudo-2ϕ** logic of this type simplifies the timing, but makes the distribution of the clock signals more complicated.

It is also possible to implement 3ϕ and 4ϕ clocking in CMOS. These will not be discussed in detail here, but can be examined more closely in the literature [19].

7.2 Clock Distribution and Skew

Controlling clock signals is critical to the operation of high-performance systems. A great deal of engineering goes into proper generation and dis-

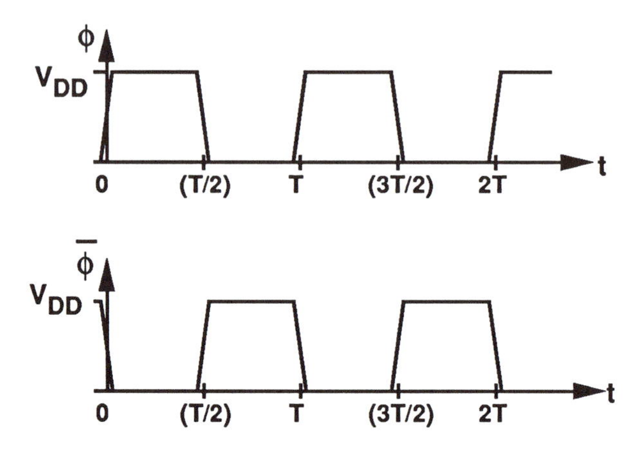

Figure 7.1: Single Clock 2-Phase Timing

tribution of the clock pulses to various points on the chip. From the perspective of chip design, the fundamental clocking signal CLK is supplied by an off-chip source. Distribution and use then become the fundamental issues of importance to the CMOS designer.

In this chapter we will study several CMOS circuit techniques which use the properties of the clock to synchronize the data movement. Since the circuits are sensitive to the behavior of the clock, it is important to understand the problems which arise when the basic signal CLK is distributed and modified for use on the chip. We will rely on a qualitative analysis for the present discussion; more details on the clock distribution problem are presented in Chapter 8.

Many clocking problems tend to originate from the distribution circuit. Consider the simplest case where we start with a single clock CLK and want to generate 2-phase signals ϕ and $\overline{\phi}$. From the logic viewpoint, all we need to do is add an inverter as shown in Figure 7.3, and the textbook solution is complete! Do we get an A+? Probably not, since this ignores the delay introduced by the inverter. The outputs ϕ and $\overline{\phi}$ are staggered in time, so that they are not really true complements. This problem of **clock skew** can cause problems in timing and data control. The **skew time** t_{skew} for this case is just the delay time t_p of the inverter.

A simple solution would be to delay ϕ by an equal amount to reestablish the timing. One way to accomplish this is to use a conducting TG as an equivalent resistor R_{TG} as shown in Figure 7.4. The actual delay depends

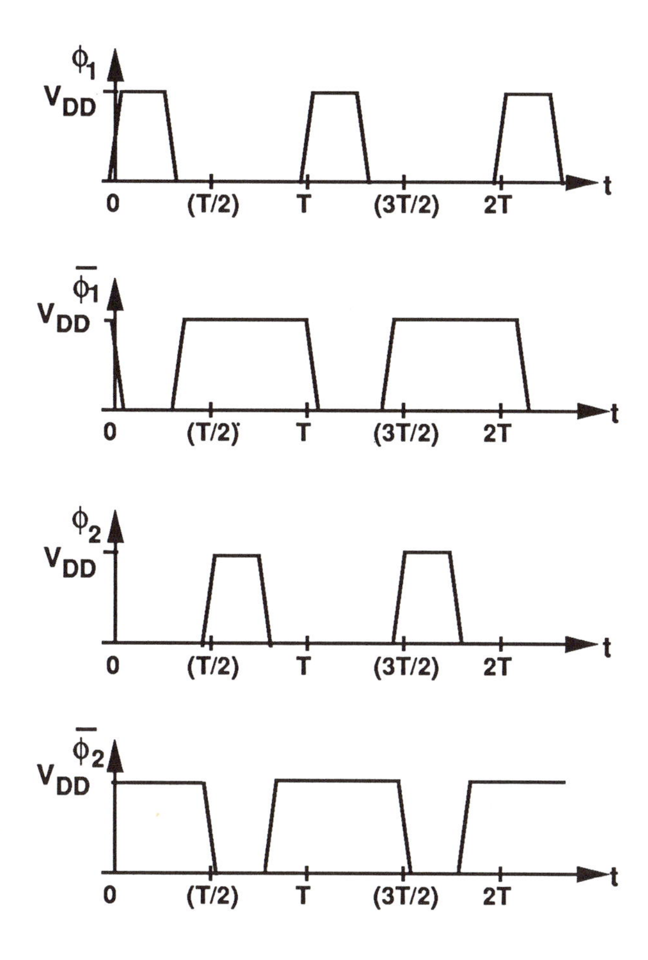

Figure 7.2: Pseudo 2-ϕ Clocking

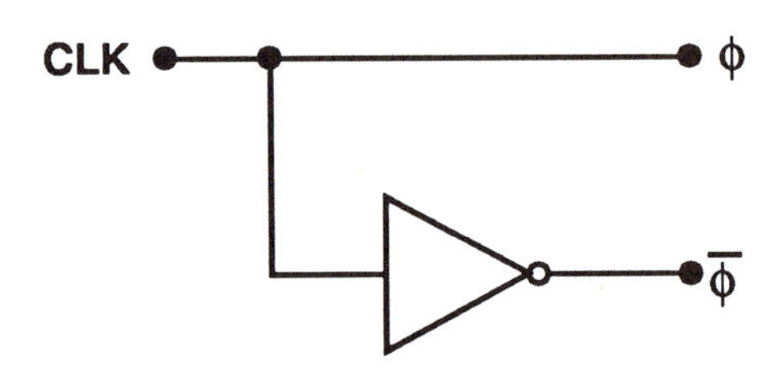

Figure 7.3: Generating ϕ and $\overline{\phi}$ from CLK

on the line capacitance of both the ϕ and $\overline{\phi}$ signal paths. Other approaches

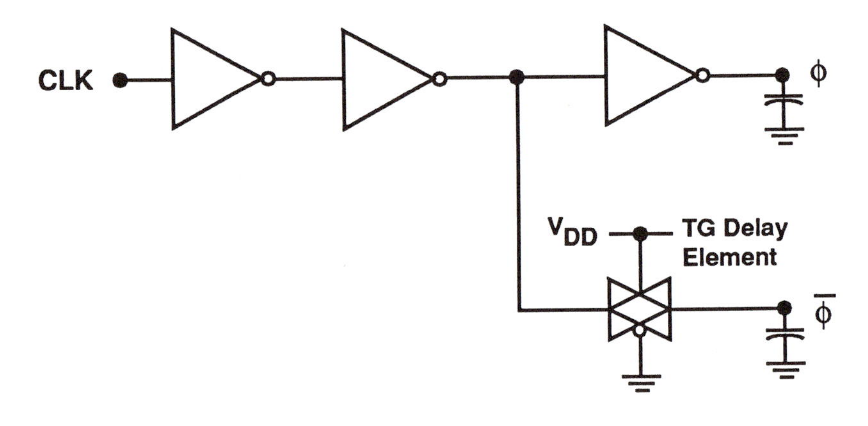

Figure 7.4: TG Delay Circuit

employ multiple gates to generate clock signals with equal delay times. Regardless of the approach, the performance of the circuit is very sensitive to the load and the interconnect parasitics.

Clock distribution circuits can be designed to provide excellent characteristics at low frequencies. However, as the data rate increases, so do the clocking problems. Some can be solved by straightforward techniques, while others, such as those due to process variations, are much more difficult to deal with. Designing circuits in a realistic environment requires careful attention to the clocking characteristics.

7.3 Clocked Static Logic

Synchronized data transfer can be achieved by using clocked transmission gates between static logic gates. Figure 7.5 illustrates how this is accomplished in a simple shift register; we have chosen single-clock 2ϕ timing for this example. When $\phi = 1$ the odd-numbered TGs transmit; during the next half-cycle where $\phi = 0$, the even-number TGs are conducting. It is

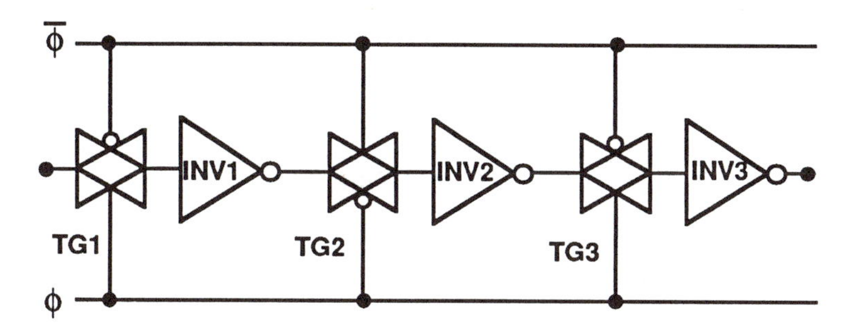

Figure 7.5: Shift Register

easy to visualize the synchronized data transfer in this case.

7.3.1 Design Factors

Designing clocked static logic is relatively straightforward. The basic problem is to insure that the pass characteristics of the TGs are sufficient to drive the gate, and vice-versa. Figure 7.6 illustrates a shift register segment with input and output transmission gates. It is assumed that the data input V_A can vary from 0 [V] to V_{DD} in a step-like manner. The primary design variables are the device geometries. To understand the design problem we will analyze the circuit requirements for the two clock states.

Data Transfer Limitations

A clock state of $\phi = 1$ allows data to move through TG1 to the inverter input. Since a CMOS TG can transmit full-rail voltage swings, there are no DC voltage level problems to contend with. Transient response thus becomes the primary issue, and the discussion of Section 5.1 can be directly applied to the present case.

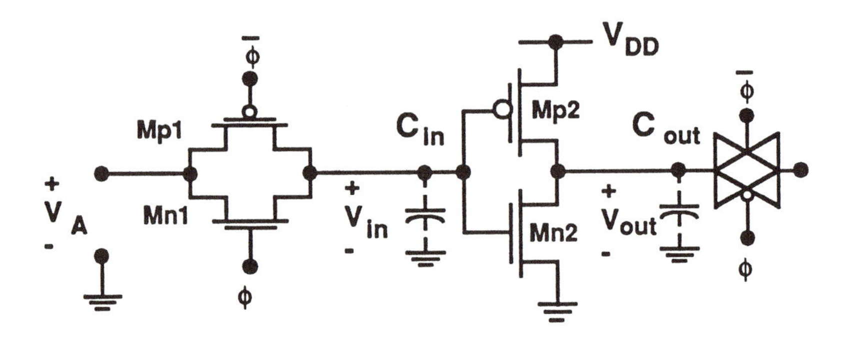

Figure 7.6: Clocked Shift Register Circuit

Modeling the TG as a linear resistor with value R_{TG} gives a charging-discharging time constant of

$$\tau_{TG} = R_{TG}C_L \tag{7.2}$$

where

$$C_L = C_{TG} + C_{in} + C_{line} \tag{7.3}$$

is the load capacitance seen by the transmission gate. The TG parasitics R_{eq} and C_{TG} are set by the aspect ratios and layout. C_{line} includes all of the interconnect contributions, while C_{in} represents the inverter input capacitance.

Case I: $V_A = V_{DD}$. A high input voltage gives an inverter input which varies as

$$V_{in}(t) \simeq V_{DD}[1 - e^{-t/\tau_{TG}}] \tag{7.4}$$

where we have assumed that the capacitor C_{in} is initially uncharged with $V_{in}(0) = 0$. The inverter interprets the input as a logic 1 when $V_{in}(t_1) = V_{IH}$ which occurs at a time

$$t_1 \simeq -\tau_{TG} \ln\left[1 - \frac{V_{IH}}{V_{DD}}\right]. \tag{7.5}$$

The inverter transistor parameters enter directly into this equation: the relative device ratio (β_p/β_n) determines V_{IH}, while the device geometries determine

$$C_{in} = C_{ox}[(WL)_n + (WL)_p]. \tag{7.6}$$

The inverter also introduces a propagation delay t_P which is required to charge the output capacitance.

Case II: $V_A = 0$. With a low voltage, the inverter input is described by

$$V_{in}(t) \simeq V_{DD} e^{-t/\tau_{TG}} \tag{7.7}$$

where we have assumed an initial condition of $V_{in}(0) = V_{DD}$. To have a logic 0 transferred into the gate requires that V_{in} must fall to a value of V_{IL}. This requires a time

$$t_0 \simeq \tau_{TG} \ln\left(\frac{V_{DD}}{V_{IL}}\right) \tag{7.8}$$

where the inverter parameters determine both V_{IL} and C_{in}. Of course, the propagation delay time t_P is always present.

The above analyses illustrates that the transient response is determined by the MOSFET dimensions of both the TG and the inverter. The clocking signal ϕ must allow for the longest delay time; this gives us the **maximum frequency** f_{max} such that $f \leq f_{max}$ must be maintained for proper operation. In system design, f_{max} is set by the slowest circuit in the network.

Charge Leakage

A clock state of $\phi = 0$ is designed to hold the data state at the inverter input. In an ideal situation the capacitance can store the charge for an indefinite amount of time. Realistic circuits do not have this property because of unavoidable charge leakage paths through the MOSFETs. Incorrect logic levels may result if the charge varies too much, so that this is an important problem to examine.

Recall that the n-type bulks of pMOSFETs are connected to V_{DD}, while the p-type bulks of nMOSFETs are grounded. The bulk connections of the TG transistors are responsible for the leakage as shown by the redrawn circuit in Figure 7.7. As discussed above,

$$C_L = C_{TG} + C_{line} + C_{in} \tag{7.9}$$

is the load capacitance seen by the TG. Both the inverter input capacitance C_{in} and the line contribution C_{line} are due to MOS layering. Silicon dioxide is an excellent insulator, so that these are able to hold charge for extremely long periods. However, the depletion capacitance contributions to C_{TG} are due to reverse-biased pn junctions, which admit leakage currents. Denoting the pMOS and nMOS values by I_{Lp} and I_{Ln}, respectively, we see that the pMOS leakage path from V_{DD} **adds charge** to the node, while the path through the nMOS transistor **removes charge** from the node. The net leakage current *off of the node* is thus

$$I_L = I_{Ln} - I_{Lp}; \tag{7.10}$$

the actual values of the leakage current are determined by the doping and geometries.

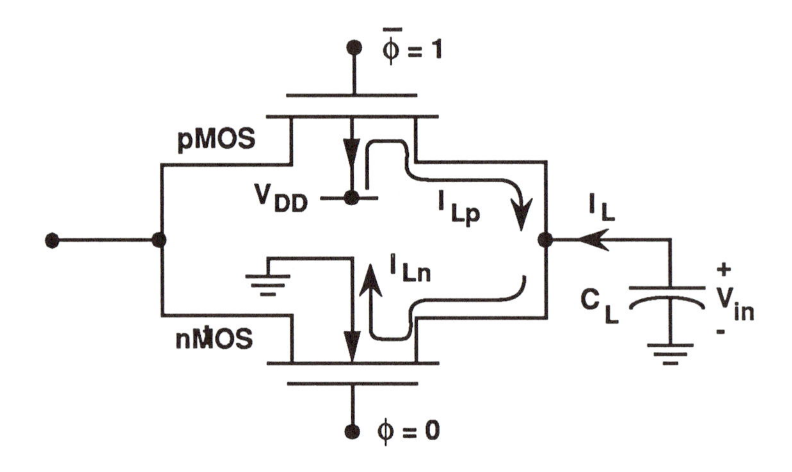

Figure 7.7: Leakage Paths in a CMOS TG

The differential equation for the inverter input voltage is

$$C_L \frac{dV_{in}}{dt} = -I_L. \tag{7.11}$$

If $I_{Ln} > I_{Lp}$, then the node can hold a logic 0 but cannot sustain a logic 1 state because the voltage decays with time. Conversely, if $I_{Lp} > I_{Ln}$, the node continuously charges; this indicates that logic 1 high voltages are easy to maintain, but that a logic 0 low voltage value will increase in time. The transition voltages are again V_{IL} and V_{IH}. Due to its importance to dynamic logic circuits, the leakage problem is analyzed in more detail in Section 7.5.

Charge leakage problems prohibit the indefinite holding of logic states in this type of circuit. To avoid problems, we define the **minimum frequency** f_{min} such that $f \geq f_{min}$ is required for reliable operation of the circuit. The value of f_{min} is set by the node which has the most severe leakage problem in the network.

7.3.2 Complex Logic Cascades

The shift register circuit is easily extended to arbitrary levels of complexity. A basic subsystem level example is shown in Figure 7.8 where we have substituted a complex logic circuit for the inverter. Clocking the TGs synchronizes the data flow through into and out of the gate. The maximum clock frequency f_{max} of the gate is determined by the transistor network. The maximum **system clock** is set by the slowest gate in the system. Using a single complicated logic gate with long propagation delay slows down the performance of the entire logic chain. Sometimes this can be avoided by choosing logic segments with approximately equal delay times; system architecture constraints may override this possibility.

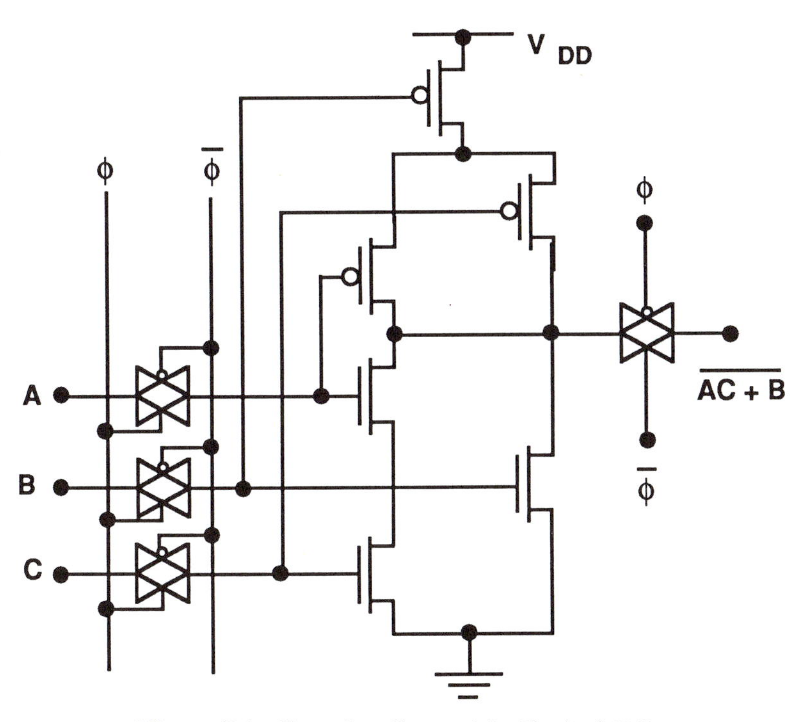

Figure 7.8: Complex Gate with Clocked TGs

7.4 Charge Storage Nodes

Dynamic logic designs are based on the synchronized movement of charge through a CMOS circuit. Capacitive **storage nodes** are used to hold charge levels between clock pulses. Typical chip capacitance values are on the order of a few femtofarads (1 [fF] = 10^{-15} [F]), so that the amount of charge $Q = CV$ stored on a dynamic node is on the order of femtocoulombs [fC]. Thus, even small perturbations from ideal behavior become important to the operation of the circuit.

In order to store a charge state on a capacitive node, the node must be isolated from both the power supply and ground. Several different types of storage nodes can be made in CMOS. The ones of interest to us here are nodes which are between two (or more) MOSFETs, as these are commonly found in switching networks. Attention is focused towards three basic combinations: nMOS–nMOS, pMOS–pMOS, and nMOS-pMOS. These are summarized in Figure 7.9 for future reference. The distinction among the three MOSFET-connected node types is made because of the difference in voltage transmission levels for nMOS and pMOS gates.

Two problems arise in maintaining the integrity of a stored logic state. First, unwanted conduction paths always exist at the chip level, leading to **charge leakage** on to, or, off of the node. Leakage currents alter the node voltage which may lead to a logic error. The second problem of **charge sharing** occurs when two isolated storage nodes become connected by a switching event and must equalize their voltages by redistributing charge. Charge sharing may result in a logic error, or, in a severe case, may block logic propagation entirely. Both problems are discussed in detail after we examine each of the node types.

nMOS–nMOS

The threshold voltage loss through an nMOS transistor limits the capacitor voltage to a value

$$V_{max} = V_{DD} - V_{Tn} \tag{7.12}$$

where

$$V_{Tn} = V_{T0n} + \gamma_n\left(\sqrt{2|\phi_{Fp}| + V_{max}} - \sqrt{2|\phi_{Fp}|}\,\right). \tag{7.13}$$

This is true regardless of the direction of current flow. The minimum voltage on the capacitor is

$$V_{min} = 0 \tag{7.14}$$

since nMOS transistors do not have any problems transmitting logic 0 levels. Charge storage on an nMOS–nMOS node is affected by the leakage paths through the p-type bulk to ground. This affects the ability of this type of node to store a logic 1 voltage[1].

[1] The charge storage properties of nMOS-nMOS nodes is discussed in detail in reference [18].

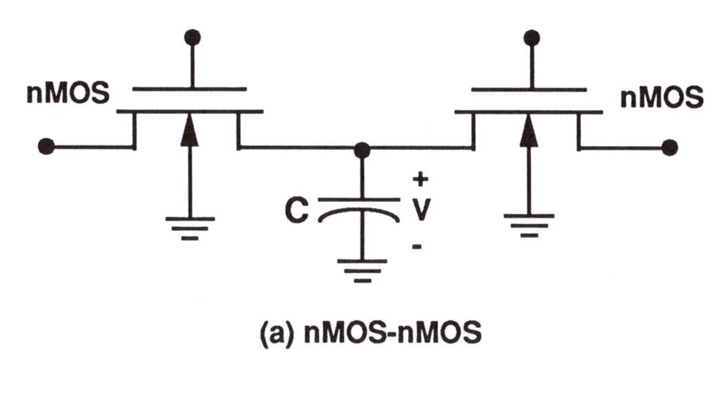

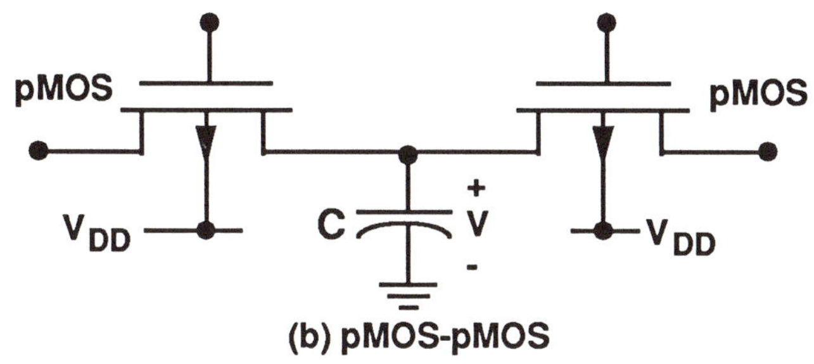

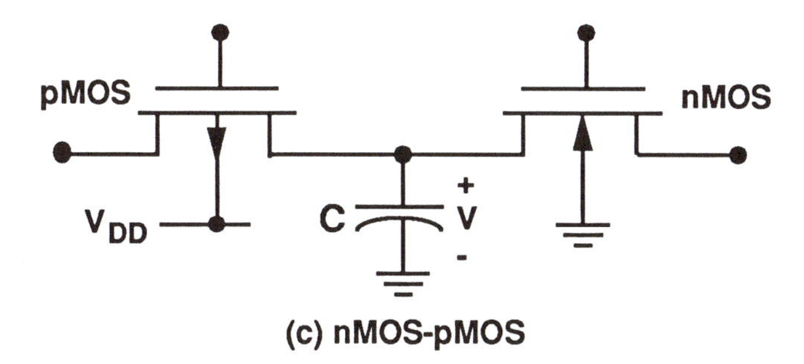

Figure 7.9: MOSFET-Connected Storage Nodes

pMOS–pMOS

A pMOS–pMOS node is the voltage complement of an nMOS–nMOS node. The maximum voltage on the storage node is

$$V_{max} = V_{DD} \tag{7.15}$$

while the smallest voltage is limited a threshold loss to a value

$$V_{min} = |V_{T0p} - \gamma_p(\sqrt{2\phi_{Fn} + (V_{DD} - V_{min})} - \sqrt{2\phi_{Fn}}\ )| \tag{7.16}$$

as discussed in Chapter 4. Since both p-channel MOSFETs have n-type bulks which are connected to V_{DD}, this type of storage node receives leakage current from the power supply. Logic 1 values can be held indefinitely, but logic 0 low voltages can only exist for short periods.

nMOS-pMOS

A complementary nMOS–pMOS storage node is usually connected to give

$$\begin{aligned} V_{max} &= V_{DD}, \\ V_{min} &= 0, \end{aligned} \tag{7.17}$$

corresponding to maximum and minimum input values of V_{DD} from the pMOS side, and 0 [V] from the nMOS side, respectively. If these are reversed so that 0 [V] is input into the pMOS side while V_{DD} is input on the nMOS side, then the range is reduced to

$$\begin{aligned} V_{max} &= V_{DD} - V_{Tn}, \\ V_{min} &= |V_{Tp}|; \end{aligned} \tag{7.18}$$

this type of operation is generally avoided since it greatly reduces the noise margins.

In a standard nMOS–pMOS storage node, leakage paths to both the power supply and ground exist. The ability to store logic 0 and logic 1 values depends on which leakage path dominates.

7.5 Charge Leakage

Charge leakage to or from a storage node occurs because of reverse leakage across a pn junction[2]. Reverse current cannot be eliminated so it must be

[2] Subthreshold leakage may also be present, but will be neglected for simplicity.

incorporated into the circuit design. Charge leakage is crucial for determining the minimum clocking frequency for a dynamic switching network.

To study charge leakage, consider the problem of a transmission gate which drives a static logic circuit; this has been redrawn in Figure 7.10. During a transmission event, the system clock goes to a value $\phi = 1$ which allows the charge state to be set. When $\phi = 0$, the node is isolated from the rest of the system. Charge leakage is important during this time interval.

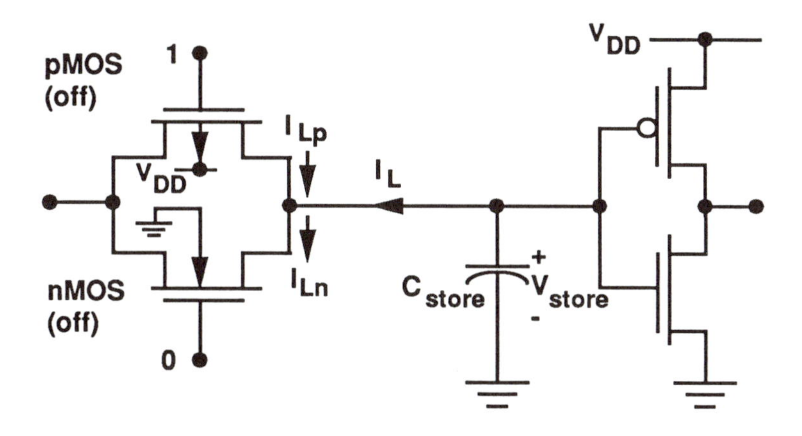

Figure 7.10: Charge Leakage Circuit

Two reverse leakage current paths exist when the TG is in a hold state. As shown in Figure 7.10, the p^+n drain-bulk connection in the pMOS transistor Mp allows leakage current to add charge from the power supply to the node. A complementary leakage path exists in the n-channel MOSFET Mn through the n^+p drain-bulk junction; this drains charge off of the node. Denoting the stored charge by Q_{store}, the leakage is described by

$$\frac{dQ_{store}}{dt} = I_{Lp} - I_{Ln} \tag{7.19}$$

where I_{Lp} and I_{Ln} respectively denote the drain/source leakage currents through the pMOS and nMOS transistors. The storage capacitance is given by

$$C_{store} = \frac{dQ_{store}}{dV} \tag{7.20}$$

with V as the voltage across the capacitor. At first sight, this appears to be a straightforward problem. However, both the charge and the reverse currents are nonlinear functions of the voltage, making the full problem quite complicated. Although a computer simulation is required to obtain an accurate analysis, simplified models can be introduced to help gain a physical understanding of the problem.

7.5.1 Constant Current and Capacitance

The simplest approximation which can be made is to assume that both leakage currents I_{Lp} and I_{Ln} are constants, and that the node charge-voltage relation is linear of the form

$$Q_{store} = C_{store} V. \tag{7.21}$$

In this case, C_{store} is a constant and the charge leakage equation is

$$C_{store} \frac{dV}{dt} = I_{Lp} - I_{Ln}. \tag{7.22}$$

Solving gives the solution

$$V(t) = \frac{(I_{Lp} - I_{Ln})}{C_{store}} t + V(0), \tag{7.23}$$

with $V(0)$ the initial voltage at time $t = 0$.

Consider the case where $V(0) = V_0$ with V_0 a small logic 0 voltage. If $I_{Lp} > I_{Ln}$, then the equation predicts a voltage which increases linearly with time. Eventually the logic 0 will change to a logic 1, inducing an error. If $I_{Lp} < I_{Ln}$, then $V(t) \to 0$, which is not a problem.

If instead we have set a logic 1 high voltage with $V(0) = V_1$, then the situation is reversed. The condition $I_{Lp} > I_{Ln}$ strengthens the logic 1 state, while $I_{Lp} < I_{Dn}$ eventually gives an error because it drains the charge off of the node.

Logic errors can be avoided by specifying a minimum clocking frequency f_{min}. To calculate the value, let ΔV be the maximum allowed change in the voltage. The node can hold charge for a maximum time interval of

$$t_{max} = \frac{C_{store}(\Delta V)}{I_L} \tag{7.24}$$

where $I_L = |I_{Lp} - I_{Ln}|$ is the net leakage current. Choosing a clock with a 50 % duty cycle, we set $t_{max} = (T_{max}/2)$ where T_{max} is the longest allowed clock period. Then

$$
\begin{aligned}
f_{min} &\simeq \frac{1}{2t_{max}} \\
&\simeq \frac{I_L}{2C_{store}(\Delta V)}
\end{aligned} \tag{7.25}
$$

gives an estimate of the minimum clock frequency. Note that f_{min} increases with increasing leakage or decreasing capacitance.

The value of C_{store} can be estimated by voltage averaging. Consider the circuit shown in Figure 7.11. Both constant MOS and nonlinear depletion capacitances contribute to the total value

$$
C_T \simeq C_G + C_{line} + C_{ols} + C_{olp} + C_{SBp}(V) + C_{DBn}(V), \tag{7.26}
$$

which explicitly shows the voltage-dependent terms. To obtain the equiv-

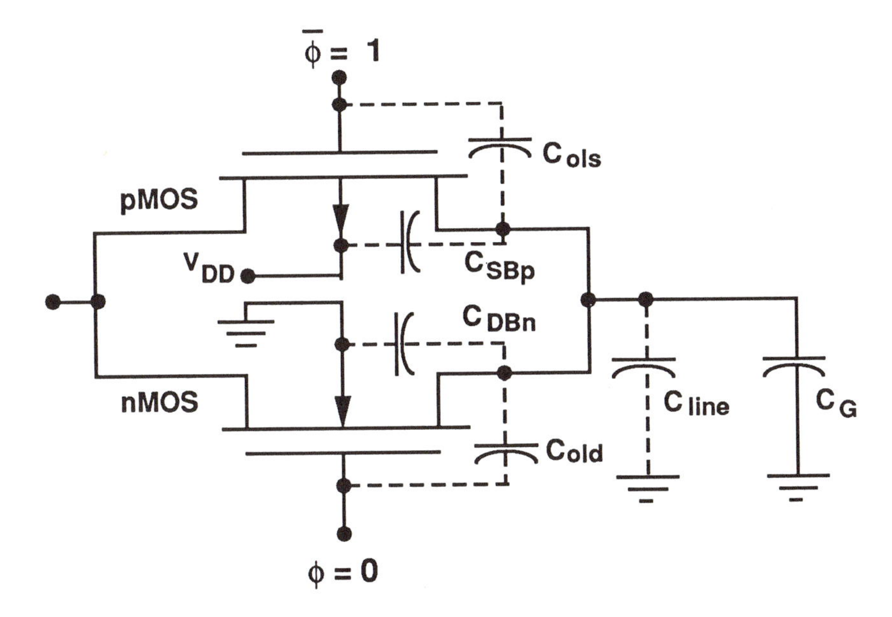

Figure 7.11: Transmission Gate Capacitance

alent storage capacitance, we average over voltage using the techniques introduced in Section 3.3. Then,

$$
C_{store} \simeq C_G + C_{line} + C_{ols} + C_{olp} + K(0, V_{DD})[C_{SBp} + C_{DBn}], \tag{7.27}
$$

where C_{SBp} and C_{DBn} are zero-bias values and $K(0, V_{DD}) < 1$ is the averaging factor.

7.5.2 Voltage-Dependent Current

A more realistic model is one where the leakage currents are dependent on the junction voltages. Figure 7.12 provides a simple circuit model. In this case, we find that an equilibrium voltage is eventually reached corresponding to balanced current flow on and off of the node. We will assume

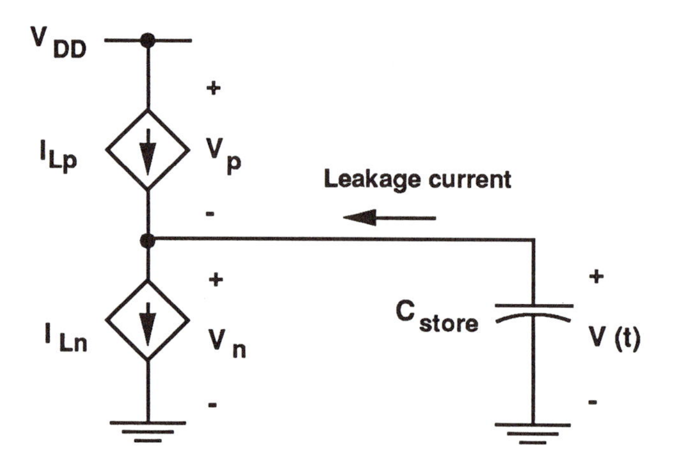

Figure 7.12: Voltage-Dependent Leakage Current

step-profile pn junctions which are characterized by the grading coefficient $m = (1/2)$; the treatment can be modified for arbitrary junction types by changing the value of m.

Recall that reverse current flow across a silicon pn junction is dominated by generation in the depletion region. For the problem at hand, the net current flow away from the storage node is described by

$$I_{Lp} - I_{Ln} = \frac{qA_p n_i x_{dp}}{2\tau_{0p}} - \frac{qA_n n_i x_{dn}}{2\tau_{0n}}, \tag{7.28}$$

where

$$x_{dp} = x_{d0p}\sqrt{1 + \frac{V_p}{V_{bi,p}}},$$

$$x_{dn} = x_{d0p}\sqrt{1+\frac{V_n}{V_{bi,n}}}, \tag{7.29}$$

are the voltage-dependent depletion widths, and x_{d0p} and x_{d0n} are the zero-bias values. Note that

$$\begin{aligned} V_n &= V, \\ V_p &= V_{DD} - V, \end{aligned} \tag{7.30}$$

relate the junction voltages V_n and V_p to the storage voltage V and the power supply V_{DD}. Assuming that C_{store} is a constant, we can write the charge leakage equation as

$$C_{store}\frac{dV}{dt} = I_{0p}\sqrt{1+\frac{(V_{DD}-V)}{V_{bi,p}}} - I_{0n}\sqrt{1+\frac{V}{V_{bi,n}}}, \tag{7.31}$$

where we have introduced the notation

$$\begin{aligned} I_{0p} &= \frac{qA_p n_i x_{d0p}}{2\tau_{0p}}, \\ I_{0n} &= \frac{qA_n n_i x_{d0n}}{2\tau_{0n}}, \end{aligned} \tag{7.32}$$

for the zero-bias leakage currents.

Direct integration of this equation is not an attractive option. Instead, we examine the physics to gain some insight into the problem. The sign of the derivative (dV/dt) is set by the leakage current flows; a positive derivative indicates an increasing voltage in time, while a negative derivative implies net charge leakage away from the node. A zero-derivative condition means that equilibrium has been reached.

Suppose that we set an initial logic 0 voltage $V(0) = 0$ and that $I_{0p} \sim I_{0n}$. When the node becomes isolated, $(dV/dt) > 0$; $V(t)$ thus increases with time. The interesting physics is seen examining the net current flow. Increasing V has the effect of *decreasing* I_{Lp} and *increasing* I_{Ln}. Eventually, the two cancel one another yielding the zero-derivative condition. Equating the two currents and rearranging gives the equilibrium value at

$$V_{eq} = \frac{I_{0p}{}^2[1+(V_{DD}/V_{bi,p})] - I_{0n}{}^2}{(I_{0n}{}^2/V_{bi,n}) + (I_{0p}{}^2/V_{bi,p})}. \tag{7.33}$$

This shows that the final equilibrium voltage is determined by the leakage current levels as expected from physical reasoning. If instead we start with $V(0) = V_{DD}$, then $(dV/dt) < 0$ indicating that charge is lost from the node. Calculating the equilibrium voltage gives the same result.

This reasoning illustrates an interesting point. Since leakage paths are formed to both V_{DD} and ground, the final equilibrium voltage V_{eq} may be either a logic 0 or a logic 1, depending on the parameters. Since I_{0n}, I_{0p}, and C_{store} all contain area factors, the behavior of each node is distinct. Generalization is possible only if one leakage current is clearly dominant.

7.5.3 Complete Solution

The complete differential equation accounts for

- Voltage-dependent capacitance
- Voltage-dependent leakage currents
- Subthreshold leakage

and assumes the general form

$$C(V)\frac{dV}{dt} = -I_L(V) \tag{7.34}$$

where both $C(V)$ and $I_L(V)$ are nonlinear functions of the voltage[3]. In general, the complexity of the functions mandates computer solutions. A circuit simulation can be used to obtain the behavior of the stored charge, or code can be generated for the problem at hand. Regardless of the approach used to study the problem, the most important result is the maximum time interval that a node can maintain a charge state without a logic error.

A simple finite-difference analysis illustrates the main points. First, we define a constant time interval Δt such that

$$\begin{aligned} t_0 &= 0 \\ t_1 &= \Delta t \\ t_2 &= t_1 + \Delta t \\ &\vdots \\ t_n &= t_{n-1} + \Delta t. \end{aligned} \tag{7.35}$$

Introducing the notation

$$V(t_n) = V_n, \tag{7.36}$$

the difference equation describing the charge leakage is given by

$$V_n = V_{n-1} - \frac{I_L(V_{n-1})}{C(V_{n-1})}\Delta t. \tag{7.37}$$

[3]The actual dependence is set by the grading factor m

To solve, we start at time $t = 0$ with $V_0 = V(0)$ the initial condition. At time t_1, the voltage is given by

$$V_1 = V_0 - \frac{I_L(V_0)}{C(V_0)}\,\Delta t; \tag{7.38}$$

at time t_2, we compute

$$V_2 = V_1 - \frac{I_L(V_1)}{C(V_1)}\,\Delta t \tag{7.39}$$

and so on. This gives a point-by-point integration as illustrated in Figure 7.13. Other effects such as subthreshold leakage can be easily included in the programming code, and we may work at arbitrary levels of complexity.

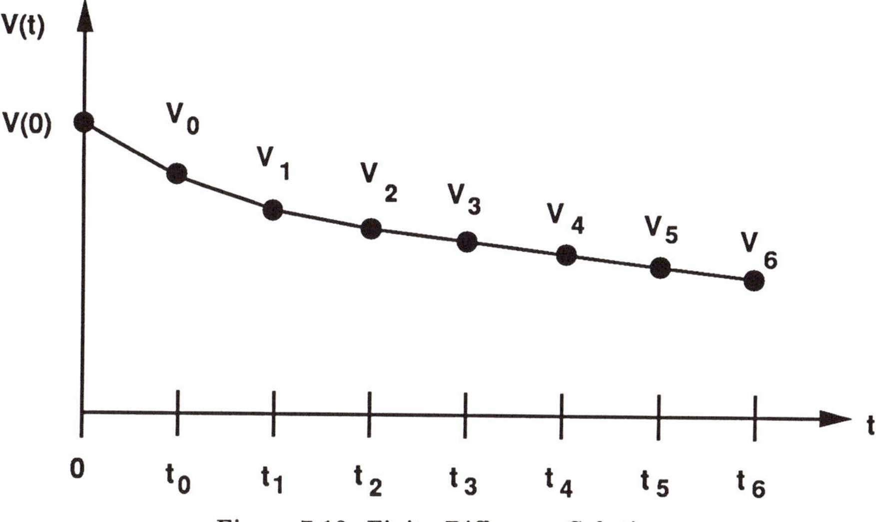

Figure 7.13: Finite-Difference Solution

The charge holding time can be directly computed by setting the extremum voltage.

7.5.4 Single-Polarity MOSFET Storage Nodes

Complex dynamic logic design introduces nMOS–nMOS and pMOS–pMOS storage nodes into the circuit analysis. Charge leakage still occurs, but as mentioned in the last section, single-polarity circuits only admit current flow in one direction.

nMOS–nMOS Leakage

Consider the drawing shown in Fig. 7.9(a). In this case, both leakage paths are to ground through n^+p junctions. Charge leakage is contained in the voltage equation

$$C_{store}\frac{dV}{dt} = -2I_{Ln}. \tag{7.40}$$

Assuming constant capacitance and current gives

$$V(t) = V(0) - \frac{2I_{Ln}}{C_{store}}t, \tag{7.41}$$

indicating that $V(t) \rightarrow 0$ as time t increases. This verifies our early comment that this type of node can store logic 0 states, but cannot hold a logic 1 state indefinitely.

pMOS-pMOS Leakage

This type of circuit uses two p-channel MOSFETs are used to form the storage node as shown in Fig. 7.9(b). Two leakage paths to the power supply are present, so we write

$$C_{store}\frac{dV}{dt} = +2I_{Lp} \tag{7.42}$$

to describe charge leakage onto the node. Solving gives

$$V(t) = V(0) + \frac{2I_{Lp}}{C_{store}}t. \tag{7.43}$$

This is valid for $V(t) \leq V_{DD}$ and shows that charge leakage increases the node voltage. A pMOS–pMOS node cannot hold a logic 0 state, but has no problem with logic 1 states.

7.6 Charge Sharing

Charge sharing occurs when a dynamic charge-storage node is used to drive another isolated node in a switching network.

A simple network which can be analyzed to illustrate the main effect is shown in Figure 7.14. Assume that for times $t < 0$, the TG is an open circuit with $(S, \overline{S}) = (0, 1)$, and that the capacitor voltages are set to $V_1(t < 0) = V_{DD}$, $V_2(t < 0) = 0$. The total charge in the system is

$$Q_T = C_1 V_{DD} \tag{7.44}$$

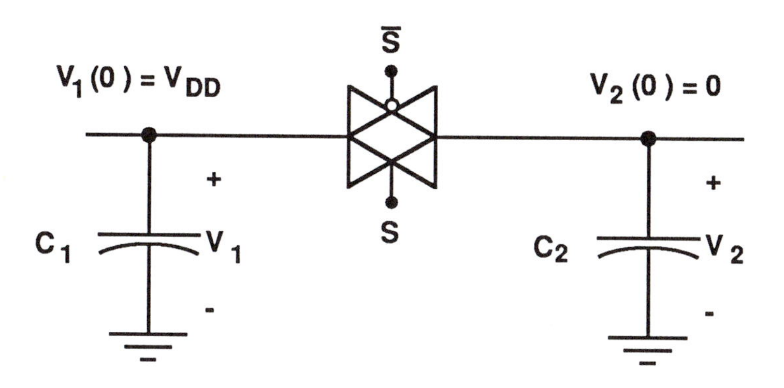

Figure 7.14: Basic Charge Sharing Circuit

and is contained entirely on capacitor C_1. At $t = 0$, we will switch the TG to a control state $(S, \overline{S}) = (1, 0)$ which allows current flow. Since this effectively places C_1 and C_2 in parallel, the charge redistributes according to

$$Q_T = (C_1 + C_2)V_f \tag{7.45}$$

where V_f is the final voltage across the capacitors. The name **charge sharing** become obvious: the total charge Q_T is shared between the two capacitors. Equating the two equations gives a final voltage of

$$\begin{aligned} V_f &= \frac{C_1}{C_1 + C_2} V_{DD} \\ &= \frac{1}{1 + (C_2/C_1)} V_{DD}, \end{aligned} \tag{7.46}$$

so that $V_f < V_{DD}$ is set by the capacitance ratio (C_2/C_1). If we design the circuit with $C_1 = C_2$, then $V_f = (V_{DD}/2)$, indicating a significant drop in voltage. A reliable forward transfer of a logic 1 state from C_1 to C_2 requires that $C_1 >> C_2$ to insure that $V_f \sim V_{DD}$. Severe voltage degradation can lead to logic errors, so that charge sharing cannot be ignored.

The general problem is shown in Figure 7.15 where we specify arbitrary initial conditions $V_1(0)$ and $V_2(0)$ on the capacitors giving the system a total charge of

$$Q_T = C_1 V_1(0) + C_2 V_2(0). \tag{7.47}$$

Applying basic circuit analysis gives the time-dependent voltages as

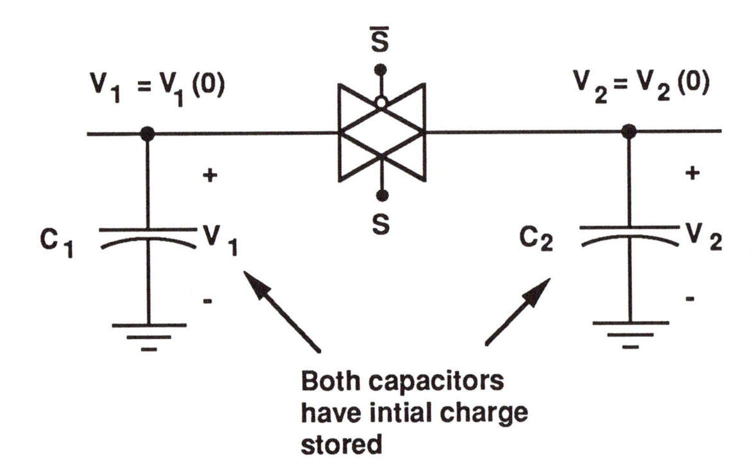

Figure 7.15: Generalized Charge Sharing Problem

$$\begin{aligned} V_1(t) &= V_2(0) + \frac{[V_1(0) - V_2(0)]}{(C_1 + C_2)}[C_1 + C_2 e^{-t/\tau}], \\ V_2(t) &= V_2(0) + [V_1(0) - V_2(0)](\frac{C_1}{C_1 + C_2})[1 - e^{-t/\tau}] \end{aligned} \quad (7.48)$$

where the time constant is given by

$$\tau = R_{TG} C_{eq} \quad (7.49)$$

with

$$C_{eq} = \frac{C_1 C_2}{C_1 + C_2} \quad (7.50)$$

as the equivalent series capacitance. In the limit $t \rightarrow \infty$, $V_1 = V_2 = V_f$ with

$$V_f = \frac{C_1}{C_1 + C_2} V_1(0) + \frac{C_2}{C_1 + C_2} V_2(0). \quad (7.51)$$

This agrees with the result from simple charge conservation by noting that the final charge distributes according to

$$Q_T = (C_1 + C_2) V_f \quad (7.52)$$

as before.

To illustrate the basic transient behavior, consider initial conditions of $V_1(0) = V_{DD}$ and $V_2(0) = 0$. Figure 7.16 shows $V_1(t)$ and $V_2(t)$ for two different values of capacitance ratios (C_2/C_1). In general, the voltage

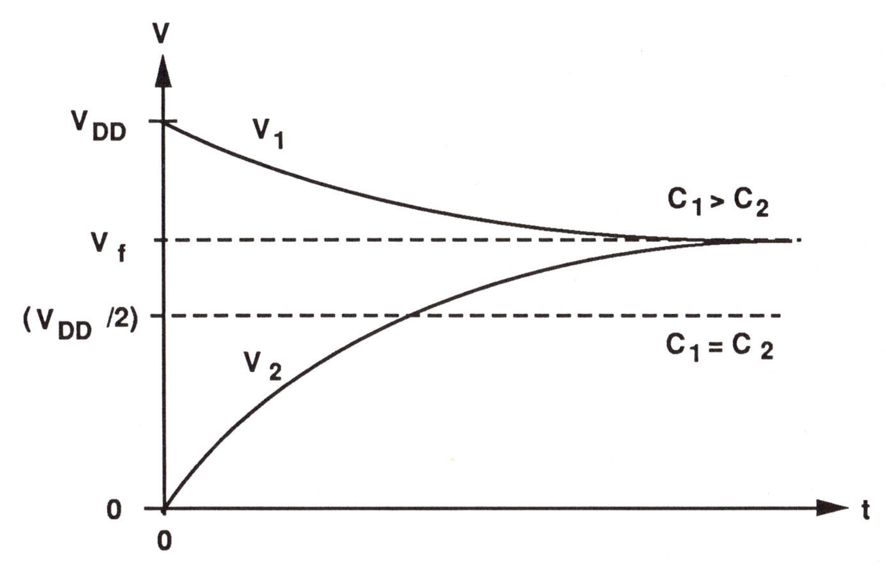

Figure 7.16: Transient Voltage Behavior

reduction from charge sharing effects depends on both the total charge and the capacitance ratios.

Charge sharing among N TG-connected capacitors is treated in the same way. The total initial charge on the system is

$$Q_T = \sum_{i=1}^{N} C_i V_i(0). \tag{7.53}$$

Assuming that the nodes are connected after a switching event, the final charge distribution is given by

$$Q_T = (\sum_{i=1}^{N} C_i) V_f \tag{7.54}$$

so that the final voltage is

$$V_f = \frac{\sum_{i=1}^{N} C_i V_i(0)}{\sum_{i=1}^{N} C_i}. \tag{7.55}$$

In general, the capacitance of the input logic node should be much larger then all secondary capacitors in the chain.

Dynamic charge sharing problems may be difficult to identify at first sight. They arise in both simple and complex networks, and usually depend on the order of switching events. Care must be taken to insure that charge sharing does not degrade the voltage noise margins or introduce glitches into the logic.

Charge sharing is also important between capacitive nodes connected by single nMOS or pMOS transistors. The analysis follows the same lines, except that the problem is complicated by threshold voltage losses through the MOSFETs. These situations are discussed in more detail in the context of specific dynamic circuit arrangements.

7.7 Dynamic Logic

Dynamic logic circuits perform logic operations using the properties of capacitive charge storage nodes. Clock signals perform a double duty: they provide logic synchronization and also allow predefined charge states to be established at a periodic rate. Outputs are defined only during a portion of the clock cycle, so that the time-varying (dynamic) properties are apparent.

The main advantages to using dynamic CMOS logic circuits is that they are generally faster and more compact than equivalent static gates. However, the operation is more sensitive to layout geometries, and the problems of charge leakage and charge sharing can become critical. Various design styles have been proposed to deal with these (and other) problems. The basic dynamic gates discussed in this section provide the foundation for more advanced design styles in this chapter.

7.7.1 Dynamic nMOS Inverter

A basic single-clock, 2-phase, dynamic inverter is shown in Figure 7.17. The NOT operation is performed by MOSFET M1. Transistors Mp and Mn are controlled by the clock signal ϕ and are respectively termed the *precharge* and *evaluate* devices; these provide dynamic control and synchronization of the charge.

The operation of the circuit is divided by the clock into two distinct phases: the **precharge** and **evaluate** intervals. A condition of $\phi = 0$ defines the precharge where Mp is conducting while Mn is in cutoff. The main purpose of the precharge is to allow C_{out} to charge to a level $V_{out} = V_{DD}$ through Mp. The input voltage V_{in} is also accepted at this time. When the clock switches to $\phi = 1$, the circuit goes into the evaluate portion of the logic cycle where Mp is in cutoff and Mn is active. If the input is at a logic 0 voltage with $V_{in} < V_{Tn}$, M1 remains in cutoff and the charge on C_{out} is held. A logic 1 input level $V_{in} > V_{Tn}$ places M1 into a conducting state, and allows C_{out} to discharge to a voltage $V_{out} \rightarrow 0$ [V]. This is

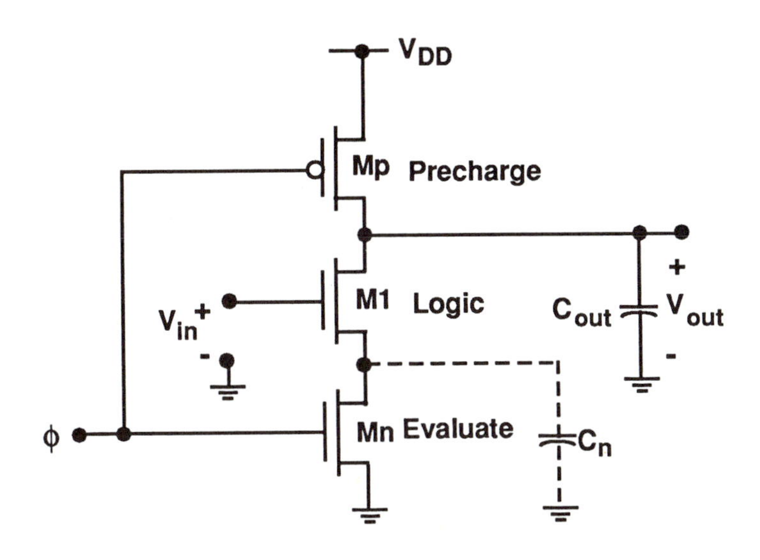

Figure 7.17: Basic Dynamic nMOS Inverter

termed a **conditional discharge**, and is characteristic of dynamic logic. The clocking diagram in Figure 7.18 illustrates the behavior of the output node during the precharge and evaluate times.

Performance benchmarks center around the transient time intervals. The precharge and evaluate (discharge) events determine both the propagation delay time and the maximum clock frequency f_{max}. In addition, charge leakage limitations establish the minimum clock frequency f_{min}.

Equivalent *RC* networks may be used to estimate the important time intervals. Consider first the precharge with $\phi = 0$. The objective of the precharge interval is to increase the voltage on C_{out} to a value V_{DD}.

The worst-case situation occurs when $V_{in} = V_{DD}$ since M1 conducts current away from C_{out} to charge C_n. Figure 7.19 provides the simplified charging circuit for this event using equivalent MOSFET resistances; the threshold voltage loss through M1 has been ignored. If M1 were off, the charging time constant would be

$$\tau_{ch} = \frac{C_{out}}{\beta_p(V_{DD} - |V_{Tp}|)} = R_p C_{out}. \tag{7.56}$$

When M1 is active, the current to C_{out} is decreased, so that τ_{ch} must be longer. The upper limit occurs when $R_1 = 0$, corresponding to having C_{out}

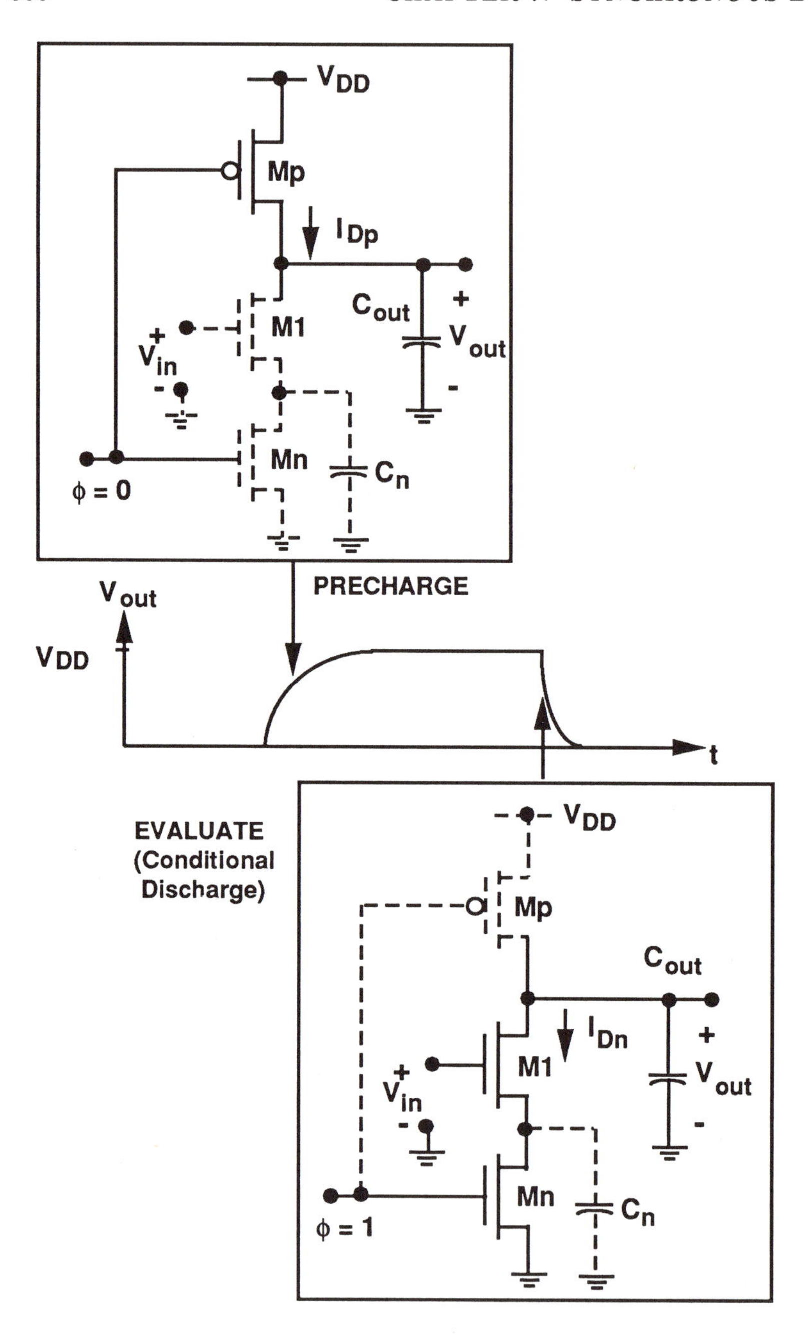

Figure 7.18: Precharge and Evaluate

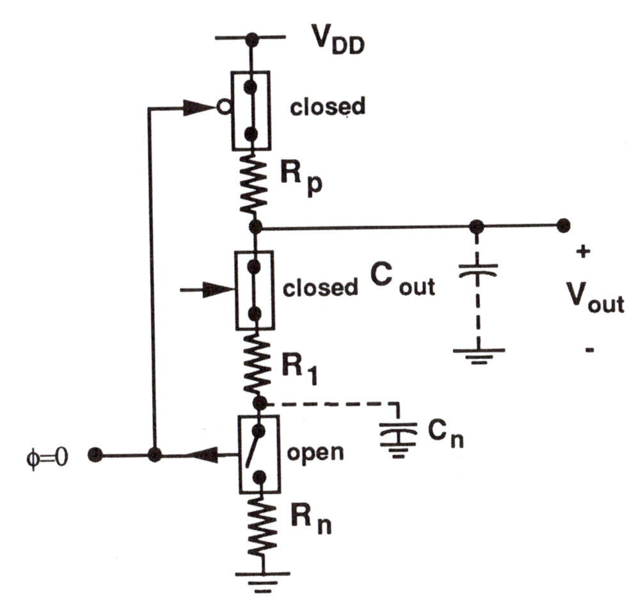

Figure 7.19: *RC* Precharge Network

in parallel with C_n. This allows us to write

$$\tau_{ch,max} = R_p(C_{out} + C_n) \tag{7.57}$$

with $R_1 = [\beta_{n1}(V_{DD} - V_{Tn})]^{-1}$ as the largest possible value. The longest precharge time t_{ch} is then

$$t_{ch,max} = \tau_{ch,max}\left[\frac{2|V_{Tp}|}{(V_{DD} - |V_{Tp}|)} + \ln\left(\frac{2(V_{DD} - |V_{Tp}|)}{V_0} - 1\right)\right] \tag{7.58}$$

by direct analogy with the static inverter analysis in Section 3.2; in this equation, $V_0 = 0.1V_{DD}$ is the 10% voltage level[4]. Due to the number of approximations, this should only be viewed as a first-order estimate.

Evaluation takes place with $\phi = 1$. If $V_{in} = V_{DD}$, then the delay time may be estimated using the equivalent circuit in Figure 7.20. The $(1/e)$ transistor chain delay is approximately

$$t_D \simeq R_nC_n + (R_n + R_1)C_{out} \tag{7.59}$$

[4]Also note that we have ignored a factor of V_0 in the first term. This is usually done for simplicity; the charge time now represents the time needed to reach V_{DD}, not the 90 % point.

which suffices for a zero-order estimate. If M1 and Mn have the same channel width W, we may estimate the discharge time by using the time constant

$$\tau_{dis} = \frac{(L_1 + L_n)C_{out}}{k_n' W(V_{DD} - V_{Tn})} \tag{7.60}$$

where L_1 and L_n are the channel lengths and C_n has been neglected. Then,

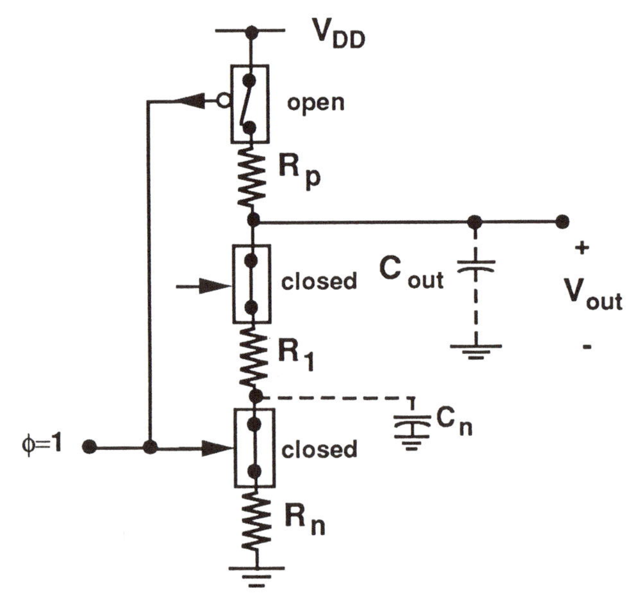

Figure 7.20: *RC* Discharge Network

$$t_{dis} = \tau_{dis}\left[\frac{2V_{Tn}}{(V_{DD} - V_{Tn})} + \ln\left(\frac{2(V_{DD} - V_{Tn})}{V_0} - 1\right)\right], \tag{7.61}$$

approximates the required discharge time.

The maximum clock frequency f_{max} can be estimated using the above calculations. Assume that the clock has a 50 % duty cycle and a period T. Since the clock must remain low for at least the charge time t_{ch}, and must remain high at least a time t_{dis}, we define

$$t_M = \max(t_{ch}, t_{dis}) \tag{7.62}$$

as the maximum time interval. To avoid problems, we must have $(T/2) = t_M$ as the shortest half-cycle allowed so that

$$f_{max} \simeq \frac{1}{2t_M} \tag{7.63}$$

gives the maximum allowed clock frequency **for this stage**. The system clock limits are set by the slowest stage in the network.

If $V_{in} = 0$, M1 is in cutoff during the precharge. When the clock changes to $\phi = 1$, the output node operates in a hold state and is subject to the charge leakage problems. This establishes the minimum clock frequency f_{min} as discussed in Section 7.4.

7.7.2 Dynamic pMOS Inverter

Dynamic logic circuits can be created using pMOS transistors as the logic devices. A pMOS-based dynamic inverter is shown in Figure 7.21. This is similar to the nMOS circuit detailed above, except that the output node is at the top of the nMOS clocking transistor. Moreover, operation of the two circuits are voltage complements of each other.

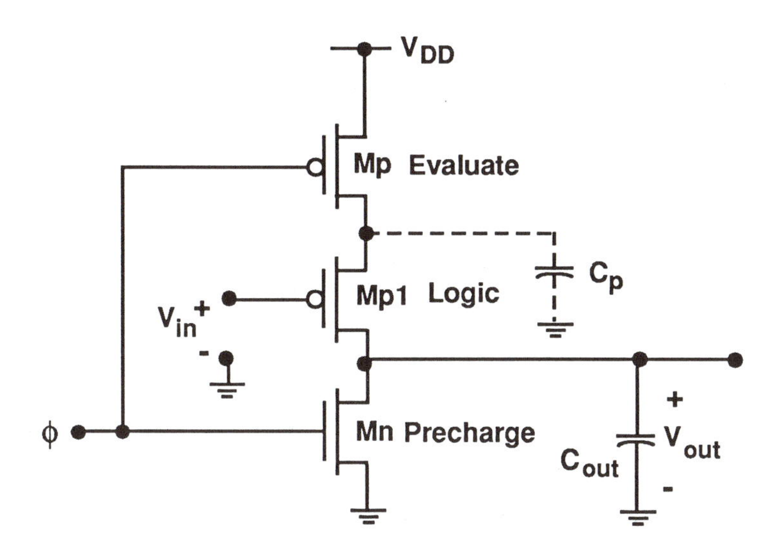

Figure 7.21: Dynamic pMOS Inverter

Consider first the clock state $\phi = 1$, placing Mp in cutoff while Mn conducts current to ground. This defines the "precharge" for the circuit during which the output capacitance C_{out} is initially set to a **zero-charge state** with $V_{out} = 0$ [V]. Although the terminology may sound confusing[5], it is easy to remember because it is exactly opposite to the nMOS operation.

[5] Some may prefer to call this the "zero-voltage precharge" or a "discharge" event.

The input is valid during this time. With a logic 1 input, $V_{in} = V_{DD}$ and Mp1 remains off. If $V_{in} = 0$ [V] (a logic 0 input), Mp1 conducts and C_p is allowed to discharge with normal threshold voltage limits.

Evaluation takes place when $\phi = 0$. Mp is active so that C_p is charged. A **conditional charge** of the output capacitance occurs during this time. If $V_{in} = V_{DD}$, Mp1 remains in cutoff and the precharge voltage $V_{out} = 0$ [V] is held at the output. On the other hand, a logic 0 input voltage of $V_{in} = 0$ [V] indicates that Mp1 is conducting so that C_{out} charges through Mp and Mp1 to a value $V_{out} \rightarrow V_{DD}$.

The precharge time is obtained from

$$t_{pre} = \tau_{pre}\left[\frac{2V_{Tn}}{(V_{DD} - V_{Tn})} + \ln\left(\frac{2(V_{DD} - V_{Tn})}{V_0} - 1\right)\right], \tag{7.64}$$

where the time constant

$$\tau_{pre} = \frac{C_{out}}{\beta_n(V_{DD} - V_{Tn})} \tag{7.65}$$

describes the discharge through Mn. Similarly, when the input is a logic 0 the delay time is approximated by

$$t_D \simeq R_{p1}C_p + (R_{p1} + R_p)C_{out} \tag{7.66}$$

corresponding to the time needed to charge V_{out} to (V_{DD}/e). If both pMOSFETs have the same width and C_p is negligible, we can estimate the evaluation by introducing the time constant

$$\tau_{ev} = \frac{(L_1 + L_p)C_{out}}{k_p' W(V_{DD} - |V_{Tp}|)} \tag{7.67}$$

such that

$$t_{ev} = \tau_{ev}\left[\frac{2|V_{Tp}|}{(V_{DD} - |V_{Tp}|)} + \ln\left(\frac{2(V_{DD} - |V_{Tp}|)}{V_0} - 1\right)\right] \tag{7.68}$$

provides a reasonable approximation. Of course, pMOS transistors have larger resistances than nMOS devices, so that the value of f_{max} will be reduced accordingly.

7.7.3 Complex Logic

Dynamic logic is easily extended to complex logic gates using the rules discussed in Section 4.7 . Figure 7.22 shows the general structure for both nMOS and pMOS dynamic blocks. The MOSFET arrays are arranged

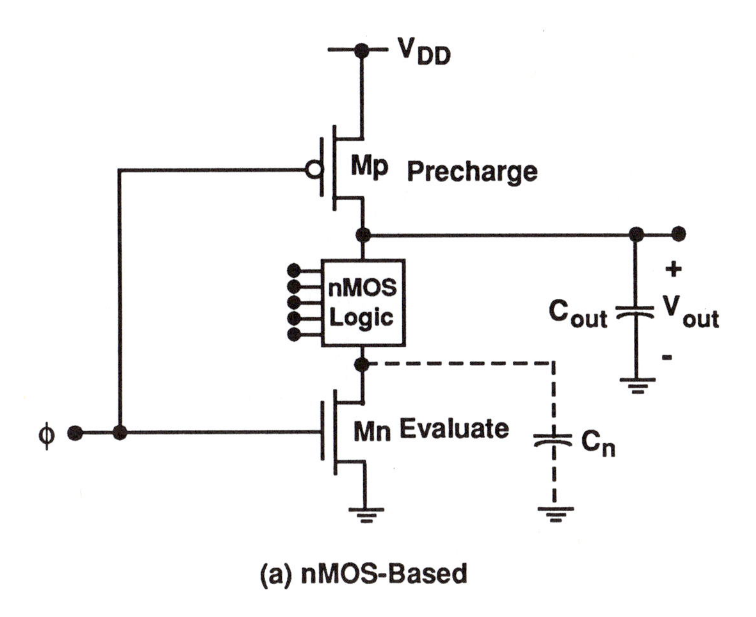

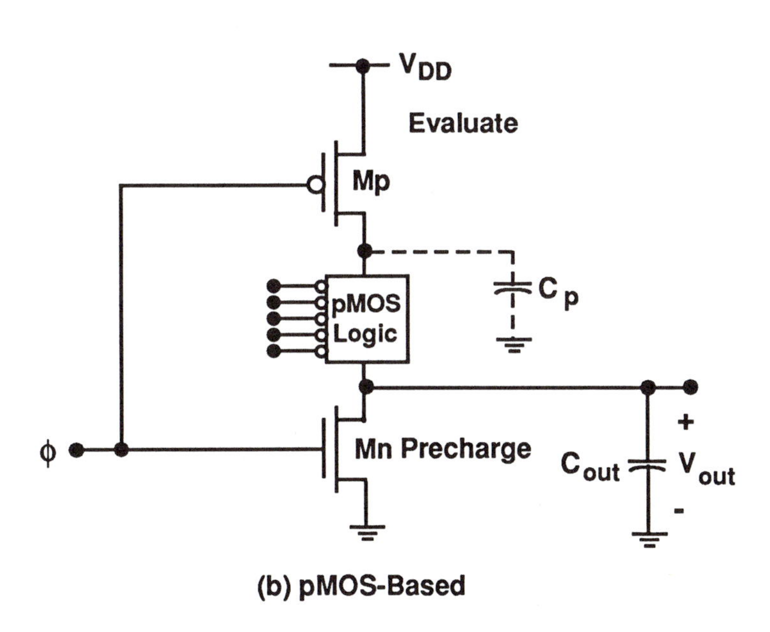

Figure 7.22: Complex Dynamic Logic

according to the desired function. Arbitrary levels of AOI and OAI logic can be implemented in either circuit.

EXAMPLE 7.1: Dynamic nMOS Gate

A half adder uses inputs (A, B) and produces the sum and carry functions

$$S = A \oplus B = A\overline{B} + \overline{A}B$$
$$C = AB.$$

Parallel dynamic nMOS circuits can be constructed as shown below to create the half-adder. A clock state of $\phi = 0$ defines the precharge for the half-adder. Outputs are valid during the evaluation time when $\phi = 1$.

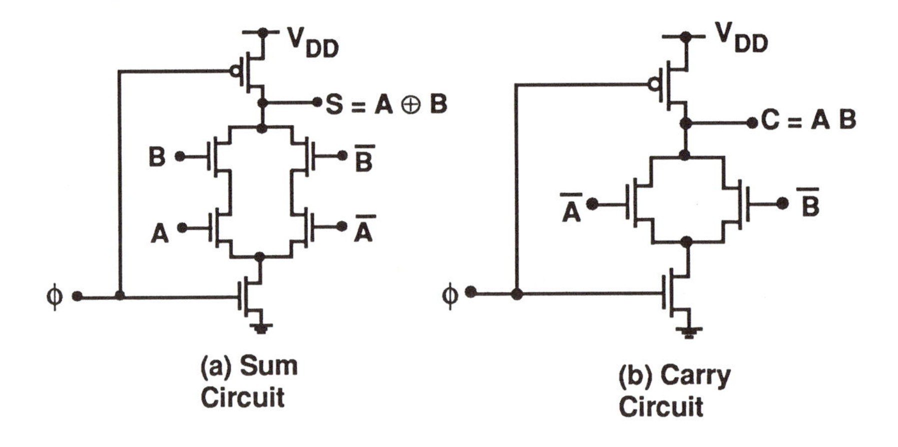

7.7.4 Dynamic Cascades

Dynamic logic gates can be cascaded so long as the logic chain is structured with alternating polarity circuits. This means that nMOS stages must drive pMOS gates and vice-versa. Violating this rule may lead to logic glitches as discussed below.

The need for alternate nMOS-pMOS gates is due to the nature of the precharge event. Consider the nMOS-nMOS cascade in Figure 7.23. During precharge ($\phi = 0$), both output capacitors $C_{out,1}$ and $C_{out,2}$ are charged to a voltage of V_{DD}. Since $V_{out,1} = V_{in,2}$, transistor M2 is conducting when the evaluation phase ($\phi = 1$) start. Charge sharing reduces the voltage on $C_{out,2}$ regardless of the logic. A glitch may occur if $V_{out,1}$ evaluates to 0, but so much charge has been lost from $C_{out,2}$ that $V_{out,2} \rightarrow 0$.

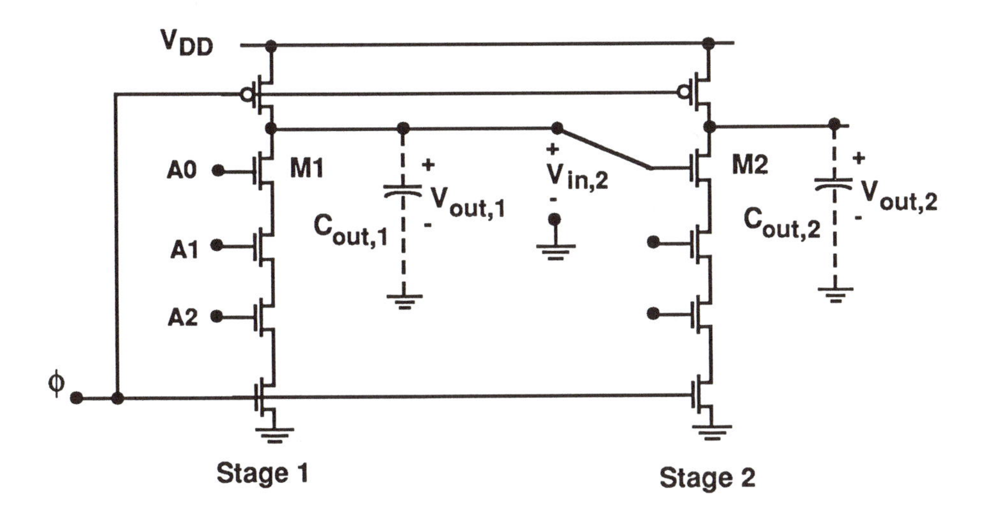

Figure 7.23: Cascaded nMOS-nMOS Glitch Problem

Glitches can be avoided by using the alternating polarity arrangement shown in Figure 7.24 [3]. Connecting nMOS outputs to the gates of pMOS transistors does not cause problems because the precharge voltage of V_{DD} forces a pMOSFET into cutoff. The same holds true for pMOS outputs connected to the gates of nMOSFETs. To avoid charge sharing problems, we structure the circuit so that dynamic outputs drive the MOSFET closest to the output node in the next stage. The main drawback of this type of circuit is that pMOS logic chains are slower than equivalent nMOS arrays, but they cannot be avoided.

7.8 Domino Logic

Domino logic (DL) is designed to provide glitch-free cascades of nMOS logic blocks by adding a static inverter to a basic dynamic circuit. This approach constitutes a **system design style**, and the name *domino* arises from the manner in which the logic propagates down a chain of gates. Let us first

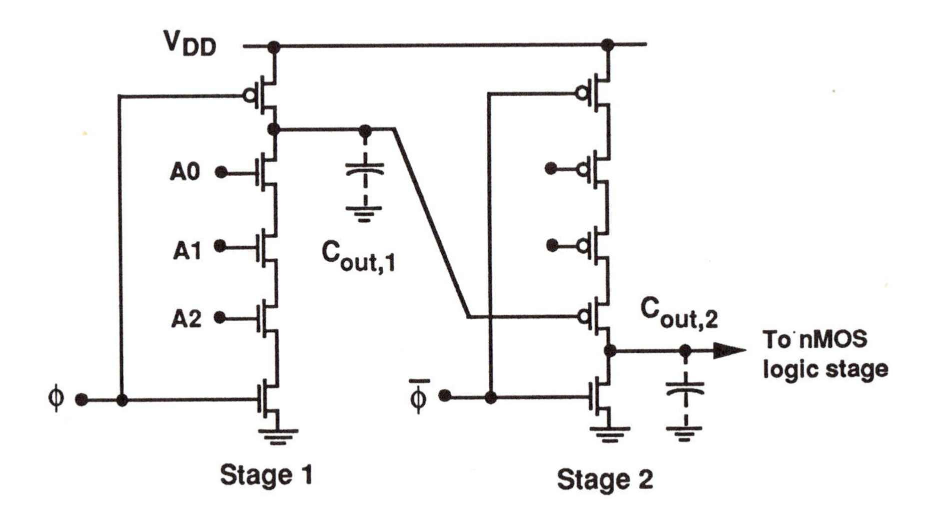

Figure 7.24: Dynamic Cascades

examine a basic logic stage, and then extend the discussion to the point where the domino effect can be understood.

A basic domino logic circuit is shown in Figure 7.25. An arbitrary domino stage consists of an nMOS dynamic logic circuit cascaded into a static inverter. We have chosen the input section to be a 4-input nMOS NAND gate for this example, but any logic function can be used. The unique feature of the domino stage is the static inverter made up of MOSFETs Mp2 and Mn2. Logically, the output from the stage is

$$F = A_0 A_1 A_2 A_3 \tag{7.69}$$

since the NOT function has been added. Aside from the fact that we have created the AND function, the most interesting aspect of the domino stage arises from the switching characteristics of the circuit.

The circuit behavior is synchronized by the clock signal ϕ. Precharge occurs when the clock is at $\phi = 0$. During this time, Mp1 conducts to charge the internal node capacitance C_X to a voltage $V_X \rightarrow V_{DD}$. Since V_X is the input to the inverter, $V_{out} \rightarrow 0$ [V], which is considered the gate output. When evaluation takes place with $\phi = 1$, Mn1 is driven into a conducting state and the circuit undergoes a conditional discharge. The NAND4 input stage drops V_X to zero only if all inputs are high: $A_0 = 1 = A_1 = A_2 = A_3$; otherwise, V_X is maintained high and V_{out} is kept at 0 [V].

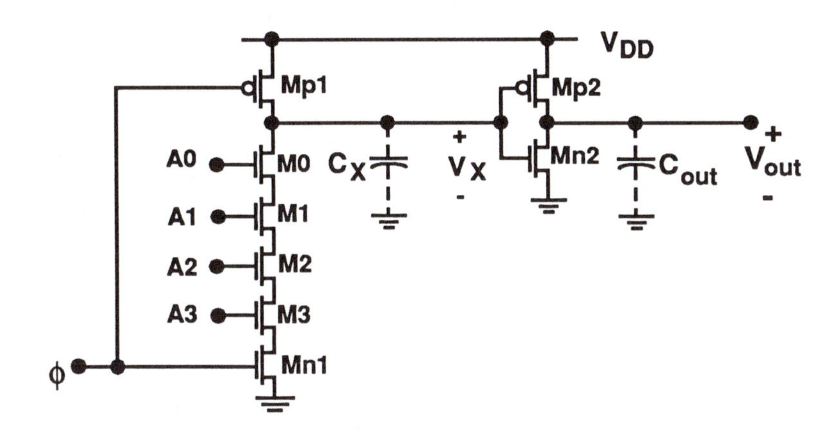

Figure 7.25: Domino AND Gate

Domino logic structuring can be understood by referring to the cascaded gates shown in Figure 7.26. Note in particular that

- Each stage is driven by ϕ
- Inverter outputs drive the nMOSFET closest to the output.

When $\phi = 0$, every stage undergoes precharging. Since the NOT function gives a gate output of $V_{out}= 0$ [V] during this time, nMOS transistors may be directly driven without worrying about glitches. Evaluation occurs when the clock switches to $\phi = 1$; during this time, the logic propagates through the chain from left to right. A gate must change its output voltage from the precharge value $V_{out} = 0$ to V_{DD} to induce a change in the following stage.

The name **domino logic** is based on making an analogy with a chain of dominos as shown in Figure 7.27. We view the precharge event as being analogous to "setting up the dominos" on ends as shown in the first drawing. When the chain undergoes evaluation, a transition in the first stage is needed to trigger the logic propagation; this is analogous to "knocking down the first domino". A change in the output of the first stage may induce a transition in the next stage and so on, so that the logic propagation is viewed as a chain reaction. Note that the logic is stopped if any stage evaluates to $V_{out} = 0$.

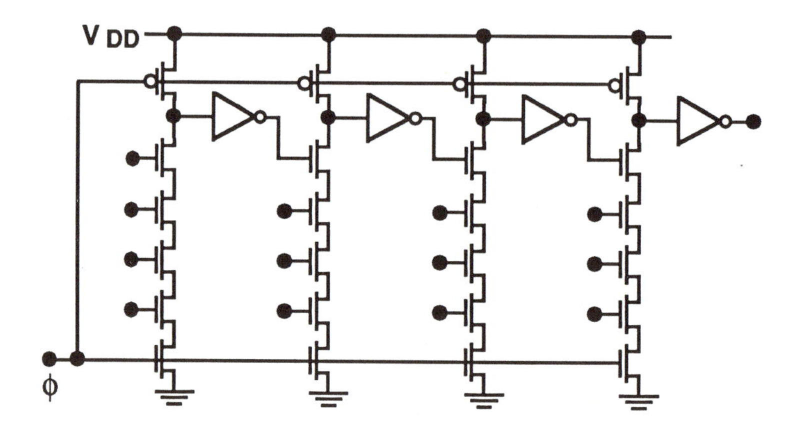

Figure 7.26: Domino Logic Chain

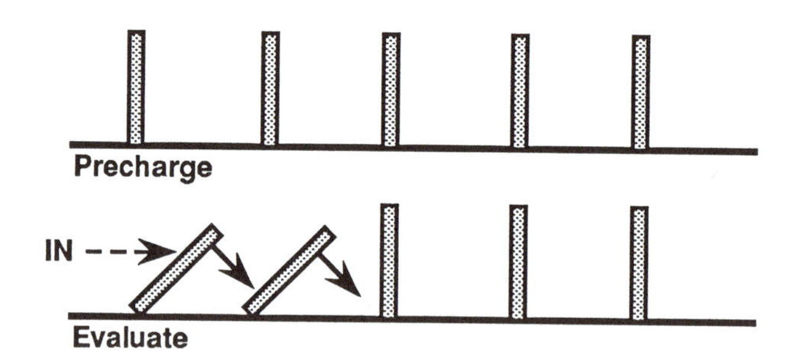

Figure 7.27: Visualization of Domino Effect

7.8.1 Analysis

Consider the AND4 circuit in Figure 7.28. The internal operation can be understood by examining the precharge and evaluate stages separately. Note that C_X has been split into two contributions such that $C_X = C_0 + C_T$. C_0 represents the capacitance due to MOSFET M0, while C_T is the total of all other contributions.

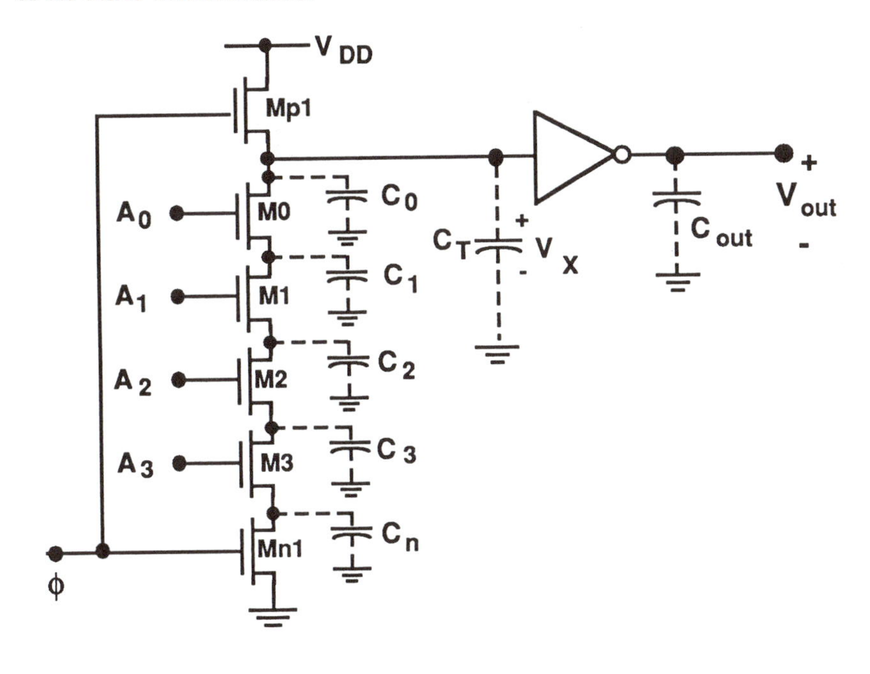

Figure 7.28: Domino AND4 Gate

Precharge

During the precharge event, $\phi = 0$ places Mp1 in conduction and Mn1 in cutoff. If we assume that the input A_0 is from a preceding domino stage (as will be the case in a chain), then $A_0 = 0$ during the precharge driving M0 into cutoff. All of the precharge current through Mp1 is thus directed to capacitor C_X, which reduces the necessary precharge time.

To calculate the minimum precharge time, note that C_X must charge to a voltage $V_X \geq V_{IH}$ to represent a logic 1 inverter input; the input high voltage V_{IH} is determined by the aspect ratio of the inverter transistors.

Modifying the results of Section 3.2, we find

$$t_{ch} \simeq \tau_{ch}\left[\frac{|V_{Tp}|}{(V_{DD}-|V_{Tp}|)} + \ln\left(\frac{2(V_{DD}-|V_{Tp}|)}{V_{DD}-V_{IH}} - 1\right)\right] \qquad (7.70)$$

where we have assumed that $V_X(0) = 0$, and have defined

$$\tau_{ch} = \frac{C_X}{\beta_p(V_{DD}-|V_{Tp}|)} \qquad (7.71)$$

as the charging time constant. Since this represents the minimum precharge time, a realistic circuit design must allow for variations, and also the normal propagation delay t_p through the inverter. Complete precharging of C_X to a voltage $V_X \to V_{DD}$ is usually desired to insure the integrity of the logic.

The value of C_X can be approximated by

$$\begin{aligned} C_X &= C_0 + C_T \\ &\simeq (C_{GDn1} + C_{DBn1}) + (C_{GDp1} + C_{DBp1}) + C_G + C_{line}, \end{aligned} \qquad (7.72)$$

where C_G is the total gate capacitance $(C_{Gn2} + C_{Gp2})$. This equation for C_X is the same as C_{out} for an inverter with a fanout of FO=1. Typically, C_X is the same order of magnitude as the smallest logic gate capacitance, so that the precharge time is comparable to the smallest value of t_{LH} in the technology.

Evaluate

Logic inputs are evaluated when $\phi = 1$. Since Mp1 is in cutoff, the internal capacitor C_X will discharge if all of the input variables are at V_{DD}, i.e., $A_0 = 1 = A_1 = A_2 = A_3$. If one or more of the inputs is zero, the conduction path to ground is severed and V_X must be maintained at a voltage greater than V_{IH}. However, both charge leakage and charge sharing may occur during the evaluate phase of the logic cycle, so the design becomes more complicated.

Consider first the case where all of the inputs are logic 1's. We may estimate the delay time t_D needed to discharge V_X to $(V_{DD}/e) \simeq 0.37\,V_{DD}$ using the analysis in Section 4.2. The worst-case value is

$$\begin{aligned} t_D &\simeq R_nC_n + (R_n+R_3)C_3 + (R_n+R_3+R_2)C_2 \\ &\quad +(R_n+R_3+R_2+R_1)C_1 + (R_n+R_3+R_2+R_1+R_0)C_X. \end{aligned} \qquad (7.73)$$

The resistances are estimated from

$$R_j = \frac{1}{k'_n(W/L)_j(V_{DD}-V_{Tn})}, \qquad (7.74)$$

where (j=0,1,2,3) or (j=n). This expression assumes that all of the internal node capacitances were initially charged in the previous clock cycle. The last term represents the important dominant factor. The evaluate portion of the clock signal must account for both the time delay t_D and the propagation delay t_P through the inverter.

7.8.2 Maximum Clock Frequency

The maximum clocking frequency f_{max} for the circuit is set by the longer of the precharge or evaluate times. In general,

$$\begin{aligned} t_{PRE} &\simeq t_{ch} + t_p, \\ t_{EV} &\simeq t_D + t_p, \end{aligned} \tag{7.75}$$

represent the minimum required time intervals. Let t_M be the larger of the two. Assuming a 50 % clock duty cycle, this then requires that the clock period T satisfy

$$(T/2)_{min} \geq t_M. \tag{7.76}$$

This sets the maximum clock frequency as

$$f_{max} = \frac{1}{2t_M}. \tag{7.77}$$

Driving the circuit with a clock frequency $f < f_{max}$ insures that the charge has sufficient time to react.

7.8.3 Transistor Sizing

Domino logic is structured around nMOS chains which can lead to long delay times t_D. Since this is generally the limiting factor in the clocking, techniques to speed up the evaluation have been investigated. A simple approach to improving performance is to vary the size of the transistors in the logic chain using a scaling technique [17].

Figure 7.29 illustrates the generic RC model used to analyze delay times. The capacitances shown represent only the intrinsic transistor contributions C_0, C_1, C_2, C_3, C_n. As noted above, R_j is inversely proportional to (W/L); however, the parasitic capacitance C_j increases with channel width, and is approximately proportional to $(W/L)_j$.

The effect of scaling the device sizes is most easily seen by rewriting the time delay expression in the form

$$\begin{aligned} t_D = R_n(C_n + C_3 + C_2 + C_1 + C_0) + R_3(C_3 + C_2 + C_1 + C_0) \\ R_2(C_2 + C_1 + C_0) + R_1(C_1 + C_0) + R_0 C_0. \end{aligned} \tag{7.78}$$

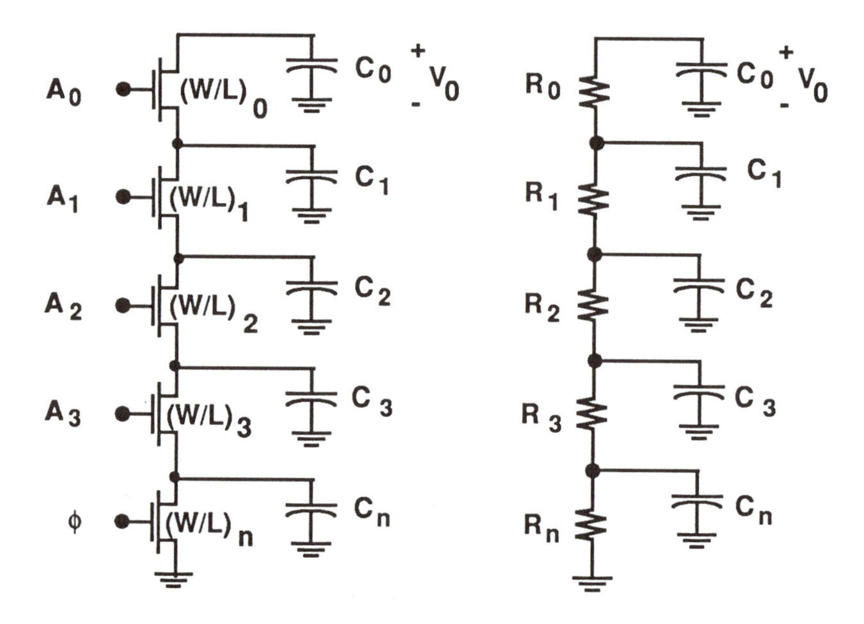

Figure 7.29: Transistor Sizing

Consider transistor M0. Suppose that we reduce the aspect ratio by a factor $\alpha < 1$ such that

$$\left(\frac{W}{L}\right)_0' = \alpha\left(\frac{W}{L}\right)_0 \tag{7.79}$$

gives the new value. The resistance and capacitance are changed according to

$$R_0' = \frac{R_0}{\alpha}, \qquad C_0' = \alpha C_0, \tag{7.80}$$

so that $R_0' > R_0$ and $C_0' < C_0$, but $R_0 C_0 = R_0' C_0'$ is unchanged. Applying these results to the time delay expression shows that the last term is invariant, but all other terms are reduced. Overall, the delay time is smaller by an amount

$$\Delta t_D = [R_n + R_3 + R_2 + R_1](1 - \alpha)C_0. \tag{7.81}$$

This is an interesting result since the increased value of R_0 has no effect on the overall time delay.

Extending the above analysis shows that we may reduce the time delay by scaling the device sizes in the entire chain. The best performance results when we successively reduce the MOSFET aspect ratios starting from the top (output) transistor working down towards ground, i.e.,

$$\left(\frac{W}{L}\right)_0 < \left(\frac{W}{L}\right)_1 < \left(\frac{W}{L}\right)_2 < \left(\frac{W}{L}\right)_3 < \left(\frac{W}{L}\right)_n. \tag{7.82}$$

Comparing this design with one which uses the same area but identical transistors shows that the scaled chain gives a smaller value of t_D. The actual value of the scaling parameter can be estimated using simple equations; however, a simple reduction of 10-30 % per level generally works quite well.

7.8.4 Charge Leakage and Charge Sharing

The minimum clock frequency f_{min} is set by charge leakage and charge sharing during the evaluate portion of the clock cycle. Charge leakage occurs whenever we attempt to hold V_X at a logic 1 voltage. This problem may be worsened if charge sharing takes place.

To understand the problem, let us assume an input pattern of

$$(A_0, A_1, A_2, A_3) = (1, 1, 0, 1) \tag{7.83}$$

and examine the charge sharing and leakage. Figure 7.30 shows the MOSFET chain for this case. During the precharge, C_X is set to $V_X = V_{DD}$; under the worst-case condition, all other capacitors will remain uncharged.

When evaluation takes place, charge is shared among $C_X = C_0 + C_T$, C_1, and C_2. Computing the total charge and applying charge conservation,

$$\begin{aligned} Q_T &= C_X V_{DD} \\ &= (C_X + C_1 + C_2) V_f \end{aligned} \tag{7.84}$$

where we have assumed that $V_f < (V_{DD} - V_{Tn})$. This gives

$$V_f = \frac{C_X}{C_X + C_1 + C_2} V_{DD} \tag{7.85}$$

as the final equilibrium voltage. If $V_f > (V_{DD} - V_{Tn})$, then

$$V_f = V_{DD} - (\frac{C_1 + C_2}{C_X})(V_{DD} - V_{Tn}) \tag{7.86}$$

since the voltages on C_1 and C_2 are subject to threshold drops. The value of V_f must be larger than V_{IH} to insure against a logic glitch.

Consider next the problem of charge leakage. After charge redistribution is completed, charge leakage paths alter the value of the stored voltage. Capacitor C_X is connected to an nMOS-pMOS node which establishes pn-junction leakage paths both from the power supply (through Mp1) and to ground (through M0). In the present example, MOSFETs M1 and M2 introduce additional leakage paths to ground through their normal drain/substrate and source/substrate junctions. Moreover, subthreshold leakage may exist through Mp1 and M3 in a short-channel process.

The simplest approach to modeling the current problem is to assume that all capacitors are constants. For the case where $V_f < (V_{DD} - V_{Tn})$ we write

$$C \frac{dV_f}{dt} = -I_L, \tag{7.87}$$

where $C = (C_X + C_1 + C_2)$ is the total capacitance, and I_L is the net leakage current off of the node to ground. This gives a linear decay of

$$V_f(t) \simeq V_f(0) - \frac{I_L}{C} t, \tag{7.88}$$

showing that $V_f(t)$ decreases in time. The minimum acceptable voltage occurs when $V_f(t_L) = V_{IH}$. Solving gives

$$t_L \simeq \frac{C}{I_L} [V_f(0) - V_{IH}] \tag{7.89}$$

as the longest holding time.

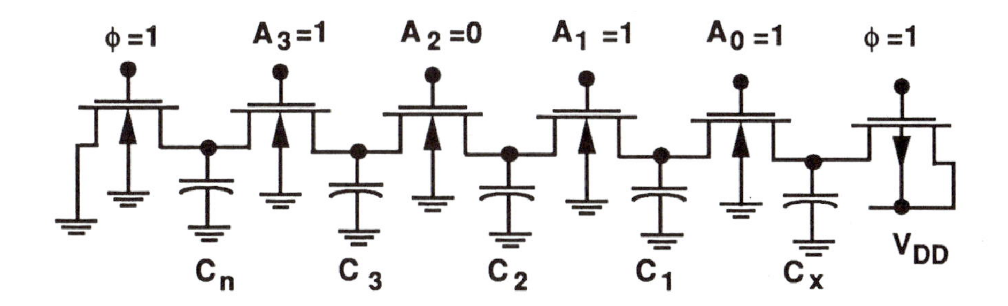

(a) Transistor Chain

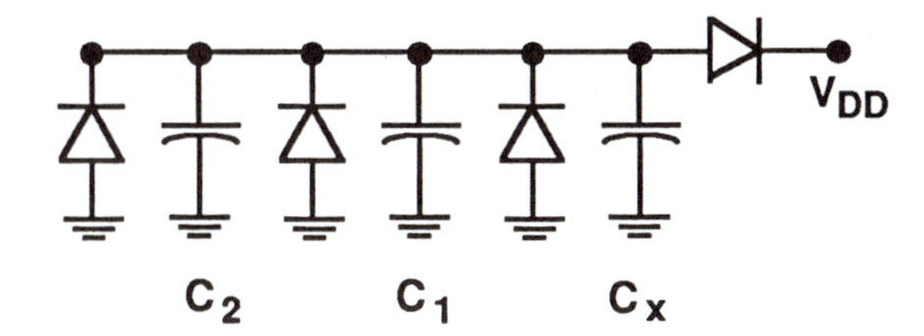

(b) Leakage Subcircuit

Figure 7.30: Charge Sharing and Leakage Example

The minimum clock frequency is set by charge leakage. Assuming a 50 % duty cycle as before, the longest half-clock period that the system can hold a charge state is

$$(T/2)_{max} = t_L. \tag{7.90}$$

This gives the general minimum frequency stipulation

$$\begin{aligned} f_{min} &= \frac{1}{2t_L} \\ &= \frac{I_L}{2C[V_f(0) - V_{IH}]}. \end{aligned} \tag{7.91}$$

The operating frequency f must be greater than f_{min} to avoid problems. In general, the worst-case value of f_{min} should be computed using the largest value of total leakage current I_L and the smallest possible capacitance C. The value depends on the number and arrangement of the logic transistors.

The stipulation of a minimum clock frequency may cause problems at the system design level. Since domino CMOS is implemented using a cascaded chain of gates, the clock frequency must be small enough to allow sufficient time for the logic to propagate all the way through the chain. There are two solutions to this problem: either restrict the designs to short chains, or compensate for the internal charge sharing and charge leakage. The latter is more attractive for implementing realistic system designs.

Additional MOSFETs may be used to partially compensate for the charge problems on the internal nodes. A simple scheme is shown in Figure 7.31. This uses a weakly-conducting p-channel transistor Mp to provide a charging path between the power supply and the internal node capacitance C_X. The gate is grounded so that $V_{SGp} = V_{DD}$ keeps the transistor in a ready-conducting state. During an evaluate phase ($\phi = 1$), $V_{SDp} = (V_{DD} - V_X)$ varies with the input combinations. To hold a high voltage at this node requires

$$\begin{aligned} I_{Dp} &= \frac{k'_p}{2}(\frac{W}{L})_p[2(V_{DD} - |V_{Tp}|)(V_{DD} - V_X) - (V_{DD} - V_X)^2] \\ &\geq I_L. \end{aligned} \tag{7.92}$$

However, I_{Dp} cannot be so large that it prohibits the transition $V_X \rightarrow 0$ when necessary. This is achieved by using a small value of $(W/L)_p$. The alternate circuit in Figure 7.32 controls the conduction state of Mp using feedback from the output. Mp accomplishes the same task, except that it is driven into cutoff when $V_X \rightarrow 0$. If long or complex logic transistor arrays are used, it may be advantageous to precharge nodes inside the chain as shown in Figure 7.33.

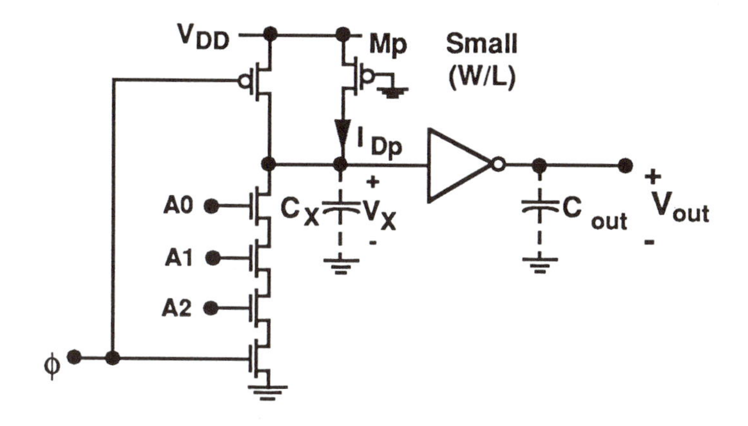

Figure 7.31: Charge Compensation Circuit

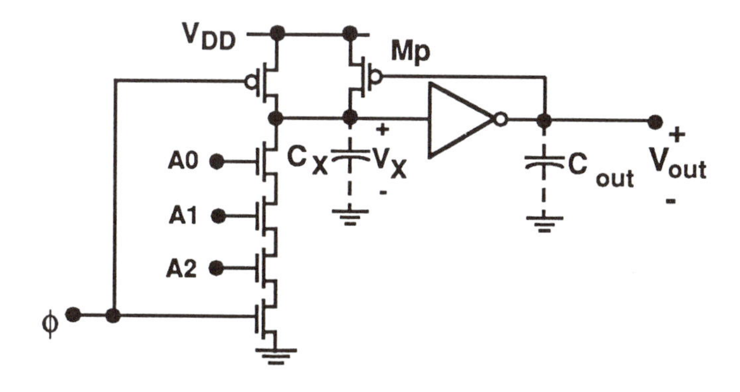

Figure 7.32: Alternate Charge Compensation Scheme

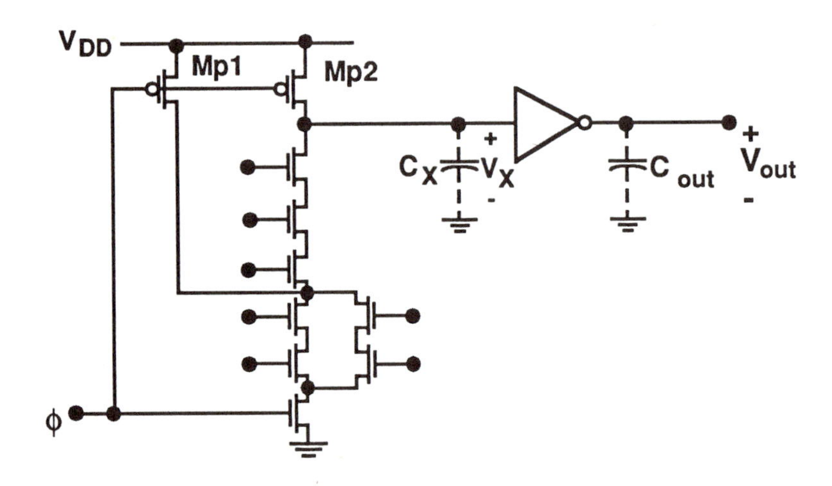

Figure 7.33: Internal Node Precharge

7.9 Multiple-Output Domino Logic

Multiple-output domino logic (MODL) extends standard domino circuits to provide both a function and one or more subfunctions using a single circuit. Suppose that we wish to implement a logical expression F such that

$$F = f_1 \cdot f_2 \tag{7.93}$$

where f_1 and f_2 are the subfunctions. MODL allows simultaneous evaluation of F and f_2 using the structure shown in Figure 7.34. This circuit splits the logic into two subblocks f_1 and f_2; adding another inverter gives the two outputs. Reversing the location of the logic transistors provides the alternate output pair F and f_1. The structure can be expanded to several subfunctions so long as they are ANDed together. For example, if $F = f_1 \cdot f_2 \cdot f_3$, then adding a logic subblock f_3 between f_2 and the evaluate MOSFET and providing another inverter gives

$$\begin{gathered} f_3 \\ f_2 \cdot f_3 \\ f_1 \cdot f_2 \cdot f_3 \end{gathered} \tag{7.94}$$

as the three possible outputs. Each subblock can be constructed with necessary level of complexity. The ability to simultaneously perform nested

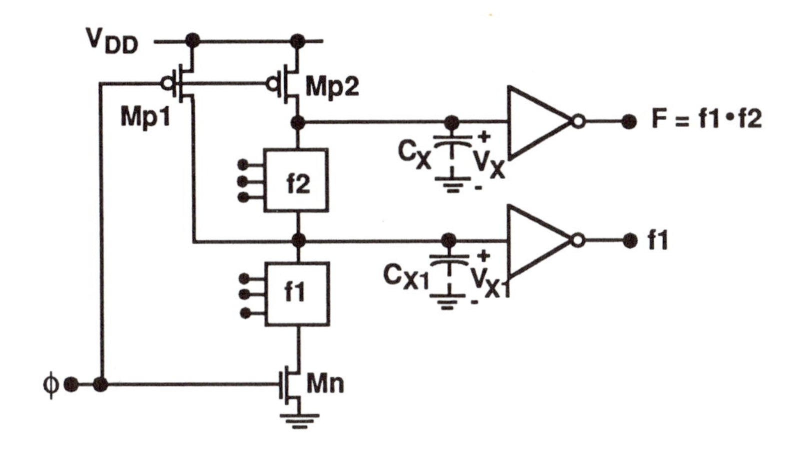

Figure 7.34: Dual-Output Domino Logic

logic functions can increase the overall system speed while also reducing the circuit complexity.

Electrically, an MODL circuit behaves just like a standard domino stage. Precharge transistors are provided for both internal nodes, so that a clock condition of $\phi = 0$ charges both C_X and C_X to V_{DD}. The time constants are respectively given by

$$\begin{aligned} \tau_{p1} &= \frac{C_{X1}}{\beta_{p1}(V_{DD} - |V_{Tp}|)} \\ \tau_{p2} &= \frac{C_{X2}}{\beta_{p2}(V_{DD} - |V_{Tp}|)}; \end{aligned} \tag{7.95}$$

the minimum precharge time is set by the larger of the two. Note that C_{X1} will have contributions from inside logic block f_1. Evaluation takes place when $\phi = 1$. Since the subfunctions are ANDed, the discharge time depends on the input combinations. When F and f_2 both provide conduction paths to ground, f_2 stablilizes before F. Note that the addition of precharge nodes makes MODL designs less susceptible to charge sharing problems than standard domino circuits.

7.9.1 MODL Carry-Look-Ahead Adder

MODL has been successfully applied to design fast parallel adders using carry-look-ahead (CLA) logic. Consider the simple "ripple-carry" 4- bit parallel adder as shown in Figure 7.35 which takes inputs $(a_3, \ldots, a_0)$ and $(b_3, \ldots, b_0)$ and produces the sum $(s_3, \ldots, s_0)$ and output carry c. The i-th stage has inputs a_i, b_i, and c_i (the carry-in) so that the basic addition algorithms for the i-th bit are given by

$$\begin{aligned} s_i &= a_i \oplus b_i \oplus c_i \\ c_{i+1} &= a_i b_i + (a_i \oplus b_i) c_i \end{aligned} \tag{7.96}$$

where c_{i+1} is the carry-out bit. The cascaded carry scheme used in connecting the half-adder circuits introduces significant delay into the circuit. For example, the output bit s_2 is not valid until s_1 is valid, which in turn needs the carry bit c_1 generated by the 0th-bit position.

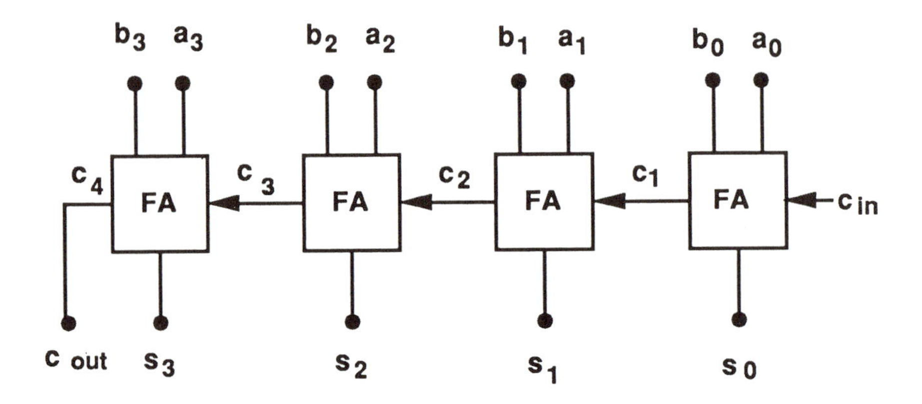

Figure 7.35: Basic 4-bit Parallel Adder

CLA algorithms are designed to give "look-ahead" capabilities which provide the carry-in bits to each stage without waiting for each bit evaluation. To see the basic idea, we define a **generate** function

$$g_i = a_i b_i, \tag{7.97}$$

and a **propagate** function

$$p_i = (a_i \oplus b_i) \tag{7.98}$$

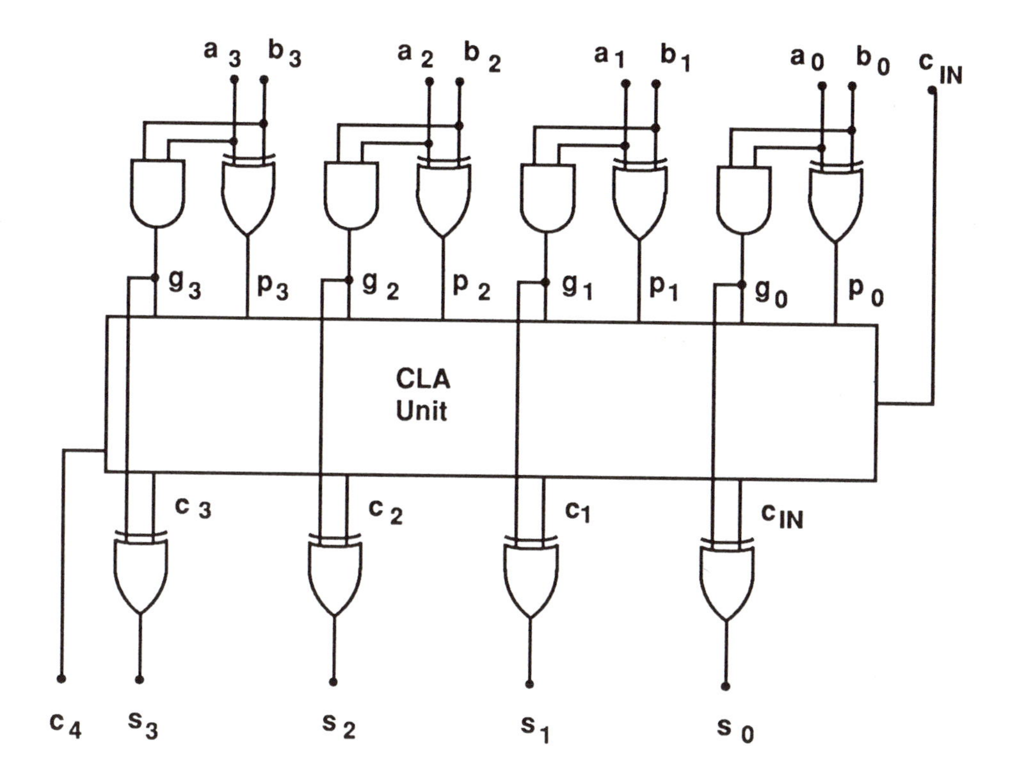

Figure 7.36: CLA 4-bit Parallel Adder

so that the carry-out bit is given by

$$c_{i+1} = g_i + p_i c_i. \tag{7.99}$$

Applying this to the 4-bit adder gives the carry-out bits as

$$\begin{aligned} c_0 &= c_{IN} \\ c_1 &= g_0 + p_0 c_{IN} \\ c_2 &= g_1 + p_1 g_0 + p_1 p_0 c_{IN} \\ c_3 &= g_2 + p_2 g_1 + p_2 p_1 g_0 + p_2 p_1 p_0 c_{IN} \\ c_4 &= g_3 + p_3 g_2 + p_3 p_2 g_1 + p_3 p_2 p_1 g_0 + p_3 p_2 p_1 p_0 c_{IN} \end{aligned} \tag{7.100}$$

by direct calculation. Adding a CLA unit to the parallel adder as shown in Figure 7.36 greatly increases the speed of the network.

The CLA algorithm can be implemented in standard CMOS, but we see that the nested functions imply that it is well-suited for MODL. The carry generator circuit shown in Figure 7.37 illustrates how a single dynamic gate can simultaneously provide all of the necessary carry bits c_1, c_2, c_3, c_4. This MODL circuit is functionally equivalent to four standard domino gates. The generate and propagate bits g_i and p_i can be obtained using single-output domino gates, and then cascaded into the MODL circuits. The speed advantage of the circuit arises from simultaneous evaluation of all carry terms; transistor count is also decreased compared to more conventional CLA implementations. The interested reader is referred to Hwang and Fisher [6] who discuss a 32-bit CLA architecture implemented using 8-bit adder segments and MODL circuits.

7.10 Latched Domino Logic

Latched domino (**Ldomino**) logic gates combine domino structuring with a cross-coupled output latch to provide many useful functions. Ldomino logic has the following characteristics:

- NOT (inversion) operations are possible;
- Ldomino circuits use **single-rail** inputs and produce **double-rail** outputs;
- Ldomino circuits can only be used as input stages to a chain of standard domino stages.

The ability to perform NOT operations overcomes the non-inverting characteristic of standard domino circuits, while double-rail outputs are useful to simplify the logic structuring of stages. However, a standard domino output cannot drive an Ldomino circuit; this restricts their placement to the input stage of a domino chain.

A basic Ldomino circuit is shown in Figure 7.38. This takes a standard domino stage and adds a dynamic cross-coupled output latch. Complementary circuitry is included to provide both the function and its complement. MOSFETs Mn1 and Mn2 provide the latching mechanism for the unbalanced output circuits, with Mp1 and Mp2 used as a load. While this will give proper operation, the performance can be improved by using a clocked dynamic latch as shown in Figure 7.39. In this circuit, capacitor C_3 is precharged to V_{DD} using pMOS transistor MLp. This drives Mn1 and Mn2 into cutoff and provides a delay time for the nMOS logic block to settle before latching action occurs. Decoupling the latch from the logic

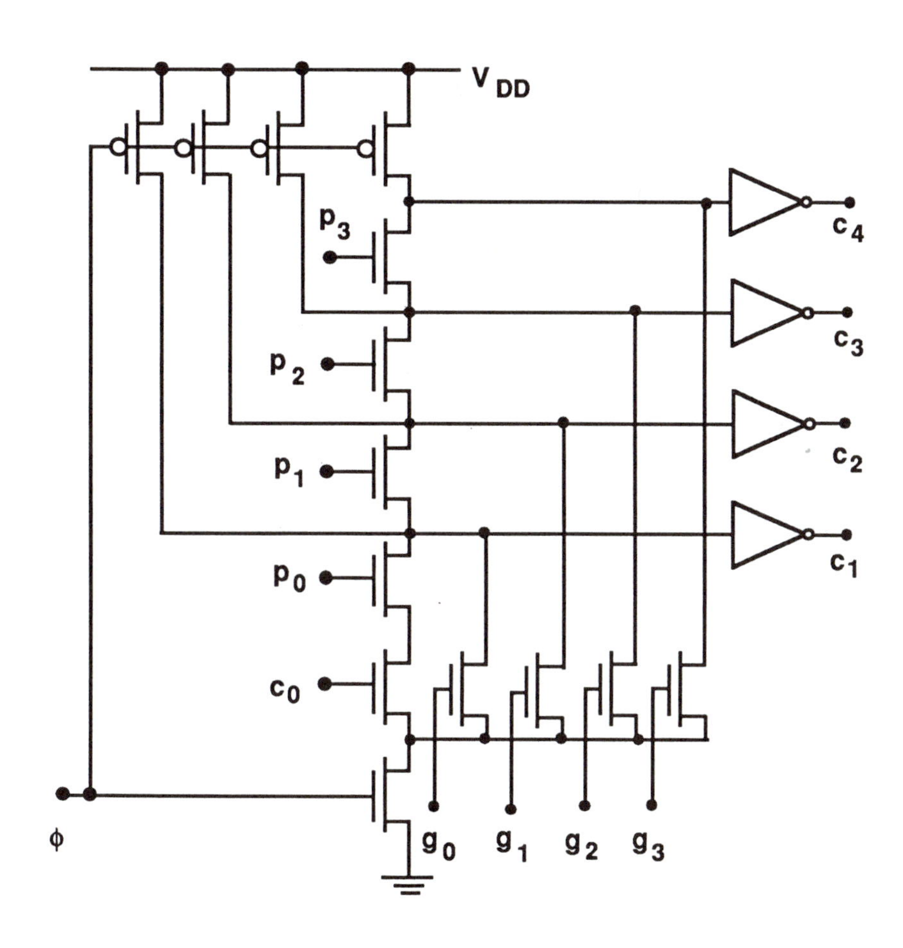

Figure 7.37: MODL CLA Carry-Generation Circuit

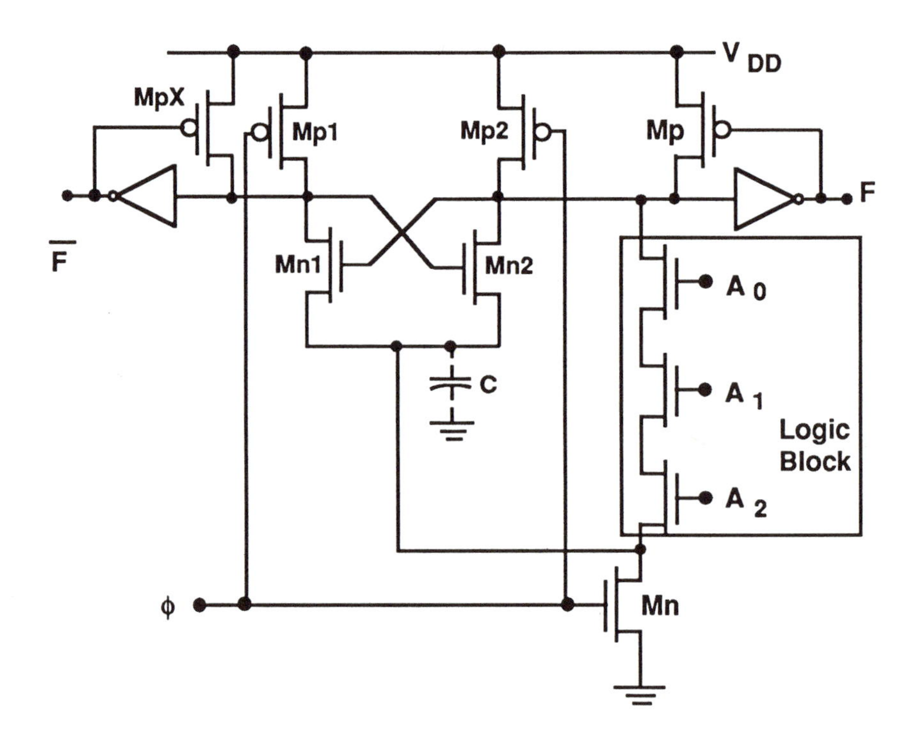

Figure 7.38: Basic Ldomino Gate

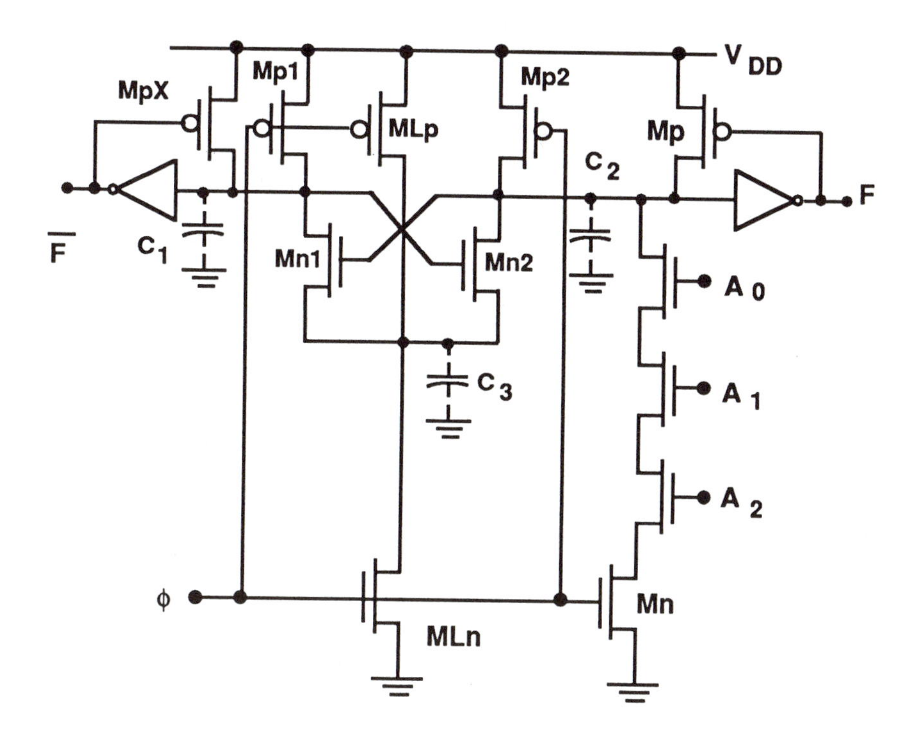

Figure 7.39: Clocked-Latch Ldomino Circuit

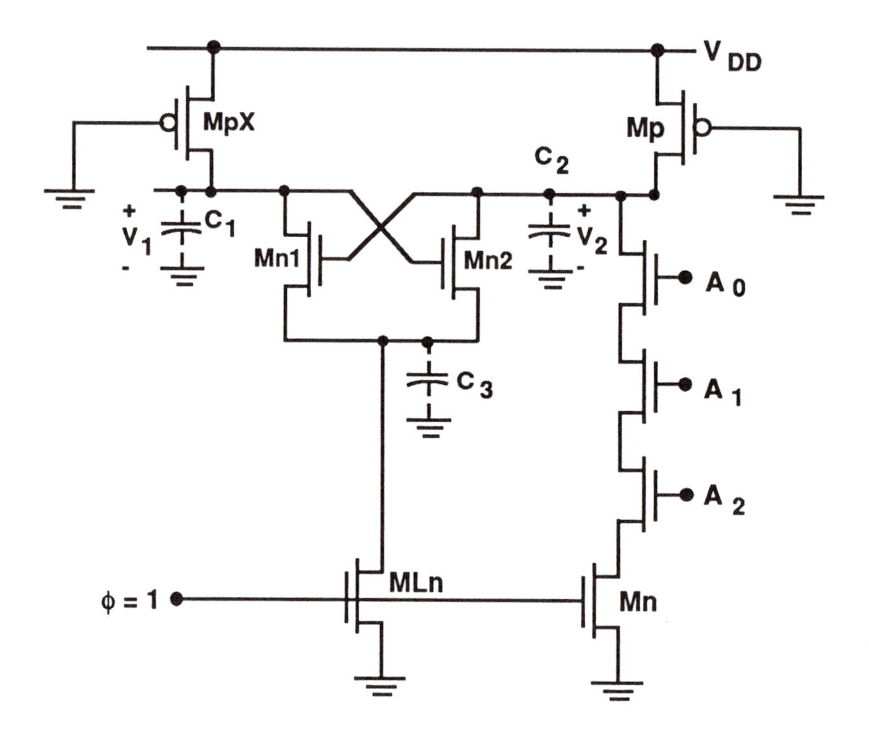

Figure 7.40: Latching Mechanism in an Ldomino Circuit

transistors using MLn allows the cross- coupled network to be designed independent of the specific logic tree.

Circuit operation is understood by tracing the clock action. Precharge occurs when $\phi = 0$. During this time, capacitors C_1, C_2, and C_3 are all charged to V_{DD}. When the clock makes an upward transition toward $\phi = V_{DD}$, evaluation transistors MLn and Mn are driven into conduction. Figure 7.40 provides the circuit for this portion of the cycle. Due to the presence of the transistor logic array, $C_2 > C_1$. Assuming that the latch is designed with $(W/L)_{Mn1} \geq (W/L)_{Mn2}$, voltage V_1 tends to fall at a faster rate than V_2. This will be reversed if the input logic evaluates to give a discharge path for C_2 through the logic array and Mn.

Ldomino logic circuits are somewhat specialized in behavior and application. They do provide dual-rail outputs with the associated inversion property and act as good interfaces with static circuit. More details of Ldomino circuits can be found in the literature [15].

7.11 NORA Logic

Transmission gates are extremely useful for controlling data flow in digital networks. Although TG-based synchronization can be effectively used to design systems of arbitrary complexity, timing problems still arise. These occur during clock transition intervals when the TG is partially transmitting, or because of **clock skew** problems when the clocking waveforms go out of phase. Either situation can induce **signal races** as described below.

No race (NORA) logic is a system design style which was introduced to overcome signal race problems introduced by transmission gate problems[4]. It is based on dynamic nMOS/pMOS CMOS logic, but uses latches instead of TGs to control signal flow. NORA logic is ideal for creating **pipelined logic** systems. In addition, it exhibits a high level of immunity against clock skew problems.

To understand NORA logic, we will first examine signal races and timing problems in standard TG-based data flow control. Once this is completed, the philosophy behind NORA circuit structuring will be discussed.

7.11.1 Signal Races

Signal races can be understood using the circuit in Figure 7.41. Data bit $D = D1$ is admitted into the circuit through TG1 when the clock is in a state $(\phi, \overline{\phi}) = (1, 0)$. After a gate propagation time t_P, the output F is ready to be transmitted to the next stage through TG2 when the clocks change to $(\phi, \overline{\phi}) \rightarrow (0, 1)$. A signal race may occur if both TG1 and TG2 conduct at the same time. In this case, a new data bit $D = D2$ is admitted through TG1 and "races" the clock to the output. If $D2$ "wins the race" by reaching TG2 while it is still transmitting the output, the information from to $D2$ is lost. Imperfect TG synchronization occurs because of normal transition intervals or clock skew.

Clock Transitions

CMOS TGs are constructed using complementary nMOS and pMOS transistors which require complementary control signals. Since it is not possible to generate a perfect square wave, the clock always has finite rise and fall times t_r and t_f. As indicated in Figure 7.42, both TG1 and TG2 conduct partially during these transition intervals. The critical voltage for turning on the nMOS transistor is V_{Tn}, while a voltage of $(V_{DD} - |V_{Tp}|)$ on the gate of the pMOSFET allows conduction.

When we use the complementary clocks to control the circuit in Figure 7.41, signal races are unavoidable. However, if $t_p >> t_r, t_f$, then no prob-

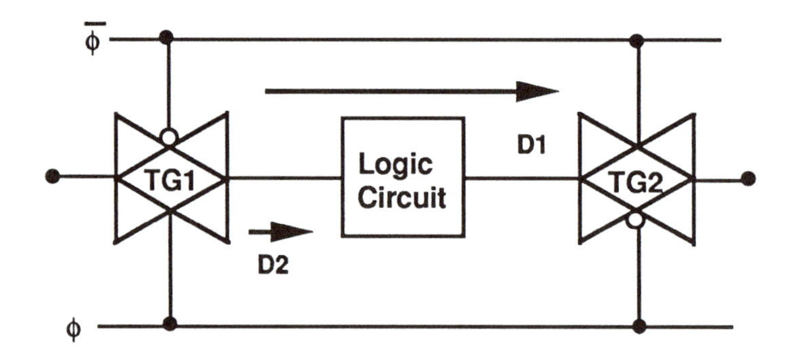

Figure 7.41: TG Signal Race Circuit

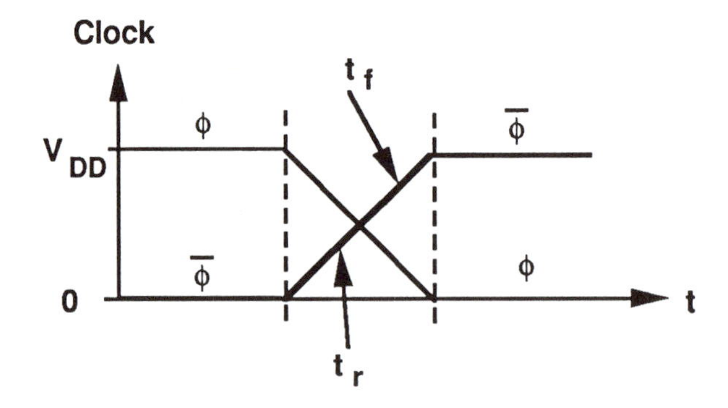

Figure 7.42: Rise and Fall Intervals

lems will occur since the proper data is always transmitted before the next data bit reaches the conducting output transmission gate.

Clock Skew

Clock skews are more difficult to deal with. Consider the situation where $\overline{\phi}$ is delayed from ϕ by a time t_{skew}; Figure 7.43 shows the signals. The overlap induced by the clock skew increases the amount of time when both TG1 and TG2 are conducting. When $t_{skew} \sim t_p$, the race results become

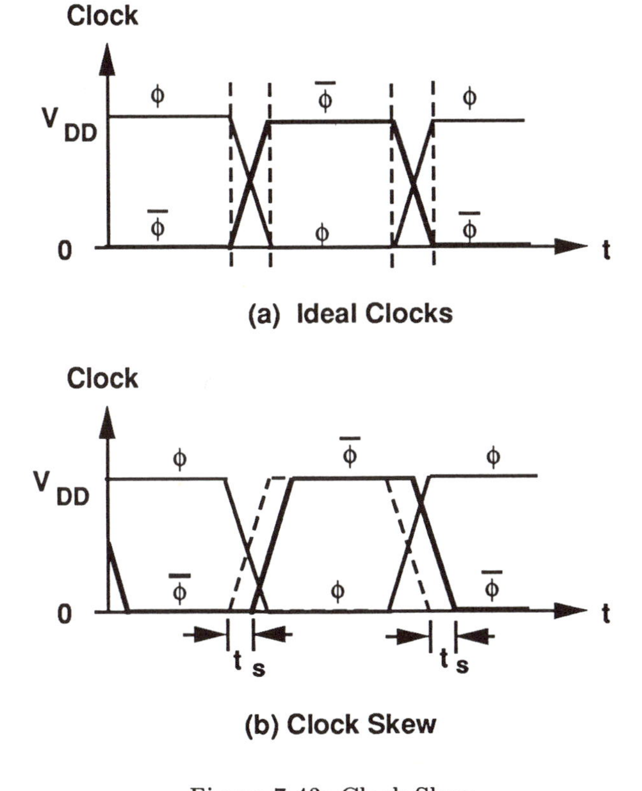

Figure 7.43: Clock Skew

critical; incorrect data may be transmitted out of the stage.

7.11.2 Data Control Using Dynamic Latches

Partial transmission problems in TG-based data control can be alleviated by using clocked latches to time the data. A block diagram is shown in Figure 7.44. Standard dynamic CMOS logic can be used for the input stages, while a clocked latch is provided at the output. The logic circuits operate using $\phi = 0$ for the precharge time and $\phi = 1$ for evaluation. The latch is designed to accept data when $\phi = 1$ and to hold the data when $\phi = 0$; no new data can be admitted during the hold time. This group of circuits constitutes what we will term a "ϕ-section ". A "$\overline{\phi}$-section " has the same structure, but reverses the clocking signal so that $\phi = 1$ is the precharge/hold time while $\phi = 0$ is used for evaluation and latch input. Cascades of alternating ϕ and $\overline{\phi}$-sections provide a basis for pipelined logic. Dynamically-latched outputs provide the basis for NORA circuits.

7.11.3 Clocked-CMOS Latches

NORA logic uses the clocked-CMOS (C^2MOS) latch shown in Figure 7.45. Mn and Mp are the logic transistors, while Mn1 and Mp1 are clock-driven to provide precharge and hold. The operation is straightforward. Precharge occurs when $\phi = 1$; data is admitted to the latch during this time. Mn1 and Mp1 are both conducting so that the charge states on C_p, C_n and C_{out} are determined by the value of V_{in}. If $V_{in} = V_{DD}$ (a logic 1), Mn is on while Mp is off; all of the capacitors discharge through the nMOS transistors. In particular, $V_{out} \rightarrow 0$ [V]. A logic 0 input voltage of $V_{in} = 0$ [V] places Mp in active conduction and Mn in cutoff. In this case, all of the capacitors charge through the pMOS transistors. C_{out} sees a conduction path to the power supply so that $V_{out} \rightarrow V_{DD}$.

When the clock changes to $\phi = 0$, the latch moves into a hold state by driving both Mn1 and Mp1 into cutoff. The output capacitance C_{out} is isolated from the rest of the circuit and acts as a charge storage node whose value was set during the precharge. Of course, the usual problem of charge leakage changes the voltage in time, so that the circuit has the normal problems associated with dynamic behavior.

7.11.4 NORA Structuring

NORA logic employs dynamic nMOS and pMOS logic circuits cascaded into a C^2MOS latch. Figure 7.46 illustrates the structure of a generic NORA ϕ-section. Static inverters are provided at the outputs of both dynamic circuits to provide for logic inversion; this overcomes the limitation of Domino structure and allows for direct implementation of arbitrary

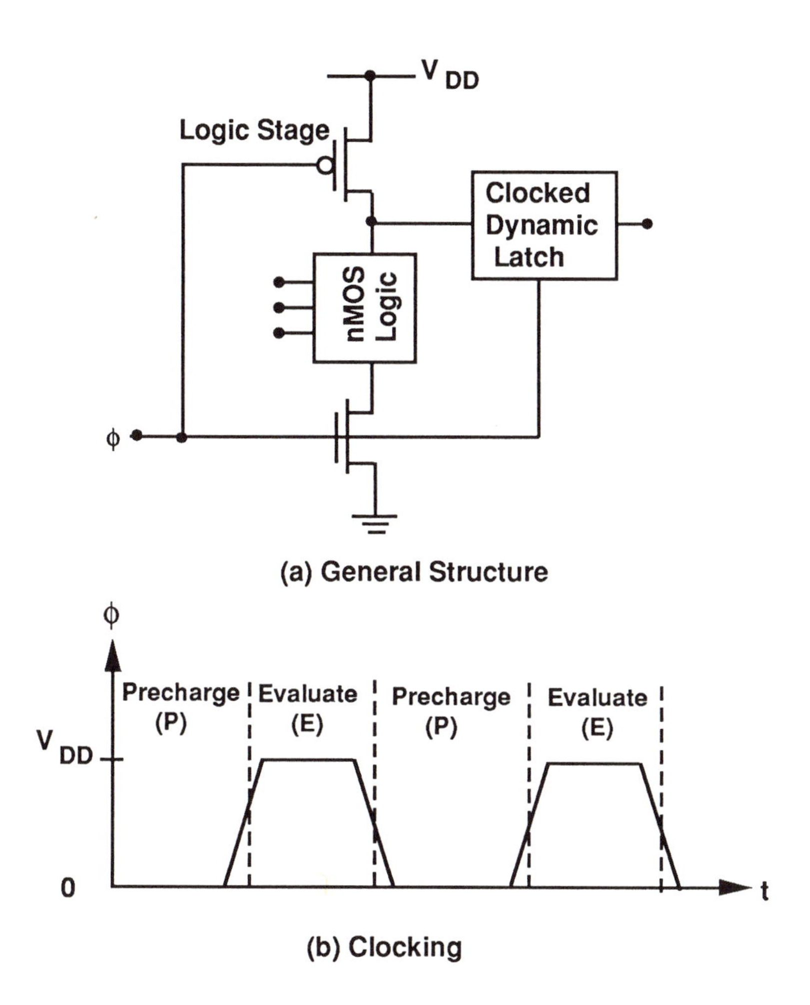

Figure 7.44: Dynamic Latch Operation

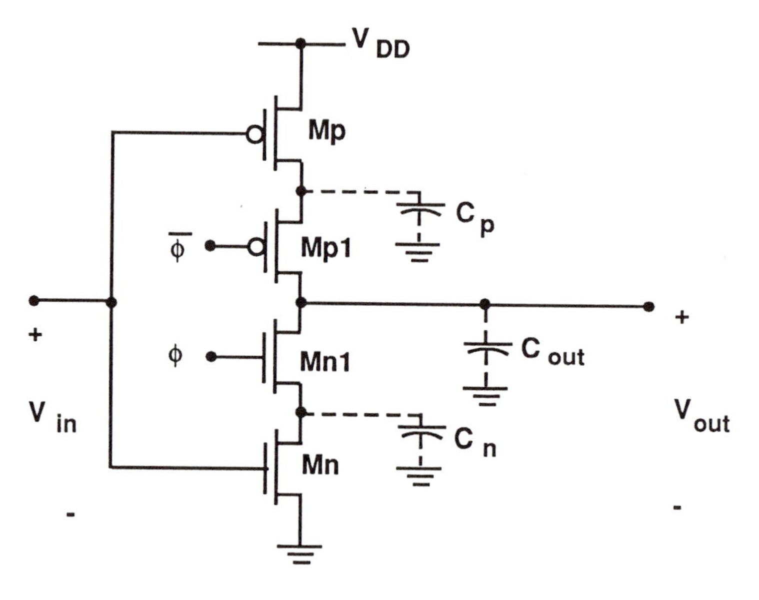

Figure 7.45: C²MOS Latch

functions without modification. The operation of the ϕ-section is easily understood. When $\phi = 0$, the logic circuits are undergoing precharge and the previous state is available at the latch output. A transition to $\phi = 1$ induces logic evaluation in both the nMOS and pMOS circuits and the result is transferred to the latch during this time. The data is latched during the next clock half-cycle when $\phi = 0$. A $\overline{\phi}$-section is also shown in the figure; this is identical in structure to the ϕ-section, but is exactly out of phase.

NORA logic cascades are designed using alternating ϕ and $\overline{\phi}$ as shown in Figure 7.47. Logic flows through the system by alternating between ϕ and $\overline{\phi}$ sections at a rate set by the clocking. Logic races do not occur by virtue of the dynamic C²MOS latching circuit, giving NO RAce properties. Alternating ϕ and $\overline{\phi}$-sections makes NORA chains well-suited for pipelined logic.

7.11.5 NORA Serial Adder

As an example of a NORA pipeline, consider the serial full adder shown in Figure 7.48. The inputs are A and b, while C is the carry bit; the clocking is designed to keep track of whether C is a carry-in or a carry-out. The

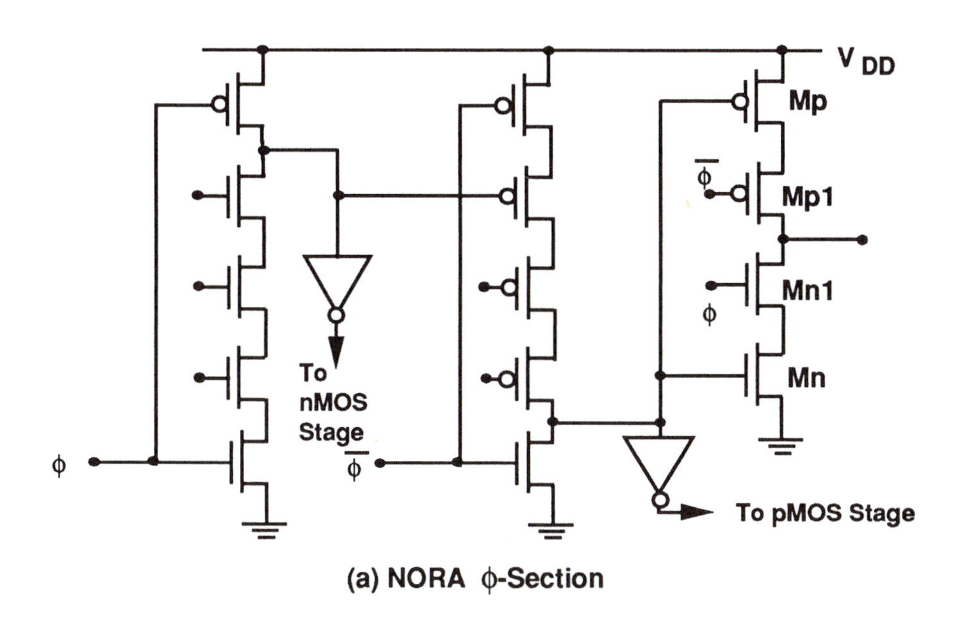

(a) NORA ϕ-Section

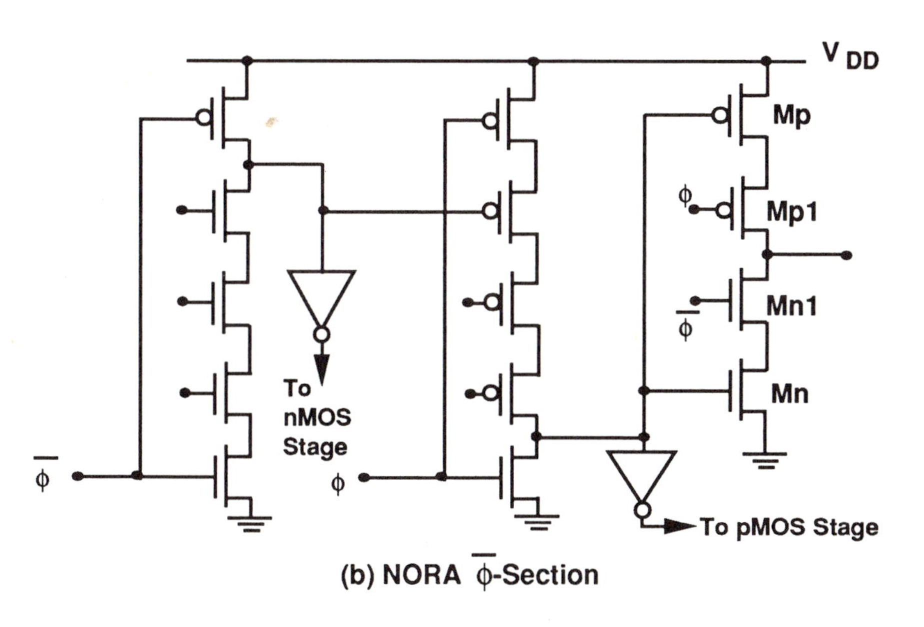

(b) NORA $\overline{\phi}$-Section

Figure 7.46: NORA ϕ and $\overline{\phi}$-Sections

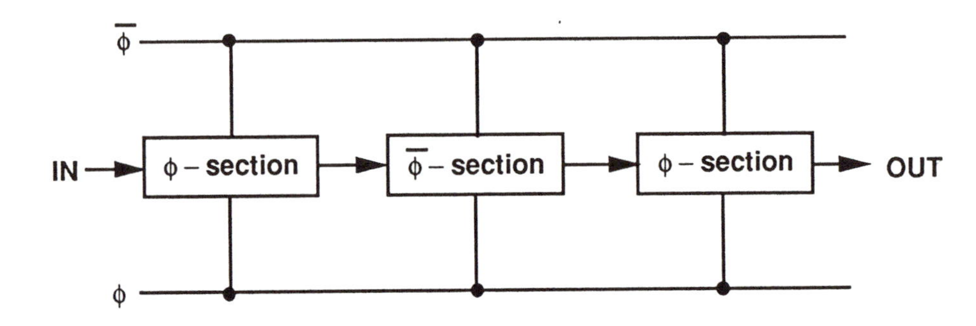

Figure 7.47: NORA Cascades

left-half of the circuit uses pMOS logic to generate a logical output of

$$\overline{F} = \overline{AB + AC + BC} \tag{7.101}$$

as verified by direct Boolean reduction; at this point C is the carry-in bit. The upper right-half of the adder uses nMOS logic and computes the sum

$$S = ABC + A\overline{B}\,\overline{C} + \overline{A}B\overline{C} + \overline{A}\,\overline{B}C. \tag{7.102}$$

The lower right-half portion of the circuit serves main two purposes. First, it has a CARRY CLEAR input which resets $C = 0$ before the LSB's of the binary words arrive. Secondly, it provides the delay period needed to distinguish between the carry-in and carry-out bits. The circuit itself is free from critical race problems, and is a relatively compact logic implementation.

7.11.6 NORA Serial-Parallel Multiplier

The serial adder circuit in the previous section can be used as a building block for constructing a multiplier circuit. The algorithm can be seen by reviewing basic binary multiplication. Consider two 4-bit words

$$\begin{aligned} \mathbf{a} &= (a_3a_2a_1a_0) \\ \mathbf{b} &= (b_3b_2b_1b_0). \end{aligned} \tag{7.103}$$

The binary product is given by

$$\mathbf{a} \times \mathbf{b} = (p_6p_5p_4p_3p_2p_1p_0) \tag{7.104}$$

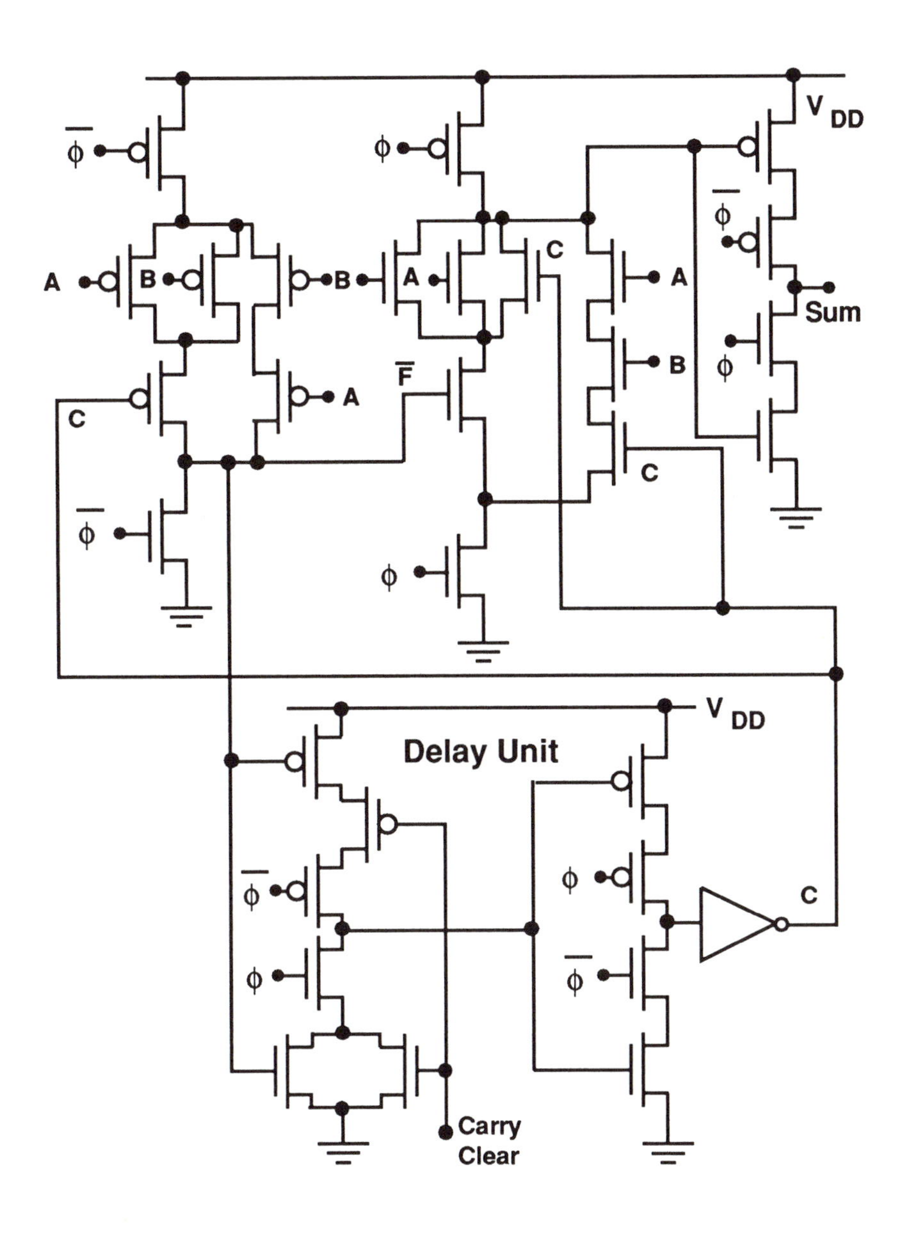

Figure 7.48: NORA Serial Adder Circuit

where the product terms are of the form

$$p_i = \sum_{\alpha+\beta=i} a_\alpha b_\beta. \tag{7.105}$$

Explicitly, the 4-bit multiplication yields

$$\begin{array}{lllll} p_0 = & & & & a_0b_0 \\ p_1 = & & & a_0b_1+ & a_1b_0 \\ p_2 = & & a_0b_2+ & a_1b_1+ & a_2b_0 \\ p_3 = & a_0b_3+ & a_1b_2+ & a_2b_1+ & a_3b_0 \\ p_4 = & a_1b_3+ & a_2b_2+ & a_3b_1 & \\ p_5 = & a_2b_3+ & a_3b_2 & & \\ p_6 = & a_3b_3 & & & \end{array} \tag{7.106}$$

with "+" denoting the binary sum.

The algorithm consists of two main operations. First, the 4-bit word **a** is multiplied by each bit of **b**, e.g.,

$$(a_3a_2a_1a_0) \times b_0 = (a_3b_0, a_2b_0, a_1b_0, a_0b_0). \tag{7.107}$$

Each term can be generated using the AND operation. The second step is simply aligning the terms and adding to produce each p_i product bit.

A serial-parallel 4-bit multiplier which is based on this sequence is shown in Figure 7.49. The **a**-bits are entered in parallel, while **b** enters the system sequentially. The AND gates provide the bit-multiplications. Delays (Δ) segments are used to align each term, and additions (Σ) are implemented using the serial NORA circuit of the previous section. The clocked network gives a sequential output of the product terms ($p_0, \ldots, p_6$) as shown.

7.12 Zipper CMOS Logic

Zipper CMOS is another example of a dynamic logic family. The main feature of zipper CMOS is the use of four clocks to control an nMOS-pMOS logic cascade as shown in Fig. 7.50. These are introduced to aid in the problems of charge leakage and charge sharing which exist in a standard circuit design. Clock signals ϕ_1 and ϕ_{1X} are in phase with each other, and are connected to the nMOS logic stages. Similarly, ϕ_2 and ϕ_{2X} are in phase, and define the timing intervals in the pMOS circuits. The clocks are defined so that

$$\phi_1 \cdot \phi_2 = 0, \tag{7.108}$$

so that this is similar to a pseudo-2ϕ system. The difference arises in the amplitudes of ϕ_{1X} and ϕ_{2X}.

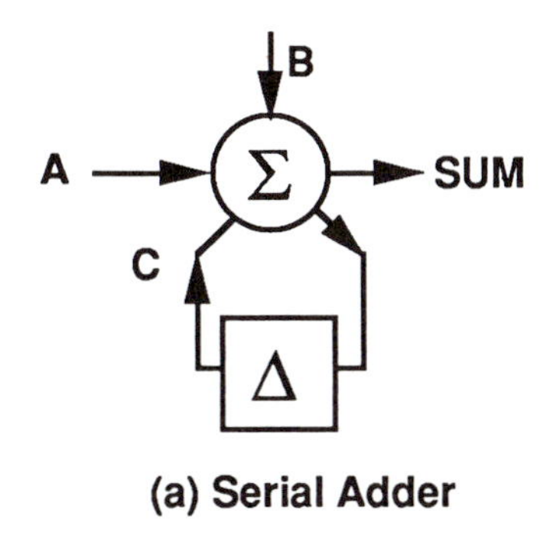

(a) Serial Adder

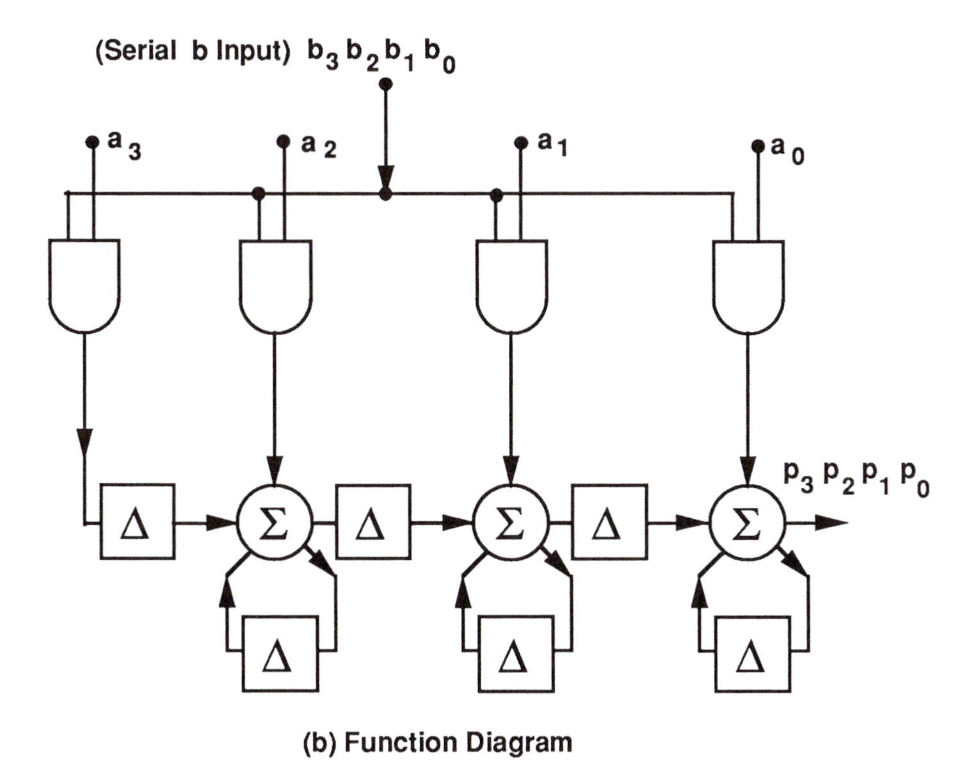

(b) Function Diagram

Figure 7.49: 4-bit Serial-Parallel Multiplier

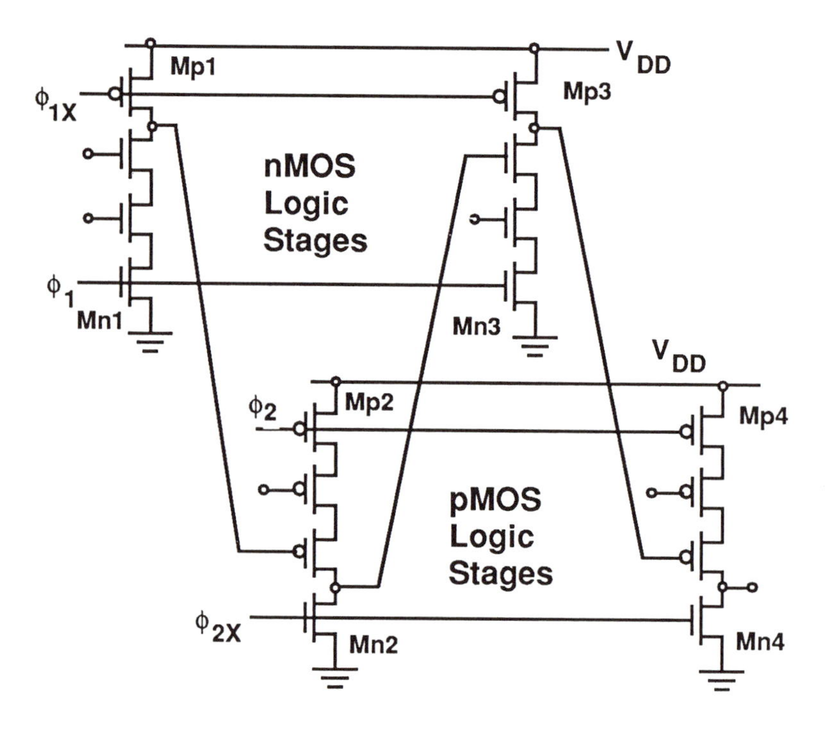

Figure 7.50: Zipper Logic Circuit

Figure 7.51 illustrates the clocking signals. Consider first the waveforms for ϕ_1 and ϕ_{1X}. We see that ϕ_1 has an amplitude which ranges from 0 to V_{DD}, but that ϕ_{1X} is restricted to the range $[0, (V_{DD} - V_T)]$, where V_T is a threshold voltage. Since ϕ_{1X} is applied to the p-channel precharge transistors Mp1 and Mp3, limiting the amplitude to a maximum of $(V_{DD} - V_T)$ keeps these devices on the edge of conduction during logic evaluation. This reduced amplitude is designed to overcome the problems of charge sharing and charge leakage in the nMOS logic stage.

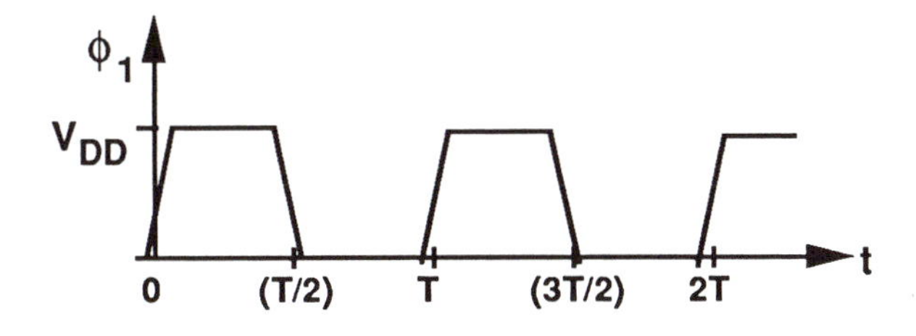

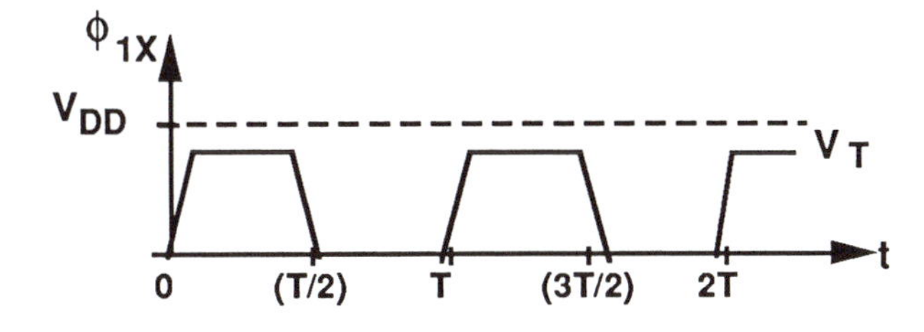

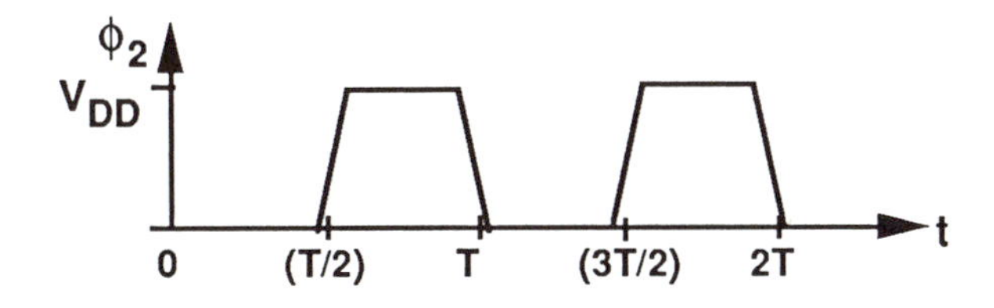

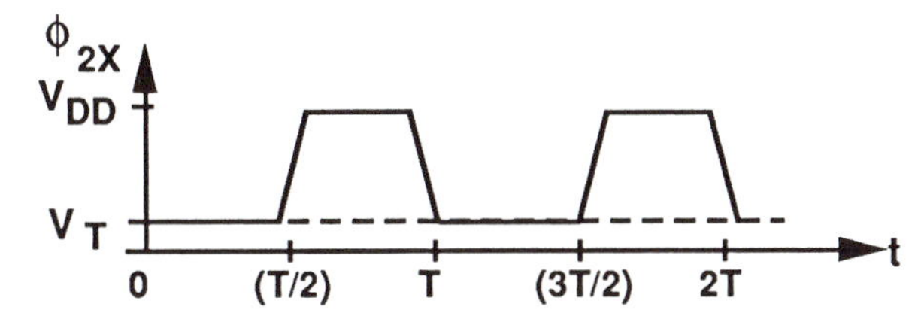

Figure 7.51: Clocks for Zipper Logic

The clock pair ϕ_2 and ϕ_{2X} are design to perform a similar function. ϕ_2 undergoes a full-rail swing, but ϕ_{2X} is restricted to the range $[V_T, V_{DD}]$. Noting that ϕ_{2X} controls the n-channel precharge MOSFETs Mn2 and Mn4, we see that the increased amplitude keeps these devices on the edge of conduction during a pMOS evaluation interval.

The name "zipper CMOS" arises from tracing the logic signal "up" and "down" as it propagates from left to right in the network; this is similar to a zipper closing. The design style itself is complicated by the need to generate and route four clocking signals. The interested reader will find more about this design style in references [12] and [18].

7.13 References

Clocked static and dynamic circuits are discussed in many textbooks and journals. The list below provides a sampling of the literature in the field.

[1] M. A. Annaratone, **Digital CMOS Circuit Design**, Kluwer Academic Publishers, Boston, 1986.

[2] M.A. Bayoumi and N. Ling, " Testing of a NORA CMOS Serial-Parallel Multiplier," IEEE J. Solid-State Circuits , vol. 24, no. 2, pp. 494-503, April, 1989.

[3] V, Friedman and S. Liu, " Dynamic Logic CMOS Circuits," IEEE J. Solid-State Circuits, vol. SC-19, no. 2, pp.263-266, April, 1984.

[4] N.F. Goncalves and H.J. De Man, " NORA: A Racefree Dynamic CMOS Technique for Pipelined Logic Structures," IEEE J. Solid-State Circuits, vol. SC-18, No. 3, pp. 261-266, June, 1983.

[5] M. Hoffmann and A.R. Newton, " A domino CMOS logic synthesis system," Proc. IEEE Int. Symp. on Circuits and Systems," pp. 411-414, 1985.

[6] I.S. Hwang and A.J. Fisher, " Ultrafast Compact 32-bit CMOS Adders in Multiple-Output Domino Logic," IEEE J. Solid-State Circuits, vol. 24, no. 2, pp. 358-369, April, 1989.

[7] N.K. Jha and Q.Tong, " Testing of Multiple-Output Domino Logic (MODL) CMOS Circuits," IEEE J. Solid-State Circuits, vol. 25, no. 3, pp. 800- 805, June, 1990.

[8] S.M. Kang, "Domino-CMOS barrel switch for 32-bit VLSI processors," IEEE Circuits and Devices Mag., vol. 3, no. 3, pp. 3-8, May, 1987.

[9] S.M. Kang and H.Y. Chen, "A global delay model for domino CMOS circuits with applications to transistor sizing," Int. J. of Circuit Theory and Applications, vol. 18, no. 3, pp. 289-306, 1990.

[10] J. Kernhof, et al., "Mixed static and domino logic on the CMOS gate forest," IEEE J. Solid-State Circuits, vol. 25, no. 2, pp. 396-401, April, 1990.

[11] R.H. Krambek, C.M. Lee, and H-F. Law, "High-speed compact circuits with CMOS," IEEE J. Solid-State Circuits, vol. SC-17, no. 3, pp. 614-618, June, 1982.

[12] C.M. Lee and E.W. Szeto, "Zipper CMOS," IEEE Circuits and Devices Mag., vol.2, no. 3, pp. 101-107, May, 1986.

[13] J.A. Pretorius, A.S. Shubat, and C.A.T. Salama, " Analysis and design optimization of domino CMOS logic with applications to standard cells," IEEE J. Solid-State Circuits, vol. SC-20, no. 2, pp. 523-530, April, 1985.

[14] J.A. Pretorius, A.S. Shubat, and C.A.T. Salama, " Charge distribution and noise margins in domino CMOS," IEEE Trans. Circuits and Systems, vol. CAS-33, no. 8, pp. 786-793, Aug, 1986.

[15] J.A. Pretorius, A.S. Shubat, and C.A.T. Salama, " Latched Domino CMOS Logic", IEEE J. Solid-State Circuits, vol. SC-21, no. 4, pp. 514-522, Aug., 1986.

[16] M. Shoji, **CMOS Digital Circuit Technology**, Prentice-Hall, Englewood Cliffs, NJ, 1988.

[17] M. Shoji, " FET Scaling in Domino CMOS Gates," IEEE J. Solid-State Circuits, vol SC-20, no. 5, pp. 1067-1071, Oct., 1985.

[18] J. P. Uyemura, **Fundamentals of MOS Digital Integrated Circuits**, Addison-Wesley, Reading, MA, 1988.

[19] N. Weste and K. Eshraghian, **CMOS VLSI Design**, Addison-Wesley, Reading, MA, 1985.

[20] J. Yuan and C. Svensson, "High-speed CMOS circuit technique," IEEE J. Solid-State Circuits, vol. SC-24, no. 1, pp. 62-70, Feb., 1989.

[21] C-Y. Wu and K-H. Cheng, "Latched CMOS Differential Logic (LCDL) for Complex High-Speed VLSI," IEEE J. Solid-State Circuits, vol 26, no. 9, pp. 1324-1328, Sept., 1991.

Chapter 8

Design of Basic Circuits

Up to this point we have only examined gate-level logic and clocked-system fundamentals. In this chapter the emphasis is placed on applying the material to a few select design examples. Space limitations prohibit covering too many circuits; besides, original design is part of the fun of working in this field in the first place. Instead, we will concentrate on circuits which have been chosen because of their overall usefulness to the CMOS circuit designer. Included in this study are input/output (I/O) circuits, chains of static gates, clocking networks, and memories.

8.1 Chip Floorplan

Integrated circuits can be studied at different levels. A convenient[1] breakdown from the bottom-up is given by

- Basic circuits
- Logic gates
- Logic units
- Subsystems
- Complete system

which starts at the basic circuit level and increases in complexity towards the finished design. Chip performance depends on the individual characteristics of each level, and also the manner in which the various levels are

[1] and arbitrary

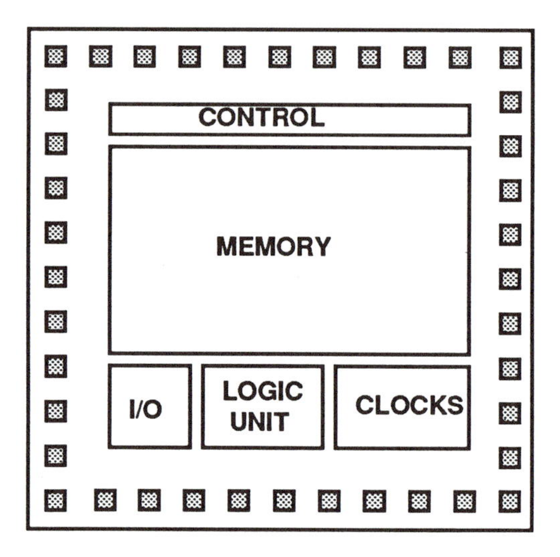

Figure 8.1: Chip Floorplan Example

linked together. Since transient switching times are heavily dependent on the layout, the overall system speed is limited by the placement of each gate, unit, and subsystem on the substrate. In addition, input and output circuits become critical connections to the outside world.

The system layout of an integrated circuit is generically termed the **floorplan**. It views the chip as interconnected functional blocks within specified geometrical boundaries. An example is shown in Figure 8.1. Input and output pads of predefined shape, location, and number surround the usable real estate. Each region of the chip has a logical function, and communications among the segments is achieved using wired interconnect lines. Technology specifications, such as the number and type of interconnect layers, directly affect the implementation. While all aspects of the floorplan are important to the chip performance, a few deserve additional comments.

The importance of the input/output pads can be seen from simple arguments. Regardless of the internal circuit design, the overall chip performance is limited by I/O interface. It is easy to design a basic CMOS inverter with a propagation delay of $t_P = 0.1$ [ns]; transmitting this switching speed to the outside world is another issue. I/O circuits also serve to protect the silicon microstructures from the outside world. Problems such as static charge buildup and short circuited outputs can easily destroy unprotected circuitry.

Another problem is that of clocking and clock distribution. VLSI-level circuits are by necessity synchronized logic systems. Once an externally-supplied clock CLK enters the chip, we usually provide circuits to generate internal timing signals ϕ_i as dictated by the architecture and system design. Moreover, the internal clocks must be distributed throughout the chip with only minimum skew permitted. This requires that we carefully structure both the circuits and the interconnects to the floorplan.

High-peformance CMOS design deals with circuits which are optimized for high data rates. Once we have addressed the issues of I/O pads and clock distribution, the primary problems are those concerned with minimizing logic delays. Critical data paths must be isolated, studied, and improved in the endless quest for circuits which operate a few megahertz faster. An isolated gate is no longer of interest; instead, we must examine the interplay among the circuits to optimize the design strategy.

8.2 Input Protection Circuits

Input pads connect data, control, or clocking signals to on-chip logic gates. When the pads are directly connected to the gate electrodes of MOSFETs, care must be taken to insure that excessive static electrical charge does not destroy the transistor. Protection circuits are designed to drain excessive charge away from the MOS capacitance to avoid static burnout.

To understand the origin of the problem, recall that a MOSFET gate is basically a capacitor of value

$$C_g = C_{ox} W L. \tag{8.1}$$

With a gate-substrate voltage V_G applied to the transistor, the internal oxide electric field is given by

$$E_{ox} \simeq \frac{V_G}{x_{ox}} \tag{8.2}$$

where we have ignored any trapped oxide or surface charge. Breakdown occurs because of the fact that silicon dioxide has a breakdown field value of approximately

$$E_{BD} \sim 7.5 \times 10^6 \quad [\text{V/cm}]. \tag{8.3}$$

If E_{ox} exceeds this value, the oxide insulating properties break down and charge is transported through the material. This usually results in destruction of the device. Since x_{ox} is usually less than about 450 [Å], the maximum gate voltage $V_{G,max} \simeq E_{BD}x_{ox}$ which can be applied to the device is a relatively small number. It is easy to build up a few kilovolts of static electricity from tribolic[2] interactions during handling, making a protection circuit very useful. Input circuit protection used to be optional (as evidenced by the number of burned-out CMOS chips typically found in old college lab cabinets), but is now a universal feature of most CMOS designs.

EXAMPLE 8.1: Oxide Breakdown Voltage

Consider a silicon dioxide layer which has a thickness of $x_{ox} = 350$ [$\mathring{A}$]. The maximum gate voltage which can be applied to the MOSFET made with this specification is

$$\begin{aligned} V_{G,max} &\simeq (7.5 \times 10^6)(3.5 \times 10^{-6}) \\ &\simeq 26.25 \text{ [V]}. \end{aligned}$$

Note that the value of $V_{G,max}$ decreases with oxide thickness.

The basic idea of an input protection circuit is to allow for alternate charge flow paths when the input voltage gets too large. Diode structures are very useful in this application since they have relatively breakdown voltages which can be controlled. Moreover, reverse breakdown in a pn junction is non- destructive, so that the protection circuit is reusable.

Consider a step profile pn junction with doping densities of N_a and N_d. The maximum internal electric field is found directly at the junction, and has a value

$$E_{pn} = \sqrt{\frac{2q(V_{bi} + V_R)}{\epsilon_{Si}(1/N_a + 1/N_d)}} \tag{8.4}$$

where V_R is the applied reverse bias. Denoting the silicon breakdown electric field by $E_{Si} \simeq 3 \times 10^5$ [V/cm], the reverse breakdown voltage of the pn junction is given by

$$V_{BD,pn} = \frac{\epsilon_{Si}\mathcal{E}_{Si}^2}{2q}\left(\frac{1}{N_a} + \frac{1}{N_d}\right). \tag{8.5}$$

[2]The field of tribology is the study of friction and similar phenomena.

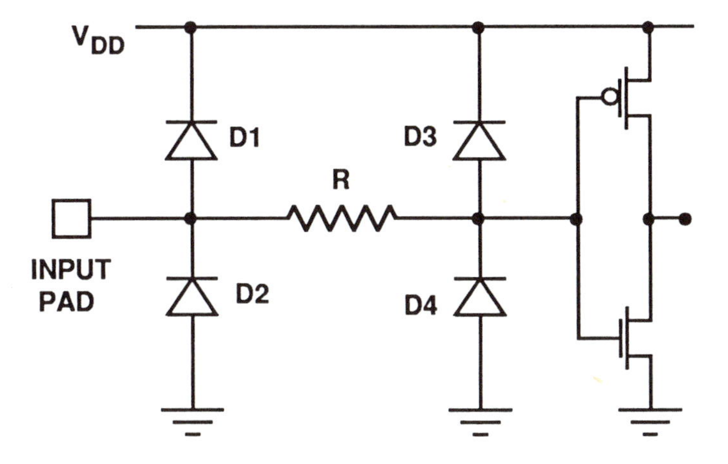

Figure 8.2: Input Protection Circuit

In a one-sided n^+p ($N_d >> N_a$) or np^+ ($N_a >> N_d$) diode, this says that

$$V_{BD,pn} \propto \frac{1}{N} \tag{8.6}$$

where N is the smaller of N_a or N_d. Junctions which are purposely used at the reverse-bias breakdown voltage are generally termed **Zener diodes**.

Figure 8.2 illustrates a simple input protection circuit for CMOS IC. Reverse-biased pn junctions are used as protection diodes, and a series-connected resistor is included to drop some of the voltage. Both diode pairs (D1,D2) and (D3,D4) are designed to undergo breakdown for positive or negative voltage surges. R is designed to reduce the voltage that reaches (D3,D4); this effectively increases the level of protection to the transistor gate.

One problem that exists with this (and other) input protection circuits is the introduction of parasitic RC time constants into the network. Figure 8.3 shows the signal equivalent circuit of the protection network using the dominant components. All of the reverse-biased diodes introduce parasitic capacitance into the input line. The RC combinations give a time delay which is approximated by

$$t_D \simeq R_{line}(C_1 + C_2) + (R_{line} + R)(C_3 + C_4 + C_{in}) \tag{8.7}$$

where C_{in} is the input capacitance of the logic gate. Depletion capacitance is proportional to area, so that C_1, C_2, C_3, and C_4 can be minimized in the layout. Noting the line resistance will generally be small on a metal

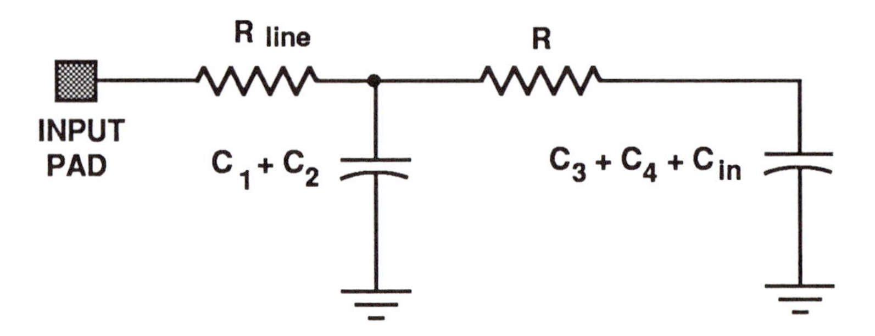

Figure 8.3: Input Delay Network

interconnect then shows that the value of R is the limiting factor in the transient response.

Other input protection schemes are used. Figure 8.4 shows a common circuit based on the properties of a thick field oxide (FOX) MOSFET. The transistor is has threshold voltage of $V_{T,F} > V_{DD}$ and is in cutoff during normal operation. A large input voltage $V > V_{T,F}$ drives the transistor into conduction, providing a path to ground to drain off the excessive charge. The breakdown voltage of the FOX MOSFET is large enough to withstand the high voltages since X_{FOX} is large.

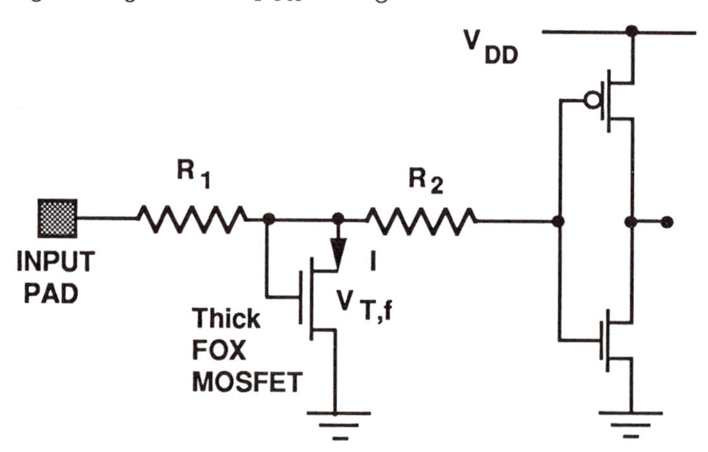

Figure 8.4: Thick Oxide MOSFET Protection Circuit

Design sets usually include recommended protection circuits which operate within the process parameters. These should be consulted whenever using a process line for the first time.

8.3 Static Gate Sizing

An interesting and useful problem is that of optimizing a chain of static gates to minimize the overall propagation delay. This type of situation arises in many different situations and is important to high-performance circuits. In particular, it is relevant to the output drivers and clocking circuits discussed in the next two sections.

A classic example is shown in Figure 8.5 where the objective is to design the fastest network for driving a large capacitance. For the problem at had, we will assume a series of inverting buffers for the driving network. At first sight, it may appear that we would want the fewest possible gates between the input and the load. This simple solution, however, ignores the effect of capacitive loading on successive stages. Accounting for these factors shows that the sizing of the transistors in the chain allows for minimization of the delay. This gives the interesting result that additional logic gates are often inserted to reduce the overall propagation delay between two points.

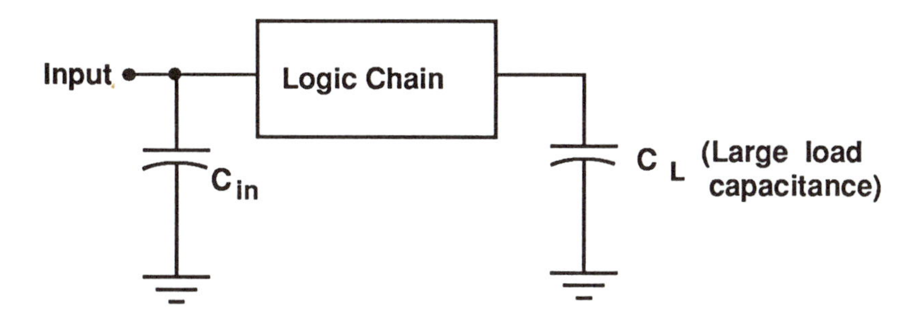

Figure 8.5: Capacitive Loading Problem

We will begin by studying the important properties of an isolated inverter. This is necessary to establish some approximations for the system analysis. Once this is completed, the propagation delay through the chain can be examined.

8.3.1 Single Inverter Model

The CMOS inverter has been examined in detail in Chapter 3. However, in order to extend the results to the system level, we need to introduce some approximations to keep the analysis tractable. Interest is directed towards simplifying the dependence of the switching delay on the device aspect ratios. Consider the transient performance of the circuit in Figure

8.6. A first-order estimate of the delay time is given by [14,20]

$$t_D = RC_o \tag{8.8}$$

where R is the equivalent MOSFET resistance. Device sizing establishes the resistance values as seen by the approximations

$$\begin{aligned} R_n &= \frac{1}{k_n'(W/L)_n(V_{DD} - V_{Tn})}, \\ R_p &= \frac{1}{k_p'(W/L)_p(V_{DD} - |V_{Tp}|)}; \end{aligned} \tag{8.9}$$

for simplicity, we assume that $R_n = R_p = R$. In addition, both the input capacitance C_i and the output capacitance C_o are affected by the channels widths W_n and W_p. In general, increasing W decreases the resistance R, but increases both C_i and C_o for the stage. This relationship provides the key to understanding the gate sizing problem and solution.

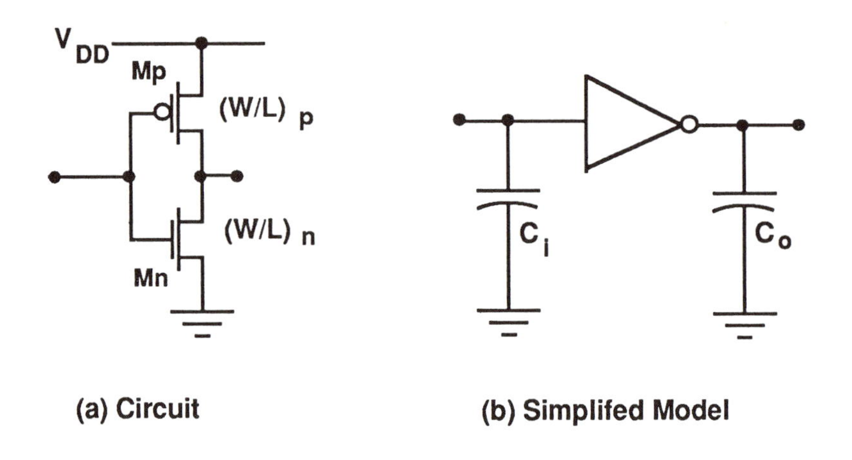

Figure 8.6: Isolated inverter

Consider first the input capacitance C_i. Assuming that this consists mainly of input gate contributions gives

$$C_i = C_{ox}(W_n L_n + W_p L_p) \tag{8.10}$$

showing a linear dependence on the channel widths. The output capacitance is more complicated. The intrinsic circuit contributions can be written as

$$C_o = C_{GDn} + C_{GDp} + C_{DBn} + C_{DBp} \tag{8.11}$$

which includes the gate-drain and drain-bulk (depletion) values. Reviewing our formulas in Chapters 2 and 3 shows that $C_{GD} \propto W$, while only half of the terms of the drain-bulk capacitances have this dependence. Nevertheless, we will approximate

$$C_o \propto W \tag{8.12}$$

for simplicity. We note in passing that these approximations limit the accuracy of the final results. Fortunately, the overall trends established in the next section remain usable even if the numbers are slightly off.

8.3.2 Inverter Chain Analysis

Attention is now directed towards the scaled inverter chain shown in Figure 8.7. Each gate is characterized by a sizing factor S_j which is normalized to the first stage such that $S_1 = 1$, while $S_j > 1$ for $(j > 1)$. By definition, the first stage has a MOSFET conduction factor

$$\beta_1 = k'\left(\frac{W}{L}\right)_1, \tag{8.13}$$

while the jth-stage is described by

$$\beta_j = S_j \beta_1. \tag{8.14}$$

The values of C_i and C_o are determined by Gate 1, and scaled for successive gates. Note than an additional capacitive component C_w has been added between stages. This represents the wiring (interconnect) contribution. We assume that the wiring capacitance is between two stages is proportional to the sizing factor of the second stage. The capacitance between the jth gate and the $(j+1)$-st gate can be summarized as follows:

- $S_j C_o$, output capacitance from Gate j
- $S_{j+1} C_i$, input capacitance to Gate $(j+1)$
- $S_{j+1} C_w$, wiring capacitance into Gate $(j+1)$.

The time delay through Gate j is thus estimated by

$$t_{D,j} = \left(\frac{R}{S_j}\right)[S_j C_o + S_{j+1}(C_i + C_w)]. \tag{8.15}$$

Our program is to compute the values of S_j for $(j = 2, \ldots)$ which minimizes the total delay through the chain.

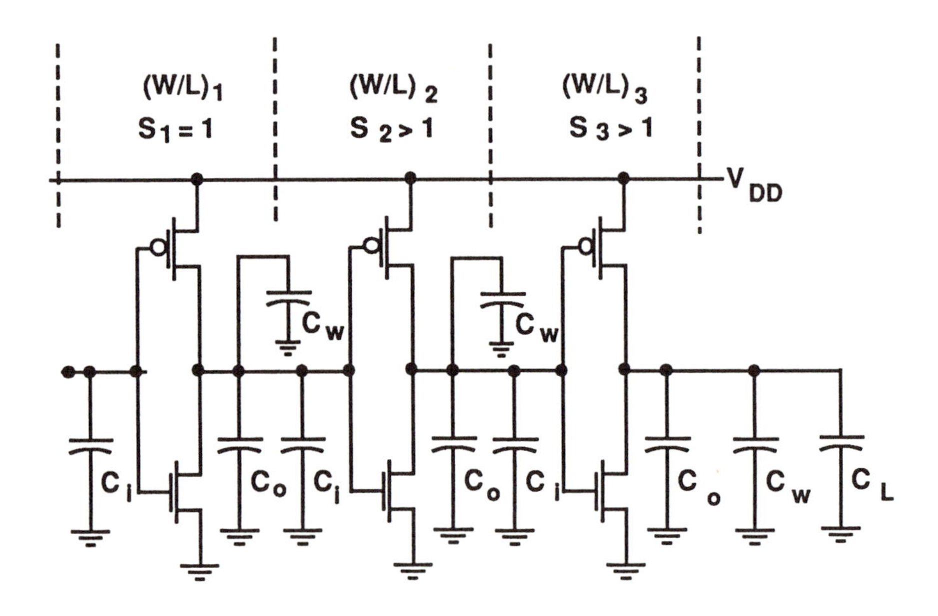

Figure 8.7: Inverter Sizing Problem

Suppose that there are N stages in the chain. The total time delay is given by

$$T_D = \sum_{j=1}^{N} \frac{R[S_j C_o + S_{j+1}(C_i + C_w)]}{S_j}. \tag{8.16}$$

To minimize T_D, we differentiate with respect to S_j and look for zero slope points via

$$\frac{\partial T_D}{\partial S_j} = 0; \tag{8.17}$$

this results in the recursion relation

$$\frac{S_{j+1}}{S_j} = \frac{S_j}{S_{j-1}} \tag{8.18}$$

for $j = 2, 3, \ldots, N$. If this is to hold for arbitrary values of j, then

$$\frac{S_{j+1}}{S_j} = K = \text{ constant} \tag{8.19}$$

must be true, Now then, the boundary conditions of the problem are

$$\begin{aligned} S_1 &= 1, \\ S_{N+1} &= \frac{C_L}{C_i}, \end{aligned} \tag{8.20}$$

at the ends of the chain. Forming the product

$$\frac{S_2}{S_1}\frac{S_3}{S_2}\frac{S_4}{S_3}\cdots\frac{S_{N+1}}{S_N} = K^N \tag{8.21}$$

and using the boundary conditions gives

$$K^N = \frac{C_L}{C_i}. \tag{8.22}$$

Thus, we obtain the scaling ratio in the form

$$K = \left(\frac{C_L}{C_i}\right)^{1/N}, \tag{8.23}$$

which is our final result. Explicitly, the scaling factors are given by

$$\begin{aligned} S_1 &= 1 \\ S_2 &= K \\ S_3 &= K^2 \\ &\vdots \\ S_N &= K^{N-1} \end{aligned} \tag{8.24}$$

as the scaling required to optimize the chain. The minimum delay is then

$$\begin{aligned} T_{D,min} &= \sum_{j=1}^{N} R[C_o + K(C_i + C_w)] \\ &= NR[C_o + K(C_i + C_w)], \end{aligned} \tag{8.25}$$

as verified by direct substitution.

One important point which is obtained from the above analysis deals with the delay time. The equation $K = (S_{j+1}/S_j)$ says physically that the minimum chain delay occurs when every stage has the same individual time delay t_D.

The final question which must be answered is the number of stages N needed to optimize the delay. To calculate this, we differentiate T_D with respect to N and set the result to 0. This gives the general equation

$$RC_o + R(C_i + C_w)\left(\frac{C_L}{C_i}\right)^{1/N}\left[1 - \frac{\ln(C_L/C_i)}{N}\right] = 0. \tag{8.26}$$

If C_o is small, this reduces to the well-publicized result [5]

$$N = \ln\left(\frac{C_L}{C_i}\right), \tag{8.27}$$

which is chosen to the nearest integer for given values of C_i and C_L.

8.3.3 Application of the Results

Although the analysis was based on a chain of inverters, the results are may be extended to any set of cascaded static gates[3]. This observation tells us a very important aspect of CMOS logic circuit design: minimum propagation speed through a logic unit does not imply a minimum number of gates. Instead, we can often speed up the overall logic path by inserting additional circuits between the input and output so long as the gates are sized to equalize the delay times. Some examples are provided in later sections of this chapter.

EXAMPLE 8.1: Static Gate Sizing

Let us apply the results of the analysis to the case where we want to match an input capacitance of $C_i = 150$ [fF] to a load capacitance of $C_L = 2$ [pF]. The number of stages we need is given by

$$\begin{aligned} N &= \ln(2000/150) \\ &\simeq 2.6, \end{aligned}$$

so we choose $N = 3$. The value of K is

$$K = \left(\frac{2000}{150}\right)^{1/3} \simeq 2.37$$

which sets the scaling at

$$\begin{aligned} S_1 &= 1 \\ S_2 &= 2.37 \\ S_3 &= 5.62 \end{aligned}$$

for the 3-gate chain. Stage 3 is designed according to the transient requirements for driving C_L; the remaining circuits are then scaled accordingly.

8.4 Off-Chip Driver Circuits

Off-chip driver (OCD) circuits are critical to the overall chip design. Much effort is put into speeding up internal switching networks. Careful output circuit design insures that the high-performance specifications apply to the

[3] In general, the fan-out and fan-in of each gate must be included in the calculation.

external characteristics as well. Some important problems which must be addressed include

- Efficient buffer circuitry between internal and off-chip drivers
- Minimization of transmission line effects
- Fast switching
- Static charge protection

as well as interface-specific items such as a CMOS-TTL level converter.

An inverter circuit can be used as a basic off-chip driver. The dominant performance factors are the transient switching times t_{LH} and t_{HL}. Transmission line effects also enter into the problem; this is complicated by the fact that the line characteristics such as Z_0 depend on the specifics of the mounting and circuit traces.

8.4.1 Basic Off-Chip Driver Design

The simplest off-chip driver circuit consists of an inverter chain which is designed to handle a large capacitive load. C_{out} includes contributions from the bonding pad, the package wiring, and the circuit board trace. Since this easily amounts to tens (or a few hundred) of picofarads depending on the interface specifications, the transistors must be relatively large.

Consider the 2-stage OCD network shown in Figure 8.8. We may use time constants to obtain first-order design estimates for the sizes of the output transistors Mn2 and Mp2 by writing

$$\begin{aligned} \left(\frac{W}{L}\right)_{n2} &= \frac{C_{out}}{\tau_n k'_n (V_{DD} - V_{Tn})}, \\ \left(\frac{W}{L}\right)_{p2} &= \frac{C_{out}}{\tau_p k'_p (V_{DD} - |V_{Tp}|)}, \end{aligned} \tag{8.28}$$

where τ_n and τ_p are the high-to-low and low-to-high time constants, respectively. Since the output capacitance seen by an OCD can be large, the MOSFET aspect ratios are also quite large. These are obtained using several parallel-connected transistors to aid in layout and parasitic control. Sizing theory may be used to determine the sizes of the first stage transistors Mn1 and Mp1.

The actual values of the fall and rise times can be estimated from

$$\begin{aligned} t_{HL} &= \tau_n \left[\frac{2V_{Tn}}{(V_{DD} - V_{Tn})} + \ln\left(\frac{2(V_{DD} - V_{Tn})}{V_0} - 1\right)\right] \\ t_{LH} &= \tau_p \left[\frac{2|V_{Tp}|}{(V_{DD} - |V_{Tp}|)} + \ln\left(\frac{2(V_{DD} - |V_{Tp}|)}{V_0} - 1\right)\right] \end{aligned} \tag{8.29}$$

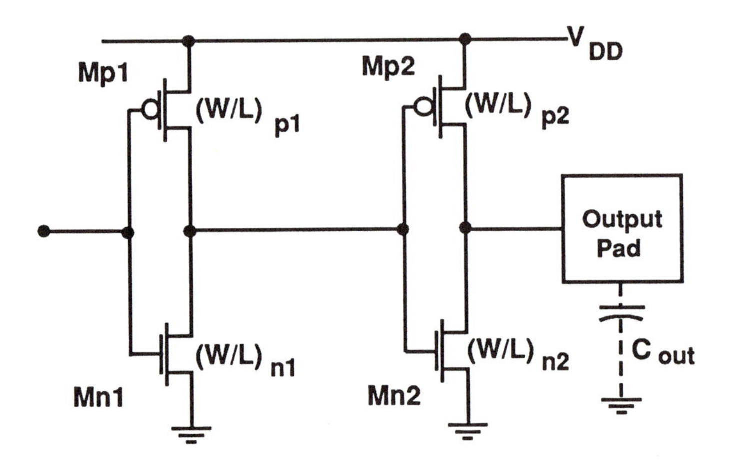

Figure 8.8: Double-Inverter OCD Circuit

where V_0 is the 10% voltage point. Note that we have simplified the expression from that given in Chapter 3. In this form, the time intervals are are referenced to a maximum value of V_{DD} (instead of the 90 % voltage V_1).

EXAMPLE 8.2: Off-Chip Driver Transistors

Suppose that a process is characterized by the nominal values

$$k'_n = 55\ [\mu\text{A/V}^2], \qquad V_{T0n} = 0.90\ [\text{V}],$$
$$k'_p = 25\ [\mu\text{A/V}^2], \qquad V_{T0p} = -0.75\ [\text{V}],$$

and the power supply is 5 [V]. We will look at the device requirements for an OCD circuit which has $t_{LH} = t_{HL} = 20$ [ns] with a maximum load of $C_{out} = 50$ [pF].

Using the equations above, we compute the time constants as

$$\tau_n \simeq \frac{(20 \times 10^{-9})}{3.10} \simeq 6.45\ [\text{ns}],$$
$$\tau_p \simeq \frac{(20 \times 10^{-9})}{3.04} \simeq 6.58\ [\text{ns}].$$

The corresponding aspect ratios are then

$$\left(\frac{W}{L}\right)_n \simeq 35, \qquad \left(\frac{W}{L}\right)_p \simeq 72,$$

to meet the minimum requirements.

Optimization of the double-inverter cascade can be accomplished using the results of Section 8.3. These values of (W/L) can be scaled by a factor $K = (C_{out}/C_i)$ to find the aspect ratios needed for the first stage.

The above analysis ignores one important large MOSFET characteristic: the **working** threshold voltage decreases as the aspect ratio increases. This is due to the somewhat ill-defined definition of V_T in the simplified square-law MOSFET model as the value where

$$I_D|_{V_{GS}=V_T} = I_X \tag{8.30}$$

with I_X a pre-set value. In the usual design environment, standard sized transistors are used to provide the electrical design parameters; the most common reference devices are relatively small to aid in the circuit design. Large transistors act differently because $I_D \propto (W/L)$ shows that large currents will flow even when the device is only weakly inverted. Applying this consideration to the design equations relaxes the transistor size requirement.

High-speed off-chip driver circuits may also be subject to transmission line effects. In this case, the analysis in Section 6.7 may be applied to minimize ringing and other problems.

8.4.2 Tri-State and Bidirectional I/O

Tri-state off-chip driver circuits are constructed by splitting the input signal to individually control each output transistor. Normal operation gives high and low voltages, while the high-impedance state is obtained by driving both the nMOS and pMOS devices into cutoff. An inverting tri-state circuit is shown in Figure 8.9. When the tri-state variable $Z = 1$, pMOSFETs Mp1 and Mp2 are off, while nMOSFET Mn conducts. This gives normal circuit operation. If $Z = 0$, then the gate voltages to output transistors are given by

$$\begin{aligned} V_p &= V_{DD}, \\ V_n &= 0, \end{aligned} \tag{8.31}$$

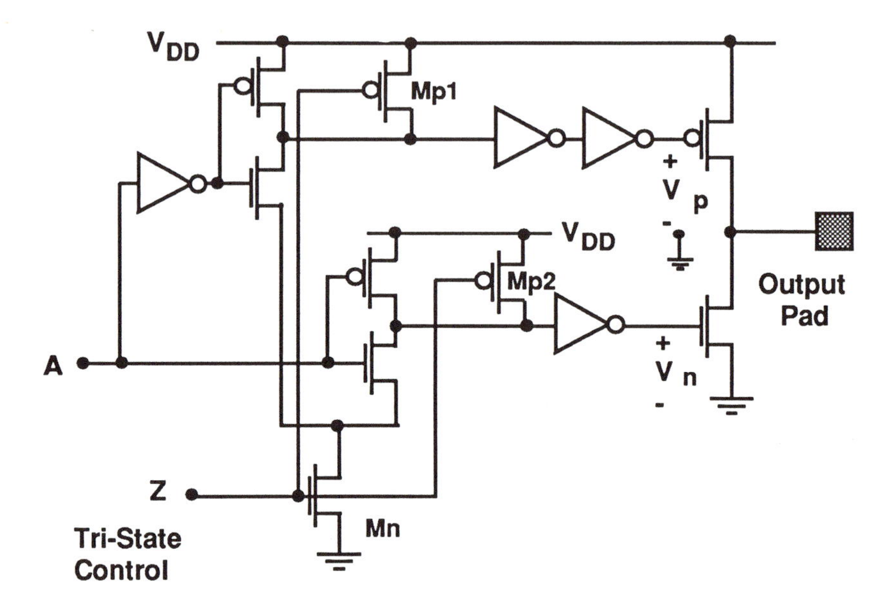

Figure 8.9: Tri-State Output Circuit

so that both are in cutoff. A condition of $Z = 0$ thus provides the necessary high-impedance state.

Bi-directional input/output (I/O) circuits are also quite useful. An example is shown in Figure 8.10. The tri-state section of the circuit is a non-inverting buffer with an enable control E, where $E = 0$ gives the High-Z state. Operation is straightforward and is easily understood by examining the circuit.

8.5 Timing and Clock Distribution

Synchronizing machine operations and data transfers with clock pulses provides us with a structured framework for dealing with the complexities of large system designs. Clocking is a global control technique which provides the "glue" for system operation. It is equally important at the circuit level, particularly in a dynamic logic stage.

8.5.1 Clocks and Timing Circles

System level timing can be described using circular timing charts [2]. Consider an ideal pseudo 2-phase (or, dual-clock) scheme with mutually- exclu-

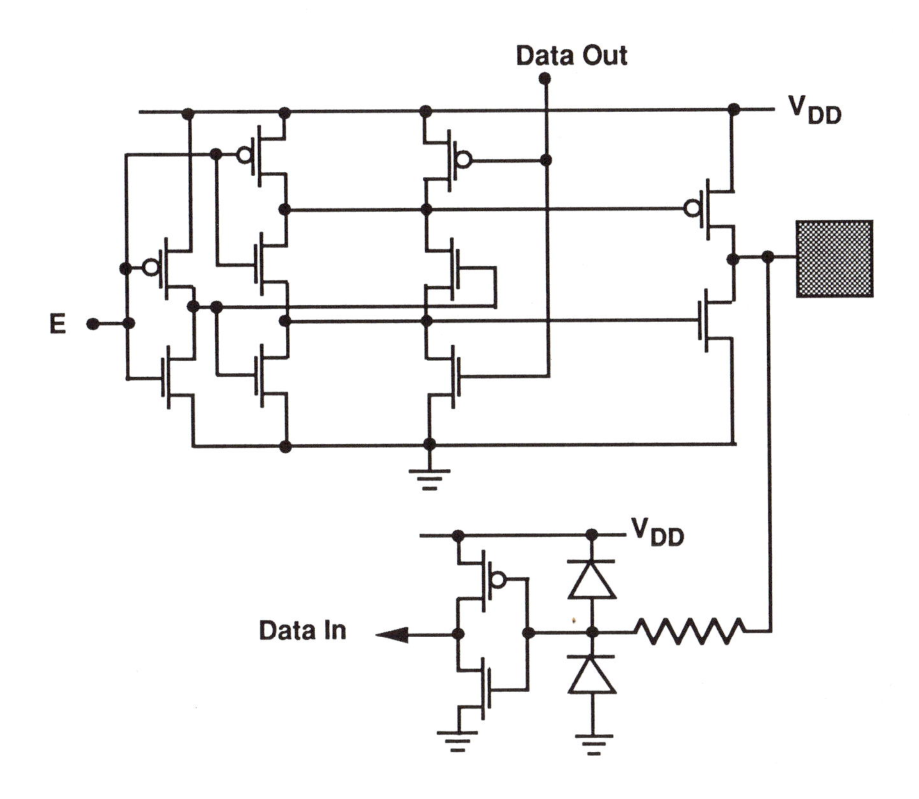

Figure 8.10: Bi-Directional I/O Circuit

sive pulses ϕ_1 and ϕ_2:

$$\phi_1(t) \cdot \phi_2(t) = 0 \quad (\forall t). \tag{8.32}$$

System timing may be described by constructing the chart shown in Figure 8.11. Time increases in a counter-clockwise direction with one full rotation corresponding to the clock period T. Segments are labeled according to time intervals when a clock signal is high (at a logic 1). In this example, $\phi_1 = 1$ during the first half-period, while $\phi_2 = 1$ during the last half-period.

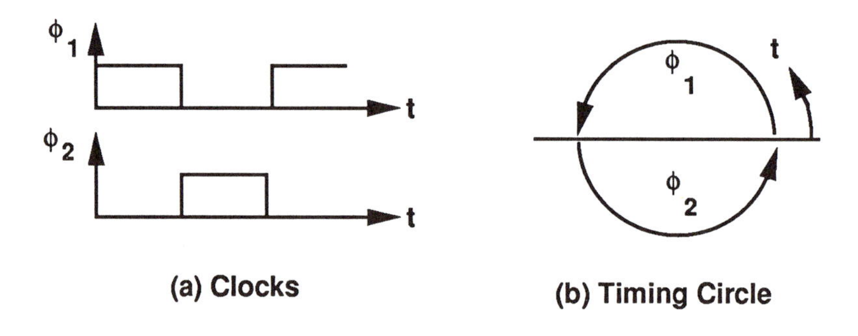

Figure 8.11: Pseudo 2-Phase Clocking Chart

Descriptions of synchronized logical operations can be added to the chart. In Figure 8.12, the loading and hold states of a TG latch are controlled by a pseudo 2-phase clocking arrangement. Data movement is described directly on the chart. At the system level, the chart corresponds to a register-transfer language (RTL) statement

$$\phi_1 : R \leftarrow D, \tag{8.33}$$

where D is the data bit and R denotes the register. Clocking charts provide a useful visualization of data movement and general communications within a synchronized digital system. They also serve to illustrate critical timing margins in circuit design. Timing charts can be constructed for arbitrary clocking schemes, including standard 2-phase (with ϕ and $\overline{\phi}$), 3-phase (ϕ_1, ϕ_2, ϕ_3), and 4-phase ($\phi_1, \phi_2, \phi_3, \phi_4$) arrangements.

A more realistic clocking arrangement is depicted by the clocking circle in Figure 8.13. If both clocks have 50% duty cycles, normal operation gives

$$\phi_1(t) \cdot \phi_2(t) = 0 \tag{8.34}$$

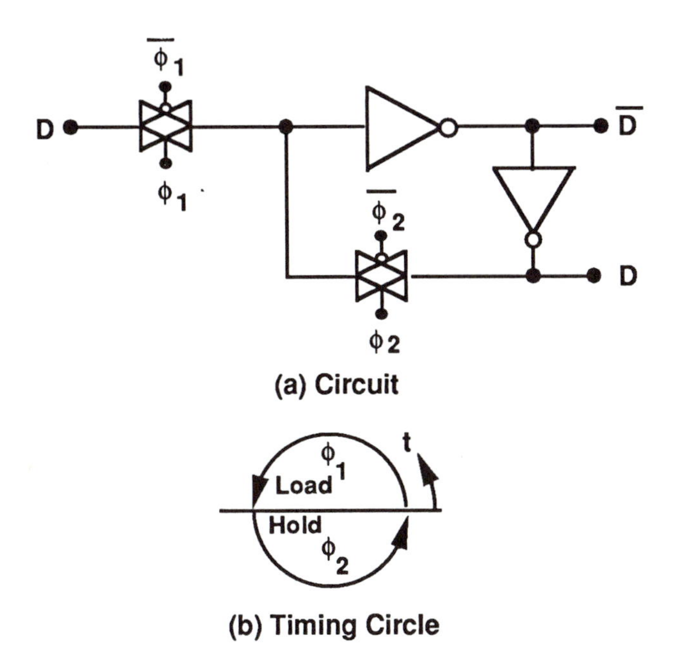

Figure 8.12: Latch Timing Chart

except during the transition times. Mutually-exclusive clock signals provide timing intervals for logical operations, and are used to allow for normal gate delay times. Overlap segments are avoided to prevent ill- defined movement of data, instructions, or control signals. Transition times can be made small

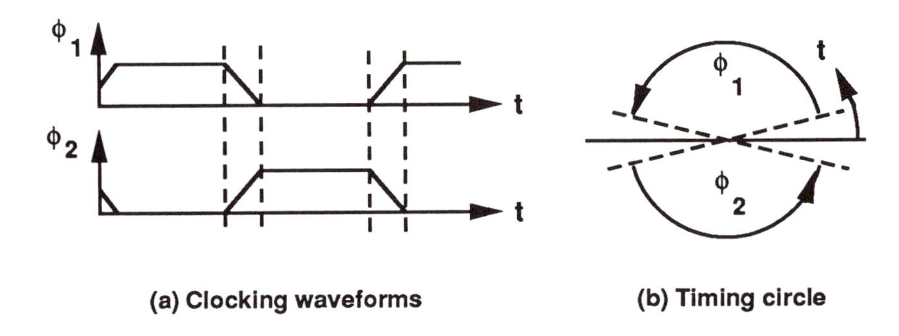

Figure 8.13: Pseudo 2-Phase Overlap Times

by proper clock generator design.

Clock skew is represented by rotating one of the clocks as shown in Figure 8.14. The skew time t_S is defined as the time interval where

$$\phi_1(t) \cdot \phi_2(t) = 1 \tag{8.35}$$

and indicates the possibility of unwanted simultaneous bit transfers. This may lead to severe conflict problems in the operation. For example, the simple latch in Figure 8.12 above will try to load new data while trying to hold on to the previous state. Signal races determine which operation wins out. Skew can originate within the clock driving circuits, or may arise from the distribution arrangement.

8.5.2 Clock Generation Circuits

A basic 2-phase clock generator circuit is designed to generate ϕ and $\overline{\phi}$ from a single input CLK signal. This is often a matter of convenience to the user: requiring only a single external clock makes the chip's usage more attractive to the board designer.

Various circuits have been developed for use in clock generation. Figure 8.15 provides a CMOS generator/driver which uses a transmission gate as a delay element. MOSFETs Mn1 and Mp1 form an inverter which acts as the first driver for the chain. The upper branch of the circuit consists of

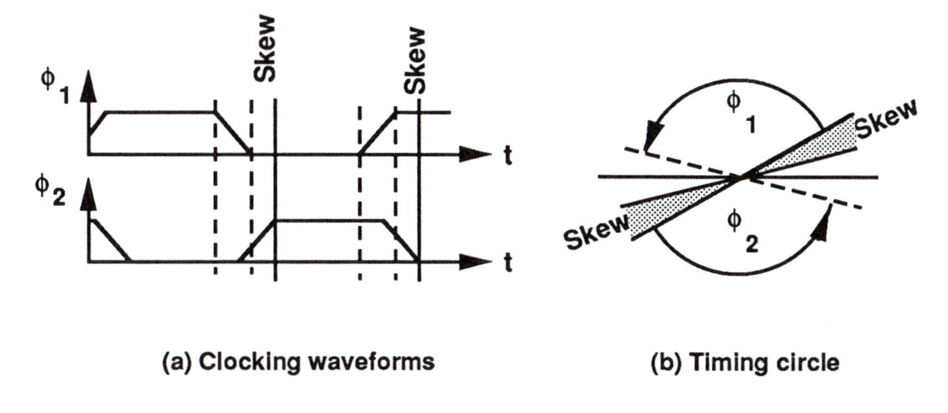

Figure 8.14: Clock Skew

two cascaded inverters and generates the signal $\overline{\phi} = \overline{\text{CLK}}$ while the lower branch only has a single inverter and gives $\phi = \text{CLK}$. Transmission gate TG is used as a delay element to minimize clock skew between ϕ and $\overline{\phi}$. Since it is biased into active conduction, we will model it using an equivalent resistance R_{TG}, and introduce the time constant

$$t_D \simeq R_{TG} C_{in}. \tag{8.36}$$

If the propagation delay through an inverter is t_P, then choosing

$$t_D \simeq t_P \tag{8.37}$$

equalizes the delay between the upper and lower branches. Recalling that the transmission gate conductance can be approximated by

$$G_{TG} \simeq \beta_n(V_{DD} - V_{Tn}) + \beta_p(V_{DD} - |V_{Tp}|), \tag{8.38}$$

we see that clocking skew can be controlled by adjusting the size of the TG transistors. The problem is complicated by the fact that C_{in} varies with the aspect ratios. A simple design procedure is to first design the $\overline{\phi}$ double-inverter chain, and then adjust the TG and ϕ inverter accordingly. A detailed discussion of this circuit can be found in reference [21].

Another straightforward approach uses an SR latch as shown in Figure 8.16. The clocking signal CLK is inverted, and CLK and $\overline{\text{CLK}}$ are used to drive the SR circuit. The 2-phase clock signals ϕ and $\overline{\phi}$ are taken from the latch outputs. This logic can also be used to generate pseudo 2-phase clocks ϕ_1 and ϕ_2 by redefining the outputs.

To insure proper operation of the circuit, two items should be checked. First, the propagation delay through the inverter must be small compared

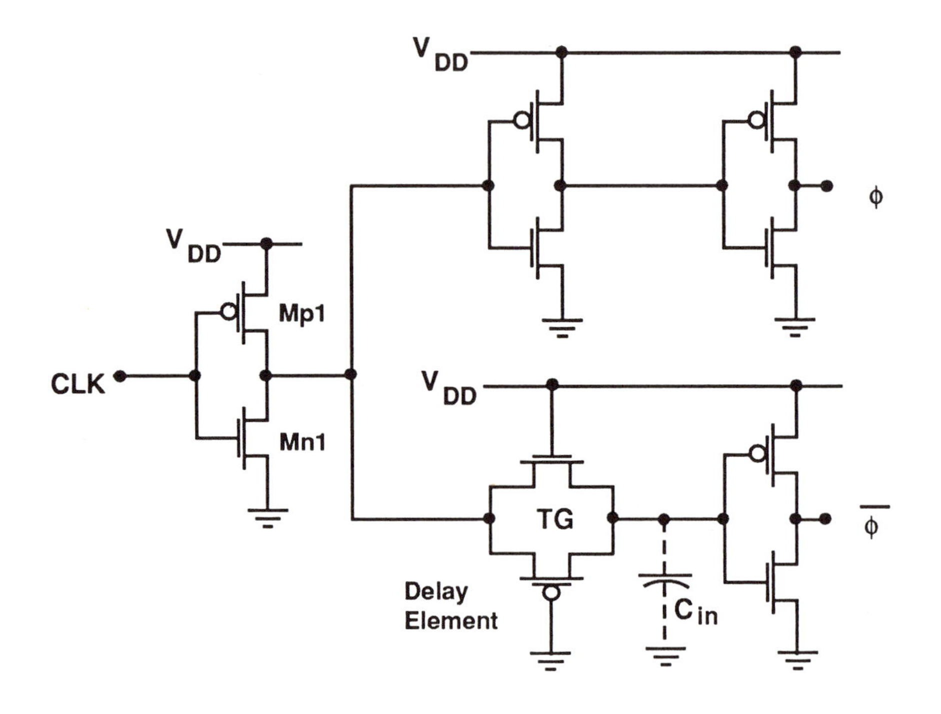

Figure 8.15: Clock Generator With a TG Delay

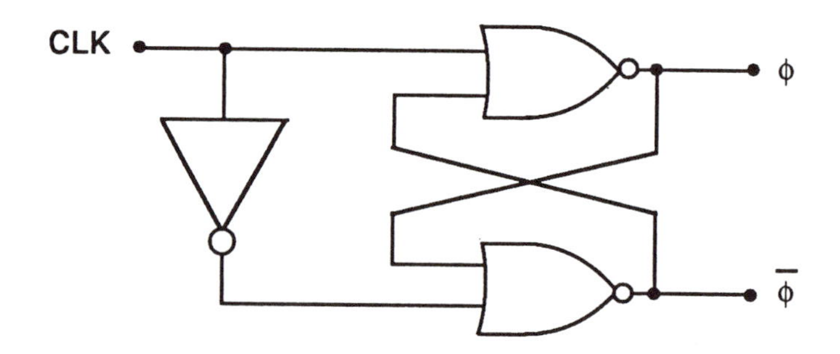

Figure 8.16: Latch-Based Clock Generator

to the clocking period so that $\overline{CLK}$ has time to enter the latch. The second problem deals with the design of the NOR2 gates. If identical circuits are used for both gates, then the output lines should have the same capacitance for equal switching delays. Since the total capacitance depends on the interconnect, the actual values seen at each output are sensitive to the layout geometry.

8.5.3 Clock Drivers and Distribution Techniques

Once the clocking pulses are generated they must be distributed throughout the chip in a manner which minimizes clock skew. Figure 8.17 illustrates the problem in a pseudo 2-phase circuit by showing timing circles at various points on a chip. Skew problems originate mostly from

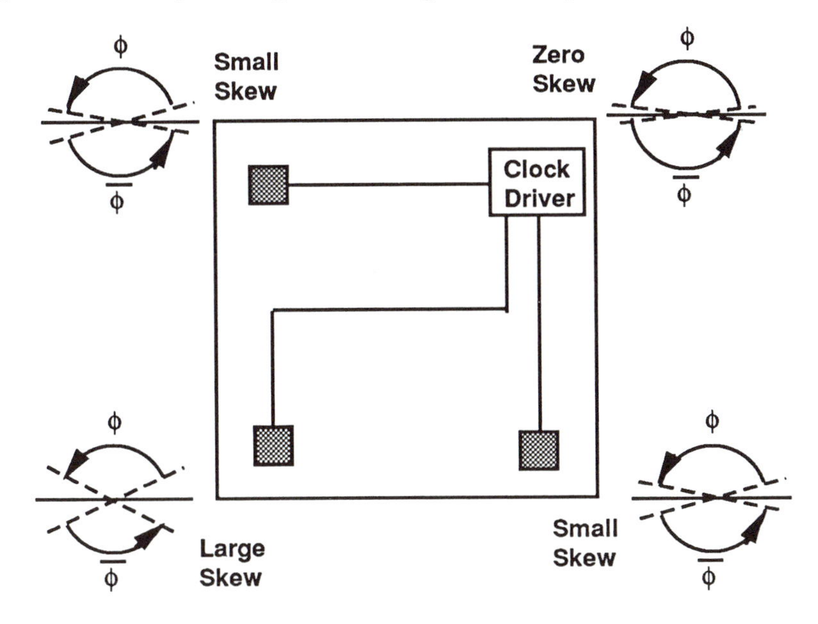

Figure 8.17: Clock Skew Due to Distribution

- Unbalanced loads at the driver,
- Unequal RC line delays,

so that the driver circuits and associated distribution schemes are important in maintaining the synchronous logic design. A related problem is that the drive capability of the circuit must be able to handle large capacitive loads at the required clock frequency.

One approach to designing a clock distribution network is to use a cascaded chain of inverting buffers that matches the clock generator to the distribution line. This type of system is shown symbolically in Figure 8.18 for a 2-phase scheme. Increasingly larger inverters are designed using the approach presented in Section 8.3. Both the drive capability and the speed are maintained by proper scaling of the stages. The load C_L represents the

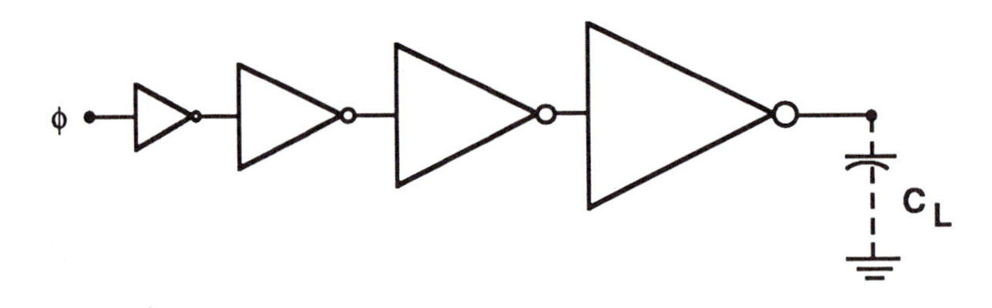

Figure 8.18: Cascaded Clock Driver Chain

capacitance seen at the input to the clock distribution lines. As illustrated in Figure 8.19, this can be expressed as

$$C_L = C_{line} + \sum_i C_{Gi}, \tag{8.39}$$

where C_{line} is the line capacitance, and

$$C_{Gi} = C_{ox} W_i L_i \tag{8.40}$$

is the gate capacitance of the i-th clocking MOSFET with dimensions $(W_i \times L_i)$. If only a single clock driver chain is used for the entire chip then the distribution geometries become critical; the total interconnect lengths for the ϕ and $\overline{\phi}$ signal distribution lines should be approximately equal. With a line interconnect of R_{line} the worst-case time constant for signals reaching their destination is approximated by the time constant

$$\tau_D = R_{line} C_L. \tag{8.41}$$

Careful global planning and structured distribution patterns can be used to solve the problem.

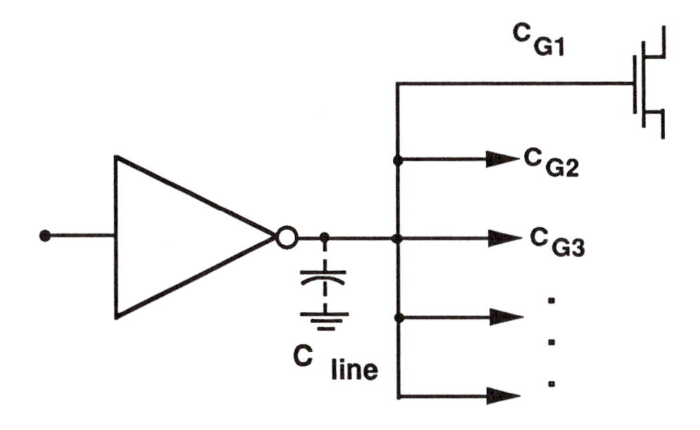

Figure 8.19: Clock Line Capacitance

Clock distribution can also be accomplished by using a balanced tree network with multiple fanouts as shown in Figure 8.20. Identical drivers can be used within a given stage. Moreover, the drive requirements of the output circuits are reduced from the single inverter design since the FO has been split into groups. Each inverter reshapes the clocking waveform, making the performance less sensitive to variations in the interconnect routing.

Clock skew problems can be minimized by using symmetrical geometries for the clock distribution lines. An example is the "H-tree" network shown in Figure 8.21 which is adaptable to Manhattan layouts. Every clock distribution point O is the same distance from the driver D, giving equal delay times. If the load capacitance is the same at every O-point, then the clocks will all be in phase with one another. Other geometrical patterns can be used so long as the general design criteria are unchanged.

8.6 Memory Circuits

Memory circuits are interesting to study for many reasons. The most obvious is that we often need on-chip data storage capabilities. They also illustrate many important aspects of the interplay between logic structuring and circuits. The complete study of even a single chip would require an entire book. Instead, we will concentrate on examining various types of memory cells, the architecture of a multiple-bit memory, and some of the associated circuits.

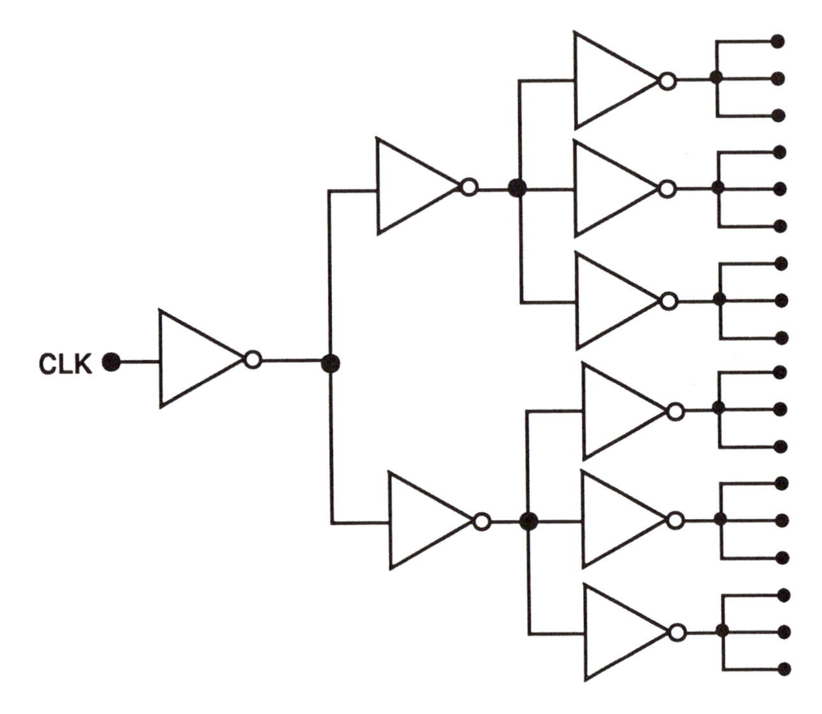

Figure 8.20: Tree Network Clock Distribution

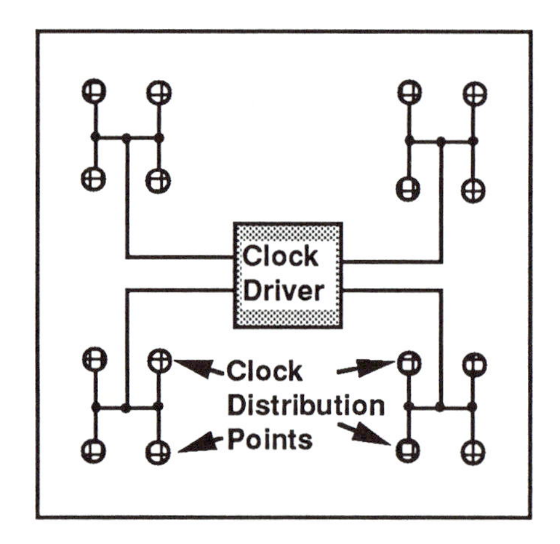

Figure 8.21: H-Tree Clock Network

Memories can be classified according to their user characteristics. The main divisions are

RAM: Random-access memory which has read-write capabilities

ROM: Read-only memory where the contents cannot be changed by the user.

Although ROMs are also randomly-accessible structures, current usage of the acronyms reserves "RAM" to imply a cell with both read and write abilities. ROMs are constructed using various types of transistor arrays. Since these are somewhat specialized, ROMs will not be discussed in detail here. Rather, we will concentrate on RAM cell operation and design.

RAM cells are divided into two main types:

- Static cells which are built from static CMOS circuits
- Dynamic cells which use charge-storage nodes.

These two cell types have very different characteristics. Static cells can hold data so long as power is applied, while a dynamic memory cell is subject to charge leakage and must be continually refreshed. One tradeoff is in density. Static memories require more transistors and have large real estate requirements, while dynamic cells are extremely simple and allow for high levels of integration. However, static RAMs (SRAMs) have traditionally been faster than dynamic RAMs (DRAMs), making them mandatory in critical system designs.

Cell density is heavily dependent on the technology. Current SRAM designs of 4-Mbit and larger are based on advanced fabrication capabilities. Dual-metal, quadruple-poly processes with channel lengths less than 0.6 [μm] are relatively common.

A generic RAM cell is illustrated in Figure 8.22. In addition to the required power supply connections, each cell must have a minimum of two connections

1. A DATA or BIT line
2. A SELECT line which chooses the cell.

The BIT line is used to transfer data in and out of the cell, while the SELECT control connects the cell to the data line. Complementary signals may also be used depending on the circuit structure.

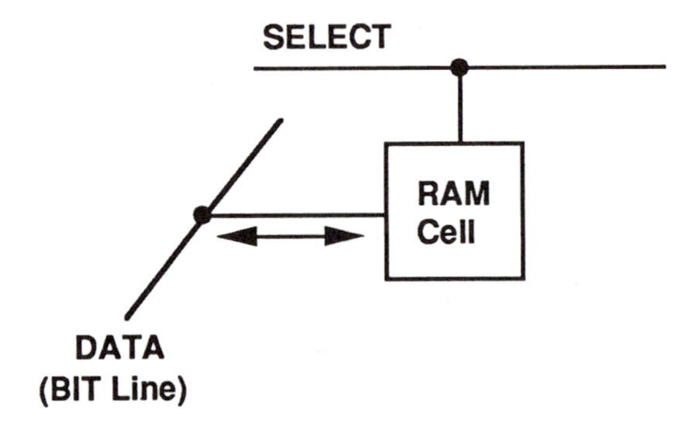

Figure 8.22: General RAM Cell

8.6.1 Static RAM Cell

A common CMOS static RAM cell is built using cross-coupled inverters to form a bistable latch as shown in Figure 8.23. Access to the cell is achieved through nMOS pass transistors Ma1 and Ma2, which are connected to BIT and $\overline{\text{BIT}}$ lines, respectively. When the voltage on the SELECT line is high, data can be written to, or read from, the cell. A condition of SELECT=0 decouples the cell from the data lines and corresponds to a hold state. Symmetrical operation is achieved by proper choice of device aspect ratios.

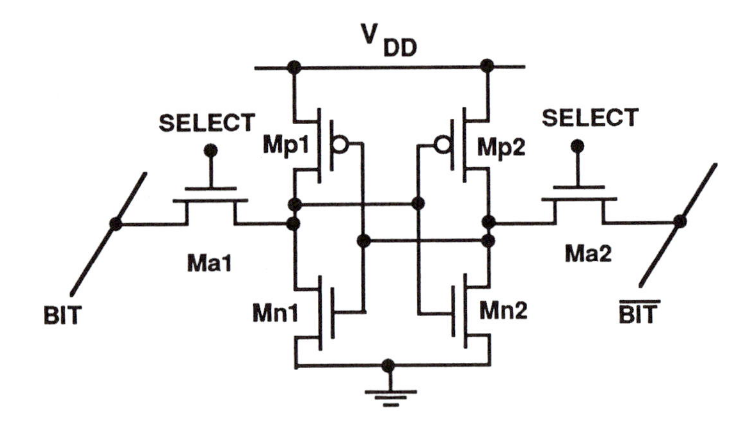

Figure 8.23: Static RAM Cell

We will assume pairs of identical MOSFETs such that (Mn1, Mn2) are both described by β_n, (Mp1, Mp2) both have β_p, and access transistors (Ma1, Ma2) have identical transconductance parameters of β_a.

Static Noise Margin

The ability of the cell to hold a stable state can be analyzed using the graph shown in Figure 8.24 [3]. Curve 1 is the ideal static transfer characteristic for the inverter made up of Mn1 and Mp1. Reflecting this around the unity gain line gives the reverse curve 2 which describes the inverter made up of Mn2 and Mp2. The existence of intersection points at the two V_{DD} axes values corresponds to the two stable states of the circuit. The separation

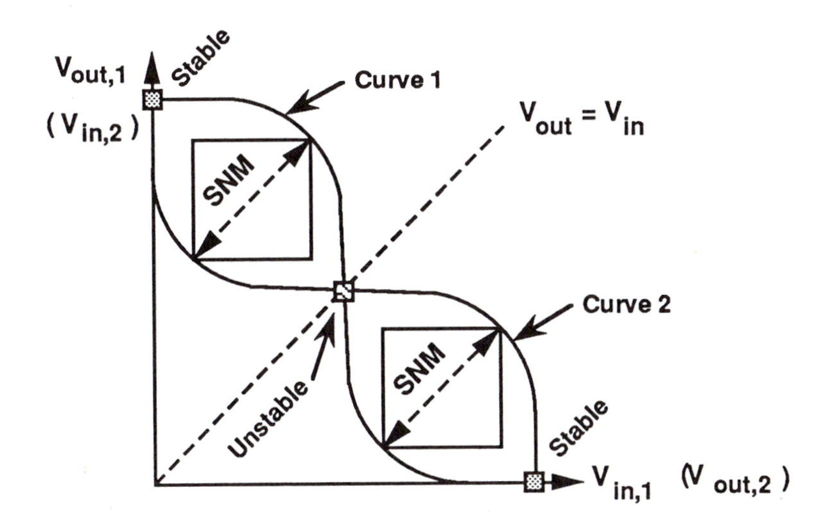

Figure 8.24: SRAM Transfer Curves

between the two curves gives an indication of how well the circuit rejects noise, and can be used to define the **static noise margins** (SNM) as shown. Qualitatively, a large "box" implies a high noise immunity.

It is possible to calculate the SNM using the circuit shown in Figure 8.25 [18]. The important sections of the circuit are the inverter made up of (Mn1, Mp1), the access transistor Ma2, MOSFET Mn2, and the noise voltage sources V_N. The noise voltages are used to represent a perturbation which tries to switch the state of the cell. By definition, the static noise margin is the largest value of V_N that can be tolerated without altering the

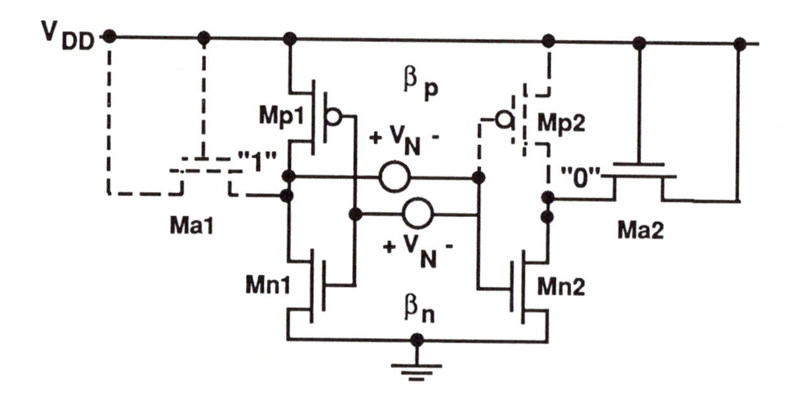

Figure 8.25: Subcircuit for SNM Analysis

state of the flip-flop: $SNM = max(V_N)$. Inspection of the circuit shows that Mn1 and Ma2 are saturated, while Mn2 and Mp1 are non-saturated. Equating currents in the inverter gives

$$\frac{\beta_n}{2}(V_{GSn1} - V_{Tn})^2 = \frac{\beta_p}{2}[2(V_{SGp1} - |V_{Tp}|)V_{SDp1} - V_{SDp1}^2], \tag{8.42}$$

while applying KCL to Mn2 and Ma2 yields

$$\frac{\beta_a}{2}(V_{GSa2} - V_{Tn})^2 = \frac{\beta_n}{2}[2(V_{GSn2} - V_{Tn})V_{DSn2} - V_{DSn2}^2]; \tag{8.43}$$

body-bias in Ma2 is ignored for simplicity. The KVL equation set for the circuit is given by

$$\begin{aligned} V_{GSn1} &= V_N + V_{DSn2}, \\ V_{SDp1} &= V_{DD} - V_N - V_{GSn2}, \\ V_{SGp1} &= V_{DD} - V_N - V_{DSn2}, \\ V_{GSa2} &= V_{DD} - V_{DSn2}, \end{aligned} \tag{8.44}$$

which can be substituted into the current equations. Performing the algebra gives

$$\begin{aligned} (V_{DSn2} + V_N - V_{Tn})^2 &= \frac{r}{q}(V_{DD} - V_N - V_{GSn2}) \\ &\quad \cdot(V_{DD} - V_N - 2V_{DSn2} - 2|V_{Tp}| + V_{GSn2}) \\ (V_{DD} - V_{Tn} - V_{DSn2})^2 &= q[2(V_{GSn2} - V_{Tn})V_{DSn2} - V_{DSn2}^2], \end{aligned} \tag{8.45}$$

where we have used

$$r = \frac{\beta_n}{\beta_a}, \qquad q = \frac{\beta_p}{\beta_a}, \tag{8.46}$$

to denote the transconductance ratios. Combining the expressions and linearizing gives the SNM as [16]

$$\begin{aligned} SNM &= V_{Tn} - \frac{1}{1+K}\Big[\frac{(V_{DD} - \frac{2r+1}{r+1}V_{Tn})}{1 + \frac{r}{K(1+r)}} \\ &\quad - \frac{(V_{DD} - V_{Tn} - |V_{Tp}|)}{1 + K(\frac{r}{q}) + \sqrt{\frac{r}{q}(1 + 2K + \frac{r}{q}K^2)}}\Big] \end{aligned} \tag{8.47}$$

where

$$K = \frac{r}{r+1}\left[\sqrt{\frac{r+1}{r+1-(V_s/V_r)^2}} - 1\right] \tag{8.48}$$

with

$$\begin{aligned} V_s &= V_{DD} - V_{Tn}, \\ V_r &= V_s - \frac{r}{r+1}V_{Tn}. \end{aligned} \tag{8.49}$$

This expression is valid for the general case where V_{Tn} is different from $|V_{Tp}|$. It is seen that the SNM is independent of the absolute values of β; rather, it is established by the ratios r and q.

Poly-Resistor Loads

Polysilicon load resistors can be used to create the SRAM cell shown in Figure 8.26. Since the resistor patterns can be placed on top of the transistor locations, this design may provide higher cell density. The SNM for this circuit is given by [18]

$$SNM = \frac{\sqrt{r}-1}{\sqrt{r}+1}V_{Tn} + \frac{r+1-\sqrt{2r^{3/2}+r+1}}{r(\sqrt{1+\sqrt{r}})}(V_{DD} - V_{Tn}). \tag{8.50}$$

This is more sensitive to the β ratio as evidenced by the fact that $SNM = 0$ when $r = 1$. In addition, it can be shown that the R-load SRAM cell is less stable when small power supply voltages V_{DD} are used.

Switching Times

The switching of the cell is determined by the output capacitance and the feedback network. Consider the basic cell shown in Figure 8.27. The time

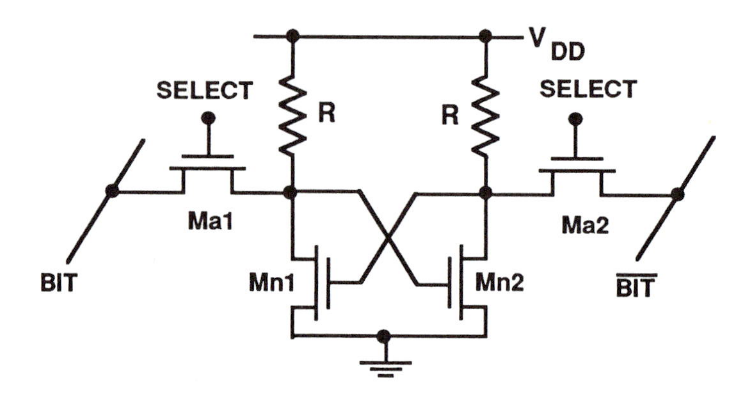

Figure 8.26: Poly Resistor Loads

constants which control the charging and discharging are

$$\begin{aligned}\tau_{ch} &= \frac{C_L}{\beta_p(V_{DD} - |V_{Tp}|)}, \\ \tau_{dis} &= \frac{C_L}{\beta_n(V_{DD} - V_{Tn})},\end{aligned} \tag{8.51}$$

with C_L the total load capacitance on the output nodes. The main contributions are seen by approximating

$$C_L \simeq C_{inv} + C_{BIT} + C_G, \tag{8.52}$$

where C_{inv} is the normal inverter output capacitance, C_{BIT} is the additional capacitance associated with the access circuitry and bit line, and C_G is the gate capacitance of the opposite inverter. Minimizing the time constants within the constraints imposed by the SNM requirements gives a reasonable criterion for the initial design.

8.6.2 Dynamic RAM Cell

A dynamic RAM cell uses dynamic charge storage nodes to hold data. Figure 8.28 shows a basic 1-Transistor (1T) cell consisting of an access nMOSFET Mn and a storage capacitor C_{store}. The operation of the circuit is straightforward. If SELECT=1, Mn is active and allows current flow between the storage capacitor and the bit line. A logic 0 state corresponds to a capacitor voltage of $V_{cap} = 0$. However, since the nMOSFET exhibits a threshold loss, the largest voltage on C_{store} is given by

$$V_{max} = V_{DD} - V_{Tn}. \tag{8.53}$$

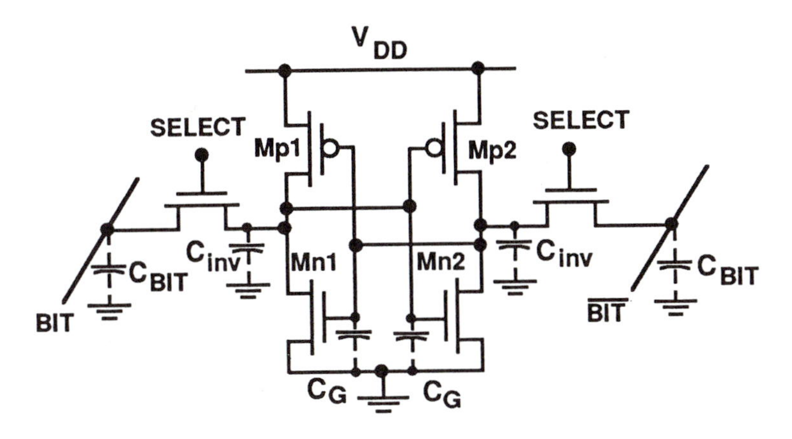

Figure 8.27: SRAM Capacitances

Including body-bias effects gives the self-iterating equation

$$V_{max} = (V_{DD} - V_{T0n}) - \gamma_n(\sqrt{2|\phi_{Fp}| + V_{max}} - \sqrt{2|\phi_{Fp}|}\ ). \qquad (8.54)$$

The logic 0 and logic 1 charge values are thus given by

$$\begin{aligned} Q_0 &= 0\ [\mathrm{V}], \\ Q_1 &= C_{store} V_{max}. \end{aligned} \qquad (8.55)$$

Since C_{store} is on the order of femtofarads [fF], the difference in charge values is on the order of femtocoulombs [fC].

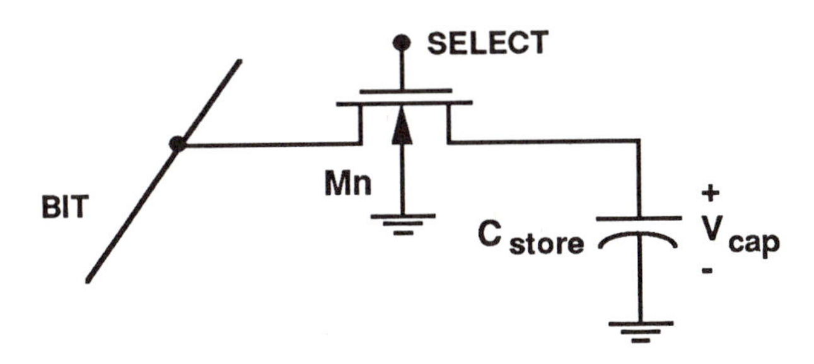

Figure 8.28: 1T Dynamic RAM Cell

EXAMPLE 8.2: Dynamic RAM Charge Storage

Consider a 1T DRAM cell with a storage capacitor of $C_{store} = 50$ [fF]. Assuming nMOS parameters of $V_{T0n} = 0.70$ [V], $\gamma_n = 0.37$ [V$^{1/2}$], and $2|\phi_{Fp}| = 0.58$ and $V_{DD} = 5$ [V] gives the equation

$$V_{max} = 4.3 - 0.37(\sqrt{0.58 + V_{max}} - \sqrt{0.58}).$$

This can be solved by self-iterations. We make an arbitrary first guess of $V_{max} = 4$[V]; substituting gives a right-hand side of 3.79 [V]. Our next guess is $V_{max} = 3.79$ [V], which gives a right-hand side of 3.81 [V]. Substituting $V_{max} = 3.81$ [V] again gives 3.81 [V], so

$$V_{max} \simeq 3.81 \text{ [V]}$$

is the solution.

The logic 1 charge level is

$$\begin{aligned} Q_1 &\simeq (50 \times 10^{-15})(3.81) \\ &\simeq 190.5 \text{ [fC]}. \end{aligned}$$

This represents the maximum stored charge. The actual value will be reduced by dynamic charge leaking into the substrate.

Loading Time

Cell loading times can be analyzed using the results of Section 5.1.1 which presented the electrical properties of nMOS pass transistors. Assume that the BIT line is at a logic 1 state and set to V_{DD}. With $V_{cap}(t = 0) = 0$ [V], the storage capacitor charges according to

$$V_{cap}(t) = V_{max}\left[\frac{(t/\tau_{ch})}{1 + (t/\tau_{ch})}\right], \tag{8.56}$$

where

$$\tau_{ch} = \frac{2C_{store}}{\beta_n V_{max}} \tag{8.57}$$

is the charging time constant. The 90% voltage point $0.9V_{max}$ is reached in a low-to-high time of

$$t_{LH} = 9\tau_{ch}, \tag{8.58}$$

and is the minimum logic 1 loading interval.

A logic 0 transfer corresponds to the BIT line being a 0 [V]. Assuming the worst-case initial condition of $V_{cap}(t = 0) = V_{max}$ gives the voltage decay as

$$V_{cap}(t) = V_{max}\left[\frac{2e^{-(t/\tau_{dis})}}{1 + e^{-(t/\tau_{dis})}}\right]. \tag{8.59}$$

The discharge time constant

$$\tau_{dis} = \frac{C_{store}}{\beta_n V_{max}} \tag{8.60}$$

characterizes the discharge event. To reach the 10% voltage of $0.1V_{max}$ requires a high-to-low time of

$$t_{HL} \simeq 2.94\tau_{dis}, \tag{8.61}$$

where we have used $\ln(19) \simeq 2.94$ in reducing the final answer. Since

$$\tau_{ch} = 2\tau_{dis}, \tag{8.62}$$

we have

$$t_{LH} \simeq 6.11 t_{HL}, \tag{8.63}$$

showing that the logic 1 transfer time is the limiting factor. In other words, it takes longer to load a logic 1 than to load a logic 0. This is due to the fact that V_{GSn} decreases during a logic 1 transfer.

Charge Sharing

The read operation for a dynamic RAM cell corresponds to a charge transfer/sharing event. Consider the circuit shown in Figure 8.29 where we will assume that $V_{cap} = V_C$ is initially on C_{store}. The bit line capacitance C_{line} is assumed to have an initial precharge voltage of V_{pre}. The total system charge is given by

$$Q_T = V_C C_{store} + V_{pre} C_{line}. \tag{8.64}$$

When the SELECT signal is set to a high voltage, Mn goes active and conducts current. After the transients decay, the capacitors are in parallel and equilibrate to the same final voltage set V_f such that

$$Q_T = (C_{store} + C_{line})V_f. \tag{8.65}$$

Applying charge conservation gives

$$V_f = \frac{V_C C_{store} + V_{pre} C_{line}}{C_{store} + C_{line}}. \tag{8.66}$$

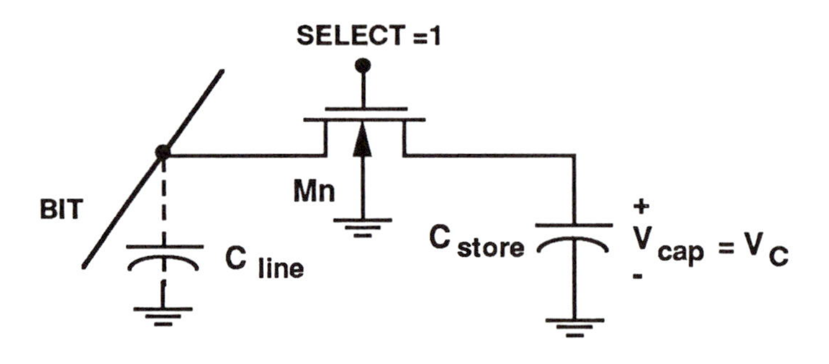

Figure 8.29: Charge Sharing Network

Defining the capacitance ratio

$$r = \frac{C_{line}}{C_{store}} \tag{8.67}$$

yields the final voltage as

$$V_f = \frac{V_C + rV_{pre}}{1+r}. \tag{8.68}$$

If a logic 1 is initially stored in the cell, $V_C = V_{max}$ and

$$V_1 = \frac{V_{max} + rV_{pre}}{1+r} \tag{8.69}$$

is the final logic 1 voltage on the line. A stored logic 0 charge corresponds to $V_C = 0$ and results in a final voltage of

$$V_0 = \frac{rV_{pre}}{1+r}. \tag{8.70}$$

The difference between the logic 1 and logic 0 voltages is

$$\Delta V = \frac{V_{max}}{1+r}, \tag{8.71}$$

which shows that a small r is desirable. Normal layout gives $C_{line} > C_{store}$, so that $r > 1$ always holds. In current 16 Mb designs, $C_{store} \sim 30$ [fF] and $C_{line} \sim 250$ [fF] are typical, giving $r \sim 8$. Careful sense amplifier design is important for proper data recovery.

Charge Leakage

Dynamic RAM cells are subject to charge leakage as shown in Figure 8.30. Suppose that the capacitor voltage is initially set to $V_{cap}(0) = V_{max}$. Assuming a constant leakage current I_L and an LTI capacitance C_{store} gives the voltage at time t_1 as

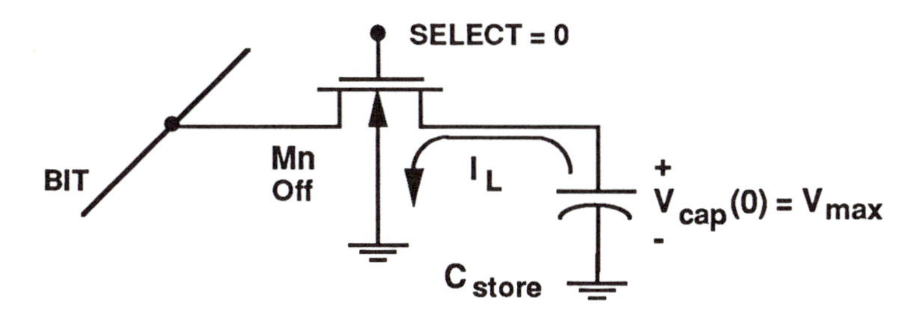

Figure 8.30: DRAM Charge Leakage

$$V_{cap}(t_1) \simeq V_{max} - \frac{I_L}{C_{store}} t_1. \tag{8.72}$$

If the cell is accessed at this time, then the final voltage on the bit line will be degraded to the value

$$V_f = \frac{V_{max} + rV_{pre}}{1+r} - \frac{I_L t_1}{C_{store}(1+r)}. \tag{8.73}$$

Care must be taken to insure that V_f does not fall to a value where it could be misread as a logic 0. In order to maintain storage of a logic 1, we must subject the cell to a dynamic refresh operation where the peripheral logic circuit reads the cell and then rewrites the bit. Periodic refreshing is required to insure the integrity of the stored data. Typical maximum refresh periods are the order of a few milliseconds.

Trench Capacitors

High-capacitor (High-C) designs use trench capacitors to increase the storage capacity while maintaining the small surface area requirements needed for high-density arrays. Various types of structures have appeared in the literature; Figure 8.31 shows the "diffusion-store" and "substrate-plate" geometries [19]. With proper bit line layout, reasonable value of r can be achieved without excessive real estate consumption.

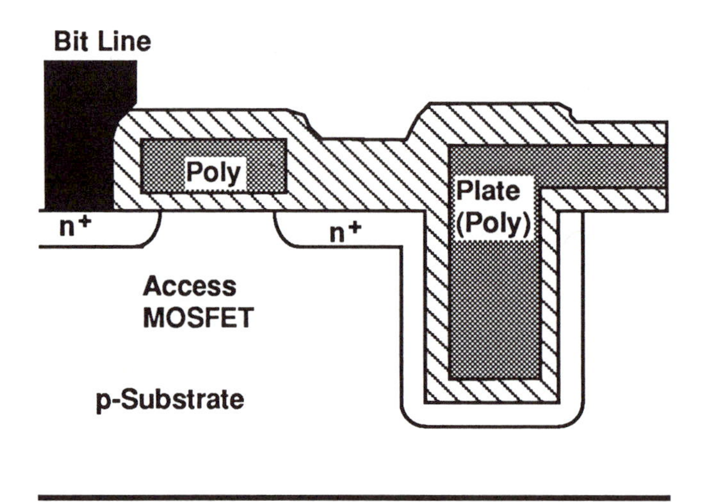

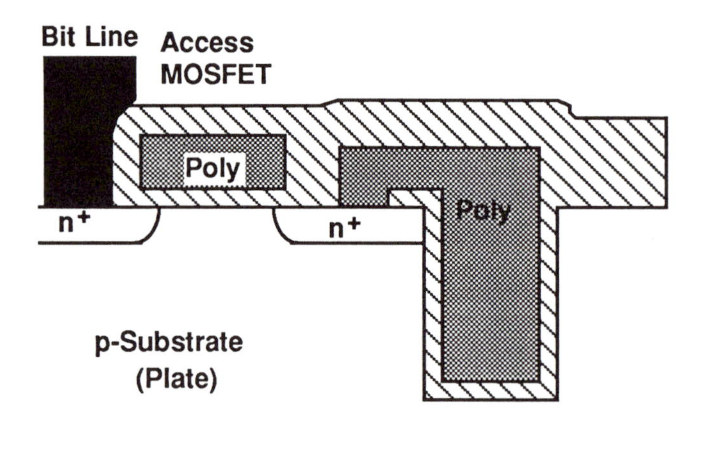

Figure 8.31: Trench Capacitor DRAM Cells

8.6.3 Architecture

Planning the architecture and layout of a memory chip requires a deep understanding of technology limits, circuit design, interconnect strategies, and several other salient aspects of CMOS VLSI. Although the specifics such as the density and cell structure form a complicated set of dynamic variables, several basic features of memory chips can be identified. We will restrict our discussion here to an overview of the architecture and building blocks in CMOS SRAM circuits as an example system.

Cell Address

Memory cells are structured into geometrical arrays to provide an orderly means for data access. Each cell is identified by a logical **address** which gives the location of the cell. The logical address is not necessarily correlated with either the relative or absolute physical location on the chip. An n-bit address word can select 2^n individual cells. Alternately, paralleling m-cells gives direct access to 2^n m—-bit words. Both cases are illustrated in Figure 8.32.

Memory Blocks

Practical addressing is accomplished using **blocks** which are rectangular cell arrays. Each location within a block is specified by a **column address** and a **row address**. The dimensions of the block are arbitrary; for example, one choice in several current CMOS SRAM designs is (1024 rows × 128 columns) which is a 128K-bit segment[4]. To access an individual cell within the block requires 10-bits for the row address and 7-bits for the column address. In most architectures, the rows constitute the word lines while the columns are the bit (or data) lines. We will employ this terminology throughout our discussion. Figure 8.33 gives the general block diagram for a basic architecture and illustrates how several cell blocks can be arranged to give a large memory array. The total storage capacity is simply the number of cells per block multiplied by the number of blocks; a 4-Mbit design can constructed by using 32 blocks which contain (1024 rows × 128 columns). Rows are selected using the **row decoder** circuits, while the output (columns) is selected using the **column decoder** sections. Sense amplifiers are used to detect the cell contents. The main peripheral circuits are discussed in more detail in the following sections.

[4]Redundant columns are usually included within the block so that the actual size is larger

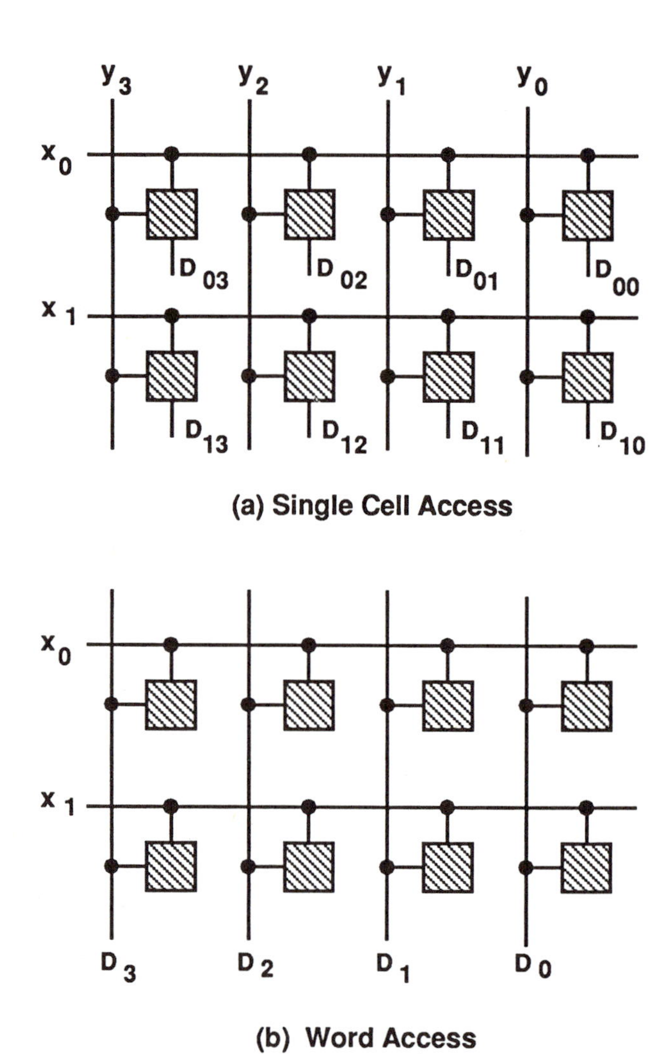

Figure 8.32: Single Cell and Word Access

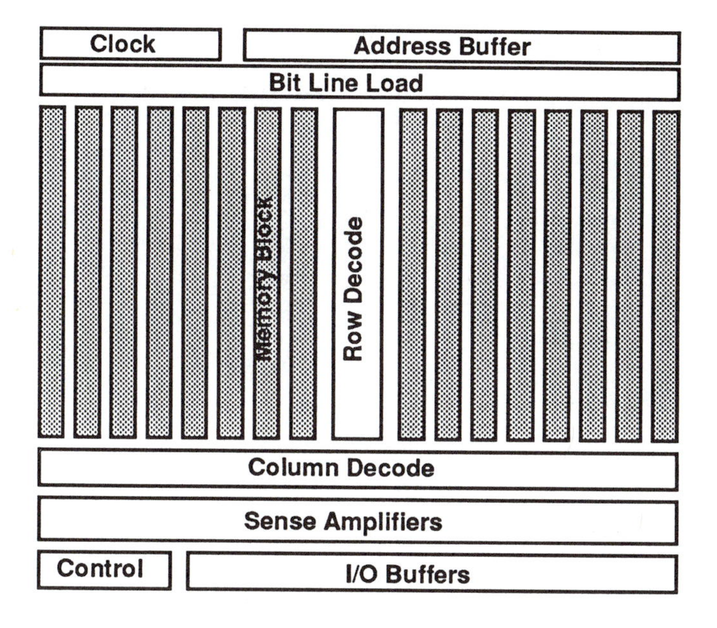

Figure 8.33: Large Memory Array

8.6.4 Address Decoders

Addressing information is captured in on-chip registers and then used as an input word to the decoder logic. Fast static decoder designs are based on OR/NOR or AND/NAND gates using standard configurations. While direct decoding can be accomplished with a single stage, double-decoder schemes provide faster access with fewer transistors [10].

To illustrate the decoder design, suppose we need to address 1024 rows using a 10-bit word $(X_9, \ldots, X_0)$. A 10-to-1024 decoder can be made using AND-logic as shown in Figure 8.34. For example, the functions $ROW0$, $ROW1$, and $ROW2$ can be defined by

$$\begin{aligned} F0 &= X_0X_1X_2X_3X_4X_5X_6X_7X_8X_9 \\ F1 &= X_0X_1X_2X_3X_4X_5X_6X_7X_8\overline{X}_9 \\ F2 &= X_0X_1X_2X_3X_4X_5X_6X_7\overline{X}_8X_9. \end{aligned} \tag{8.74}$$

This single-stage scheme requires 1024 10-input NAND gates, with each gate containing 20 MOSFETs. Even if we ignore the buffer circuits, a total of 20,480 transistors are needed. The switching response of a gate is limited

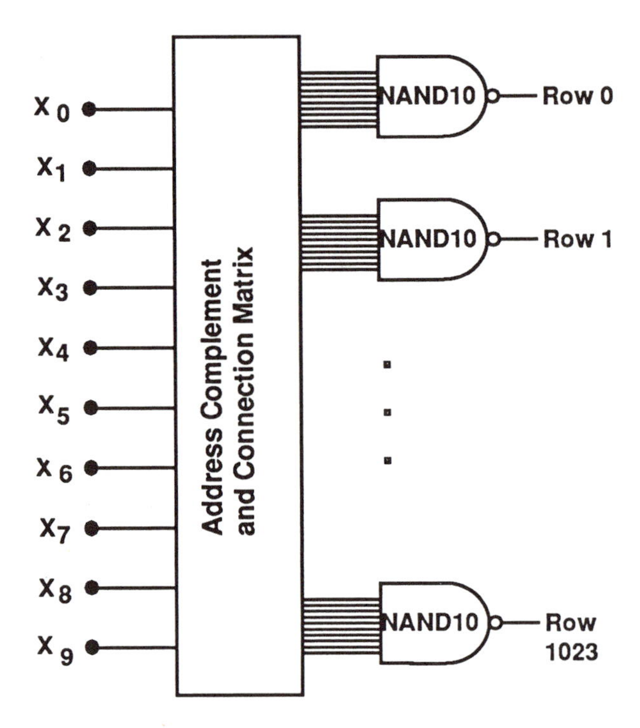

Figure 8.34: Single Stage 10-to-1024 Decoder

by the discharge time through the series-connected nMOS transistor chain. Another problem is the large fanout requirements on the buffers generating $(X_9, \ldots, X_0)$.

Double-stage decoders are used to improve performance, reduce the device count, and ease the drive requirements on the input buffers. For our example of a 10-to-1024 decoder, we start with the 10-bit address word $(X_9, \ldots, X_0)$ and **predecode** the word using 2-bit segments

$$(X_9, X_8),\ (X_7, X_6),\ (X_5, X_4),\ (X_3, X_2),\ (X_1, X_0) \tag{8.75}$$

as shown in Figure 8.35. The chip select signal CS has been added for completeness; this is defined such that $CS = 1$ activates the chip circuitry. The first stage NAND3-INV circuits produce 4 predecoded lines for every pair of inputs. For example, (X_1, X_0) gives the output combinations

$$(\overline{X_1}\ \overline{X_0}),\quad (X_1\ \overline{X_0}),\quad (\overline{X_1}\ X_0),\quad (X_1\ X_0). \tag{8.76}$$

Predecoding the 1-bit word requires 5 separate circuits. Each NAND3 gate

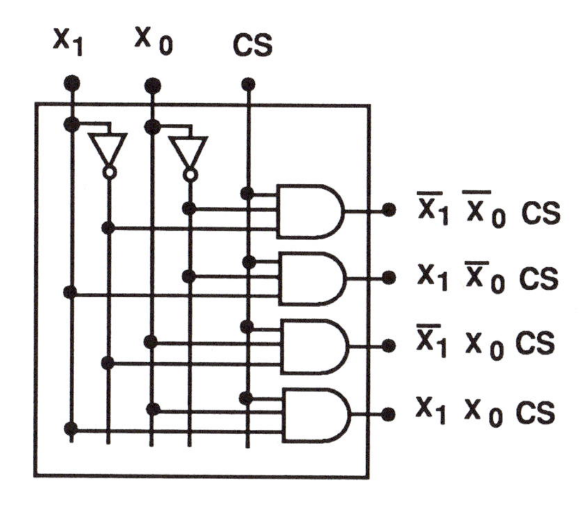

Figure 8.35: Predecoded Address Logic

has 6 MOSFETs. Including the inverters gives a total of 40 transistors in the first-stage decoders.

The second stage in this design uses 5-input NAND gates as shown in Figure 8.36. The inputs consist of the 2-bit groupings generated by the first stage. The complete decoder requires 1024 circuits. Since each NAND5 gate has 10 MOSFETs, including the inverter gives a transistor count of 12,288. Overall, the double-decoder system uses 12,329 MOSFETs, which is a substantial reduction from the single-state number. Also note that the FO of the first-stage circuits is reduced to 5, which is one-half the number found in the single-stage design. This eases the device size requirements on the large-capacitance word lines. Optimizing the gate sizes using the results of Section 8.3 further improves the performance.

Access time is the most critical performance parameter in a memory design. While predecoding circuits help reduce this time interval, further improvements are possible by changing the block and word-line structures. A modified architecture which employs a divided-word line (DWL) wiring is shown in Figure 8.37 [24]. With this layout, row selection is accomplished at two levels: the **global** and the **local**. The global word line selects a block, while the local line is used to activate a word line within the selected block. Both word line delay and power consumption are reduced using this approach. Consider our example of a 4-Mb memory made up of 32 blocks. Since each block has 1024 rows, we use the address word $(X_9, \ldots, X_0)$ at the

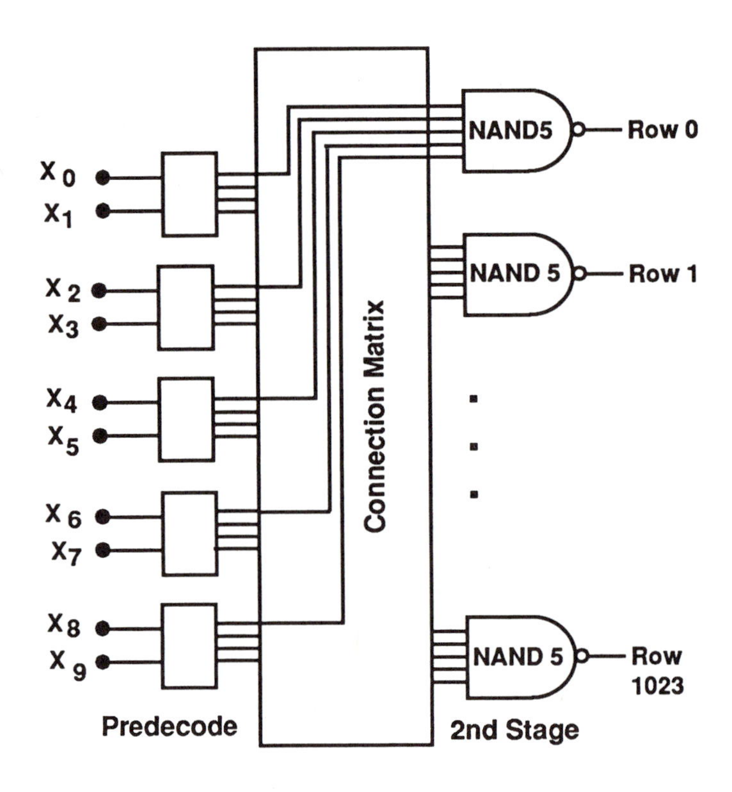

Figure 8.36: Second-Stage Decoder Logic

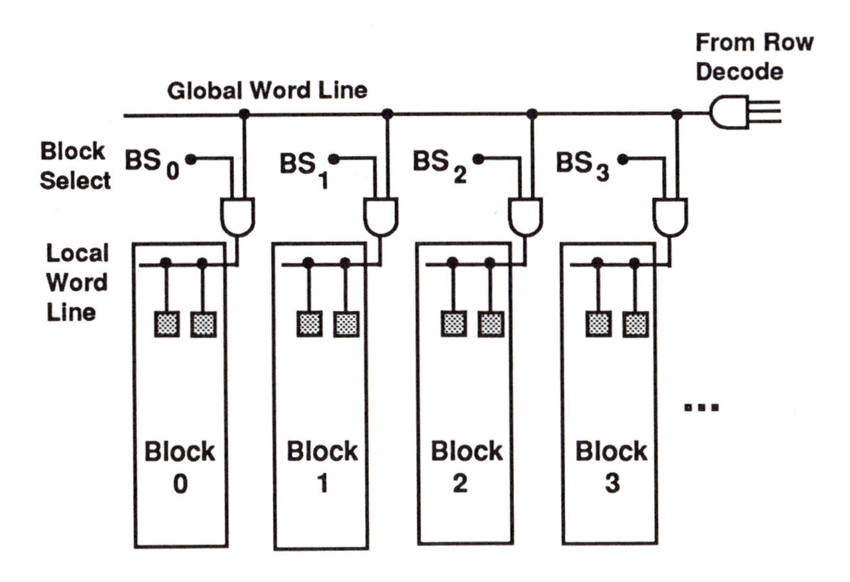

Figure 8.37: Divided-word Line Architecture

local level. Global block selection is accomplished by adding a 5-bit address $(Z_4, \ldots, Z_0)$ which specifies the block. This concept can be extended to create the hierarchical word decoding (HWD) architecture using the logic shown in Figure 8.38 [8]. Three levels of decoding are used: global, sub-global, and local. Nested word-line architectures of this type are required to maintain small access times in large memory chips.

8.6.5 Column Selector and Sense Amplifiers

A data read operation is accomplished by selecting the column and transferring the data from the cell to the bit line. Figure 8.39 shows the circuit sections involved in the event. Column decoder logic provides the Y-select signal which activates the transmission gates. The complementary BIT and $\overline{\text{BIT}}$ signals are sent to the I/O lines and the sense amplifiers for detection.

Many variations on sense amplifier design have been reported in the literature. We will concentrate on the simple current-mirror design shown in Figure 8.40 as an example. This consists of two dynamically switched differential amplifiers with active pMOS (current source) loads[5]. Modeling the operation using comparators shows that the circuit is sensitive to

[5] Differential amplifiers are analyzed in Section 9.6 of the next chapter.

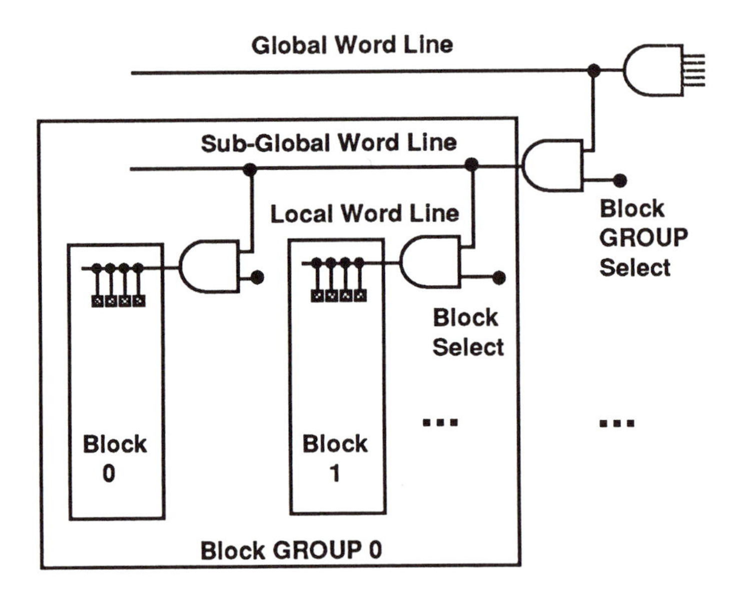

Figure 8.38: Hierarchical Word Decoding Logic

differences between the input signals. Although the gain of the circuit is usually low ($\sim$ 15 for a basic design), multiple stages can be cascaded to improve the system performance.

8.6.6 Summary

In this section we have briefly examined the major circuit segments involved in a memory array. An important aspect of the discussion was the use of small, individual circuits in an integrated environment. Even in our short analysis, we encountered the problems of

- Logic design
- Circuit performance
- Charge sharing
- Charge leakage
- Organization
- Layout

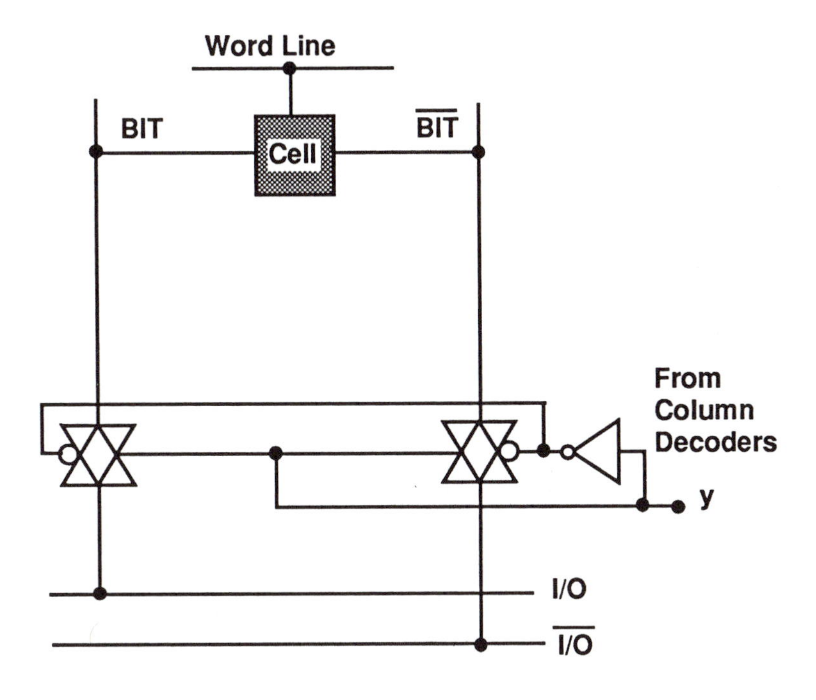

Figure 8.39: Column Select Circuit

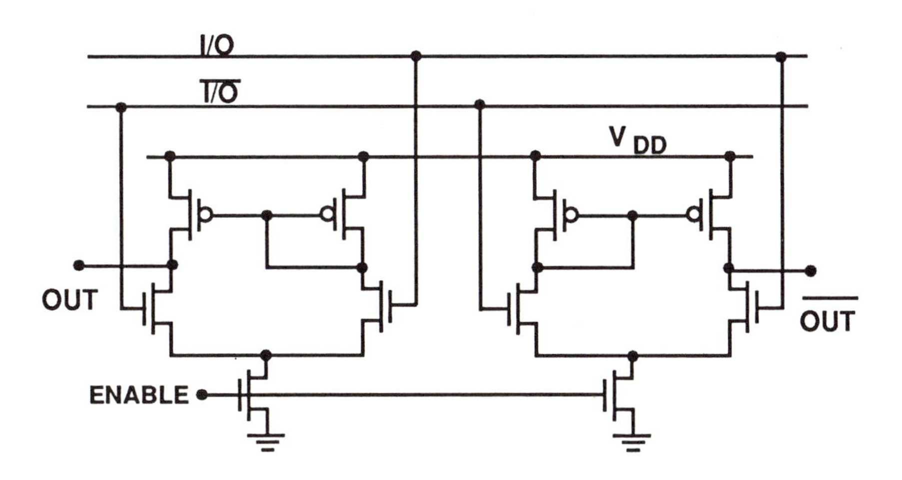

Figure 8.40: Current-Mirror Sense Amplifier

when studying the circuit. This, of course, is what we might expect of any system design in CMOS VLSI. The design puzzle can be quite complicated, with many twists and turns needed to find a solution.

8.7 References

The list below provides further reading on the circuits discussed in this chapter.

[1] S. Aizaki, et. al, " A 15-ns 4-Mb CMOS SRAM", IEEE J. Solid-State Circuits, vol 25, no. 5, pp. 1063-1067, October, 1990.

[2] F. Anceau, **The Architecture of Microprocessors**, Addison-Wesley, Wokingham, England, 1986.

[3] K. Anami, M. Yoshimoto, H. Shinohara, Y. Hirata, and T. Nakano, "Design Considerations of a Static Memory Cell", IEEE J. Solid- State Circuits, vol. SC-18, no. 4, pp. 414-417, August, 1983.

[4] H.B. Bakoglu, **Circuits, Interconnects, and Packaging for VLSI**, Addison-Wesley, Reading, MA, 1990.

[5] E.G. Fabricius, **Introduction to VLSI Design**, McGraw-Hill, New York, 1990.

[6] L. A. Glasser and D.W. Dobberpuhl, **The Design and Analysis of VLSI Circuits**, Addison-Wesley, Reading, MA, 1985.

[7] D.V. Heinbuch (ed.), **CMOS3 Cell Library**, Addison-Wesley, Reading, MA., 1988.

[8] T. Hirose, et. al, " A 20-ns 4-Mb CMOS SRAM with Hierarchical Word Decoding Architecture", IEEE J. Solid-State Circuits, vol. 25, no. 5, pp. 1068-1074, October, 1990.

[9] R.W. Hunt, "Memory design and technology," Chapter 6 in **Large Scale Integration**, M.J. Howes and D.V. Morgan (eds.), John Wiley & Sons, Ltd., Chichester, U.K., 1981.

[10] Y. Kobayashi, et al., " A 10-μW Standby Power 256K CMOS SRAM", IEEE J. Solid-State Circuits, vol. SC-20, no. 5, pp. 935-940, October, 1985.

[11] C. M. Lee and H. Soukup, " An Algorithm for CMOS Timing and Area Optimization", IEEE J. Solid-State Circuits, vol. SC-19, no. 5, pp. 781-787, October, 1984.

[12] **CMOS Macrocell Manual**, LSI Logic Corporation, 1985.

[13] R.P. Masleid, "High-Density I/O Circuits for CMOS", IEEE J. Solid-State Circuits, vol. 26, no. 3, pp. 431-435, March, 1991.

[14] C.A. Mead and L. Conway, **Introduction to VLSI Systems**, Addison-Wesley, Reading, MA, 1980.

[15] W. Noble and W. Walker, "Fundamental Limitations on DRAM Storage Capacitors", IEEE Circuits and Dev. Mag., vol. 1, no. 1, pp. 45-51, January, 1985.

[16] A. H. Sayles, "Design of Integrated CMOS Circuits for Parallel Detection and Storage of Optical Data", Ph.D. Dissertation, Georgia Institute of Technology, August, 1990.

[17] R.E. Scheuerlein and J.D. Meindl, "Offset Word-Line Architecture for scaling DRAM's to the Gigabit Level", IEEE J. Solid-State Circuits, vol. 23, no. 1., pp. 41-47, February, 1988.

[18] E. Seevinck, F. List, and J. Lohstrogh, "Static-Noise Margin Analysis of MOS SRAM Cells", IEEE J. Solid-State Circuits, vol. SC- 22, no. 5., pp. 748-754, October, 1987.

[19] H. Sunami, et al., "A Corrugated Capacitor Cell (CCC)," IEEE Trans. Electron Dev., vol. ED-31, pp. 746-753, June, 1984.

[20] C. J. Terman, "Simulation Tools for Digital LSI Design, " Ph.D. Dissertation, Massachusetts Institute of Technology, September, 1983.

[21] J.P. Uyemura, **Fundamentals of MOS Digital Integrated Circuits**, Addison-Wesley, Reading, MA, 1988.

[22] N. H. Weste and K. Eshraghian, **CMOS VLSI Design**, Addison-Wesley, Reading, MA, 1985.

[23] T. Yamanaka, et al,, " A 25-ns 64K Static RAM", IEEE J. Solid-State Circuits, vol. SC-19, no. 5, pp. 572-577, October, 1984.

[24] M. Yoshimoto, et al., " A Divided Word-Line Structure in the Static RAM and Its application to a 64K Full CMOS RAM," IEEE J. Solid-State Circuits, vol. SC-18, no. 5, pp.479-485, October, 1983.

Chapter 9

Analog CMOS Circuits

Analog CMOS circuits provide for direct on-chip interfacing with digital networks. The most obvious application would be in D/A and A/D converters. However, the field of analog CMOS opens up new areas which are being rapidly studied and implemented.

This chapter provides an introduction to the basic circuit techniques used in analog CMOS circuit design. The wide range of possible analog circuit functions prohibits an exhaustive treatment here. Instead, we will concentrate on fundamental amplifier circuits to gain an understanding of the principles. Several textbooks are listed in the References at the end of the chapter. These should be consulted if more details and expanded discussions are needed.

Notation in this chapter will follow standard analog treatments. Upper case symbols such as V_{GS} and I_D indicate large-signal DC parameters. Small-signal quantities will be denoted by lower-case letters; for example, v_{GS} and i_D represent the small-signal gate-source voltage and drain current, respectively. Elements in the small-signal model are represented using lower-case letters with lower-case subscripts, for example, g_m and r_{ds}.

9.1 MOSFET Equations

Square-law MOSFET models are sufficient to perform hand analysis of analog circuits. However, it is common to alter the basic equation set in an effort to simplify the calculations. To understand the modifications, recall that the non-saturated current is given by

$$I_D = \frac{\beta}{2}[2(V_{GS} - V_T)V_{DS} - V_{DS}^2] \tag{9.1}$$

while a saturated transistor conducts according to

$$I_D = \frac{\beta}{2}(V_{GS} - V_T)^2[1 + \lambda(V_{DS} - V_{DS,sat})] \tag{9.2}$$

with

$$V_{DS,sat} = (V_{GS} - V_T) \tag{9.3}$$

the saturation voltage. Channel-length modulation effects are described using the parameter λ [V^{-1}]. Although $\lambda = 0$ was assumed in our treatment of digital circuits, it is often important in analyzing analog networks.

The difficulty with the above equation set lies in the factor

$$[1 + \lambda(V_{DS} - V_{DS,sat})] \tag{9.4}$$

which is somewhat cumbersome to carry through long calculations . Some treatments simplify this to $(1 + \lambda V_{DS})$, but this introduces a discontinuity into the equation set as the device moves between saturated and non-saturated behavior. The convention used in SPICE gets around this problem by modifying the non-saturated current for $V_{DS} \leq V_{DS,sat}$ to

$$I_D = \frac{\beta}{2}[2(V_{GS} - V_T)V_{DS} - V_{DS}^2](1 + \lambda V_{DS}) \tag{9.5}$$

and insuring continuity by writing

$$I_D = \frac{\beta}{2}(V_{GS} - V_T)^2(1 + \lambda V_{DS}) \tag{9.6}$$

for the saturated current flow equation ($V_{DS} \geq V_{DS,sat}$). These equations are commonly used to perform basic calculations on analog circuits, and are adopted here.

It is important to point out that channel-length modulation occurs **only** in a saturated MOSFET by its very definition. Incorporating the effect into the non-saturated transistor is completely unphysical. How then, can we justify the modified equation set? Well, there are two reasons. First, the important dependences are still maintained so that the simplified analysis still gives reasonable results. Second, the square-law models represent the lower-order approximations to the actual MOSFET behavior, so that we must always be careful in interpreting the results regardless of the modifications. Computer simulations and verifications are mandatory, and provide a safety net. Owing to these considerations, we will employ the modified equations in our discussion.

9.2 Small-Signal MOSFET Model

Analog circuit analysis and design is based on the small-signal behavior of transistors. One difficulty of working with MOSFETs is that they are intrinsically 4-terminal devices, which increases the overall complexity of the model. We will base our analysis on the general small-signal MOSFET model shown in Figure 9.1. Table 9.1 provides a summary of the elements used in the model.

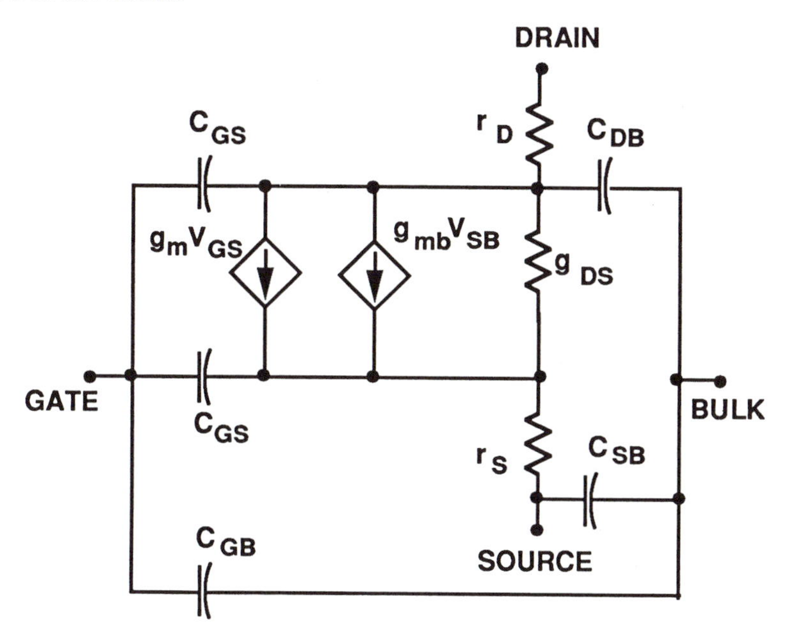

Figure 9.1: Small-Signal MOSFET Model

Analog circuits tend to be designed with the MOSFETs operating in saturation since the current is approximately proportional to the drain-source voltage V_{DS}. As discussed above, we assume an $I - V$ relation of the form

$$I_D = \frac{\beta}{2}(V_{GS} - V_T)^2(1 + \lambda V_{DS}) \tag{9.7}$$

where channel-length modulation effects are included through the factor λ.

The short discussion below provides an introduction to the analog small signal MOSFET model. Applications will be presented by using various circuits as examples.

Symbol	Name
$g_m = (\partial i_D/\partial v_{GS})$	Transconductance
$g_{mb} = (\partial i_D/\partial v_{BS})$	Bulk Transconductance
$g_{ds} = (\partial i_D/\partial v_{DS})$	Channel conductance
r_S	Source resistance
r_D	Drain resistance
C_{GD}	Gate-drain capacitance
C_{GS}	Gate-source capacitance
C_{GB}	Gate-bulk capacitance
C_{DB}	Drain-bulk capacitance
C_{SB}	Source-bulk capacitance

Table 9.1: Small-Signal MOSFET Parameters

Transconductance

The most important gain parameter is the **small-signal transconductance**

$$g_m = \left(\frac{\partial I_D}{\partial V_{GS}}\right)_Q \quad [\mho] \tag{9.8}$$

which is evaluated at the DC quiescent point (Q-point) drain current I_D. The transconductance controls the primary drain current source with a value of $g_m v_{GS}$. Differentiating gives directly that

$$\begin{aligned} g_m &= \beta(V_{GS} - V_T)(1 + \lambda V_{DS}) \\ &= \sqrt{2\beta I_D}(1 + \lambda V_{DS}), \end{aligned} \tag{9.9}$$

which is usually approximated as

$$g_m \simeq \sqrt{2\beta I_D}. \tag{9.10}$$

The value of g_m is important since it relates a change in gate-source voltage v_{GS} to a change in the current flow i_D by means of

$$i_D = g_m v_{GS}; \tag{9.11}$$

this, of course, is the main mechanism for amplification.

Bulk Transconductance

The bulk transconductance provides the change in drain current with respect to variations in the body-bias voltage v_{SB}. Consider the threshold voltage

$$V_T = V_{T0} + \gamma(\sqrt{2|\phi_F| + V_{SB}} - \sqrt{2|\phi_F|}). \tag{9.12}$$

Using the chain rule, we write

$$\begin{aligned} g_{mb} &= \left(\frac{\partial I_D}{\partial V_{BS}}\right) \\ &= -\left(\frac{\partial I_D}{\partial V_T}\right)\left(\frac{\partial V_T}{\partial V_{BS}}\right) \end{aligned} \tag{9.13}$$

and noting that the drain current satisfies $(\partial I_D/\partial V_{GS}) = -(\partial I_D/\partial V_T)$ gives us

$$g_{mb} = \eta g_m, \tag{9.14}$$

where the parameter

$$\eta = \frac{\partial V_T}{\partial V_{BS}} = \frac{-\gamma}{2\sqrt{2|\phi_F| + V_{SB}}} \tag{9.15}$$

describes the variations of the threshold voltage due to body-bias[1]. Small-signal body-bias effects are included by using a controlled current source with a value of

$$i_D = g_{mb} v_{BS} \tag{9.16}$$

in the MOSFET model.

Channel Conductance

The channel conductance g_{ds} represents the amount of conductance between the drain and source terminals with the quiescent voltages. Using the definition

$$g_{ds} = \frac{\partial I_D}{\partial V_{DS}} \tag{9.17}$$

gives us

$$\begin{aligned} g_{ds} &= \frac{\lambda I_D}{1 + \lambda V_{DS}} \\ &\simeq \lambda I_D \end{aligned} \tag{9.18}$$

for the saturated MOSFET; the drain-source resistance is the inverse $r_{ds} = (1/g_{ds})$. If channel-length modulation is ignored, this predicts that $g_{ds} \simeq 0$ and $r_{ds} \rightarrow \infty$. Although $\lambda \simeq 0$ is a reasonable approximation for most calculations in a digital network, this gives overly optimistic results when applied to analog circuits. In general, channel-length modulation should be included in a small-signal analysis.

[1]Note that $V_{BS} = -V_{SB}$.

Resistances

The resistance parameters r_S and r_D represent parasitics from the source and drain diffusion (n^+ or p^+) regions. They are layout-dependent and can be calculated using

$$R = R_{sh} n \tag{9.19}$$

where R_s is the sheet resistance in ohms/square and n is the number of squares. The model is identical to that introduced in Chapter 2.

9.3 Basic Amplifier

A simple inverter circuit can be biased to create a basic CMOS amplifier. This is evident from the voltage transfer characteristics shown in Figure 9.2 since the magnitude of the slope (dV_{out}/dV_{in}) is large in the central transition region. Voltage gain is possible if the circuit is properly biased. The slope of the voltage transfer curve can be determined using large signal models by writing the general functional dependence

$$I_{Dn}(V_{in}, V_{out}) = I_{Dp}(V_{in}, V_{out}) \tag{9.20}$$

and taking total differentials of both sides. Rearranging,

$$\frac{dV_{out}}{dV_{in}} = \frac{(\partial I_{Dp}/\partial V_{in}) - (\partial I_{Dn}/\partial V_{in})}{(\partial I_{Dn}/\partial V_{out}) - (\partial I_{Dp}/\partial V_{out})} \tag{9.21}$$

so that the gain can be calculated from the current flow equations.

9.3.1 Small-Signal Gain

Small-signal analysis is generally easier and more useful if the input voltage swings are small. To calculate the small-signal gain we apply the basic MOSFET model and analyze the network. The general rules are (1) short all DC voltage sources, and (2) replace the device by the small-signal model. In addition, we will initially concentrate on low frequencies which allows us to open all capacitors. The resulting circuit is shown in Figure 9.3. It is also necessary to specify the bias. We will assume that the Q-point is chosen as $V_{in} = V_{th} = V_{out}$, corresponding to the point where both MOSFETs are saturated and the gain is maximized.

The analysis is straightforward. We see by inspection that the output voltage is given by

$$v_{out} = -(g_{mn} + g_{mp}) v_{in} r_o \tag{9.22}$$

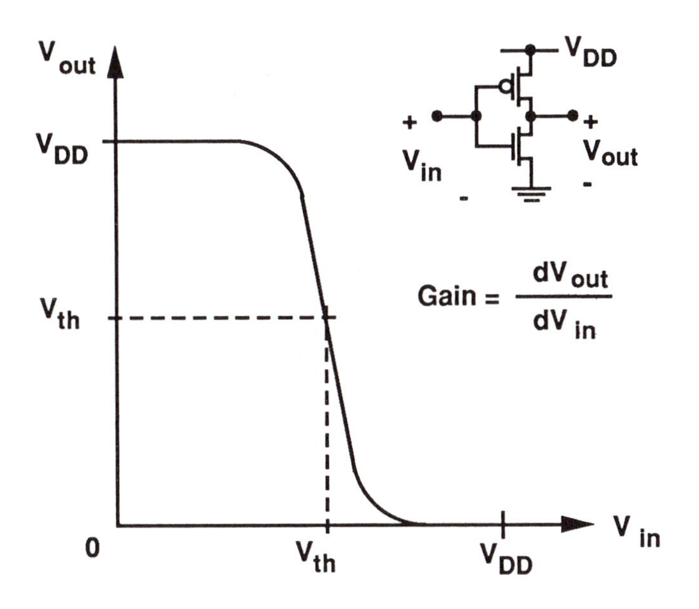

Figure 9.2: Basic Inverting CMOS Amplifier

where

$$r_o = \frac{1}{g_{dsn} + g_{dsp}} \tag{9.23}$$

is the total drain-to-source resistance. The low-frequency small-signal voltage gain is thus calculated by

$$a_v = \frac{v_{out}}{v_{in}} = -\frac{g_{mn} + g_{mp}}{g_{dsn} + g_{dsp}}. \tag{9.24}$$

Since both transistors are saturated, this can be written in the form

$$a_v = -\sqrt{\frac{2}{I_D}}\left(\frac{\sqrt{\beta_n} + \sqrt{\beta_p}}{\lambda_n + \lambda_p}\right) \tag{9.25}$$

where I_D is the bias current.

9.3.2 Frequency Response

The frequency response of the circuit is determined using the capacitances illustrated in Figure 9.4. The capacitance at the output node is

$$C_o = (C_{GDn} + C_{GDp}) + (C_{DBn} + C_{DBp}) + C_{line} + C_{in} \tag{9.26}$$

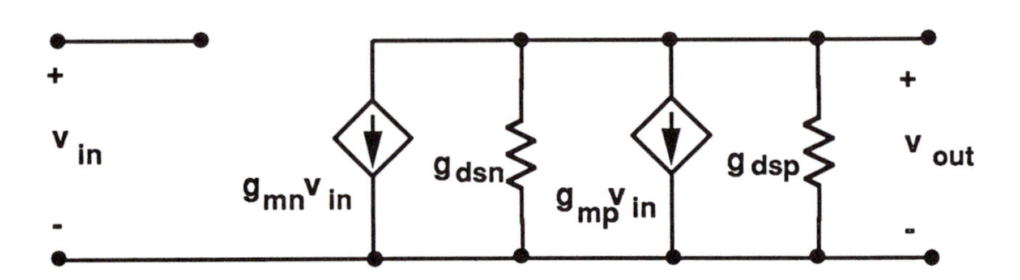

Figure 9.3: Amplifier Small-Signal Circuit

where the depletion capacitances are calculated at the bias point. This allows us to construct the single-pole network of Figure 9.4 where

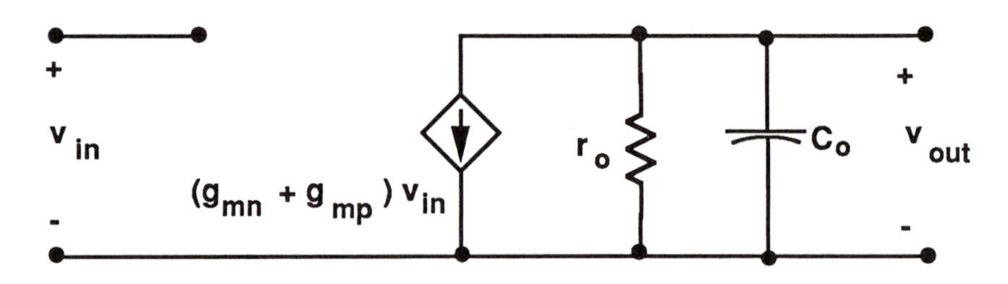

Figure 9.4: Circuit for Calculating Frequency Response

$$r_o = \frac{1}{g_{dsn} + g_{dsp}} \tag{9.27}$$

is the output resistance. By inspection we see that the s-domain output voltage is

$$V_{out}(s) = -(g_{mn} + g_{mp})\left[\frac{(r_o/sC_o)}{r_o + (1/sC_o)}\right] V_{in}(s) \tag{9.28}$$

giving the frequency-domain transfer function

$$a_v(j\omega) = \frac{-(g_{mn} + g_{mp})r_o}{1 + j(\omega/\omega_1)} \tag{9.29}$$

where

$$\omega_1 = \frac{1}{r_o C_o} \tag{9.30}$$

is the -3 dB frequency.

The analysis above illustrates that the operation of a MOSFET amplifier is very similar to any other transistor circuit.

9.4 Voltage References

Analog circuits often require reference voltages for use in the biasing. Since we tend to avoid resistors on IC designs due to large area requirements[2], voltages are generated using various types of MOSFET stacks.

Consider the simple two transistor circuit shown in Figure 9.5. Both MOSFETs are biased into saturation. Equating currents gives

$$\frac{\beta_p}{2}(V_{DD} - V_R - |V_{Tp}|)^2 = \frac{\beta_n}{2}(V_R - V_{Tn})^2 \tag{9.31}$$

where we have ignored channel-length modulation for simplicity. Rearranging gives directly

$$V_R \simeq \frac{\sqrt{\beta_n/\beta_p}V_{Tn} + (V_{DD} - |V_{Tp}|)}{1 + \sqrt{\beta_n/\beta_p}}. \tag{9.32}$$

There is no body-bias in either transistor, so that this is a closed form

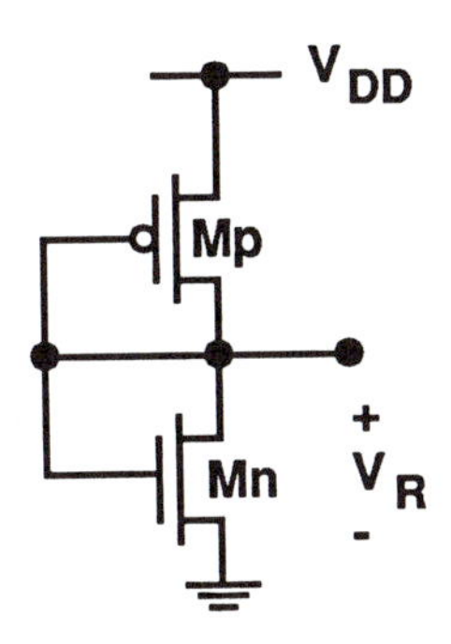

Figure 9.5: Voltage Divider Circuit

expression.

The MOSFET transconductance ratio (β_n/β_p) is the primary design variable. Rearranging the current flow expression gives

$$\frac{\beta_n}{\beta_p} = \left(\frac{V_{DD} - V_R - |V_{Tp}|}{V_R - V_{Tn}}\right)^2 \tag{9.33}$$

which can be used to compute the transconductance ratio needed for specific values of V_R.

[2] Although poly resistors are found in many analog circuits

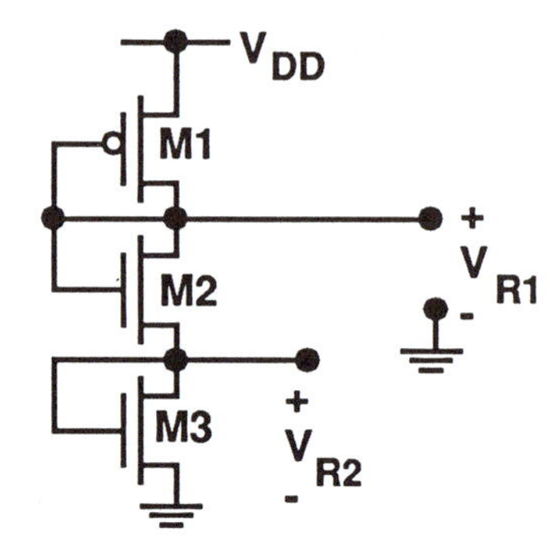

Figure 9.6: Dual Reference Voltage Circuit

Since both transistors are biased into active conducting states, we should be concerned with the power dissipation

$$P = \frac{\beta_n}{2}(V_R - V_{Tn})^2 V_{DD}. \tag{9.34}$$

Small values of V_R reduce the dissipated power corresponding to smaller DC currents.

Additional transistors can be added to the stack to generate multiple reference voltages. Figure 9.6 uses one pMOS transistor (M1) and two nMOS transistors (M2, M3) to produce two reference levels, V_{R1} and V_{R2}. To calculate the voltages, note that all of the MOSFETs are saturated. Ignoring channel length modulation gives the current as

$$\begin{aligned} I &= \frac{\beta_1}{2}(V_{DD} - V_{R1} - |V_{T1}|)^2 \\ &= \frac{\beta_2}{2}(V_{R1} - V_{R2} - V_{T2})^2 \\ &= \frac{\beta_1}{2}(V_{R2} - V_{T3})^2. \end{aligned} \tag{9.35}$$

Body-bias effects (in M2) will be ignored for simplicity. Combining the equations for M2 and M3 gives

$$V_{R2} = \frac{\sqrt{\beta_a}(V_{R1} - V_{T2}) + V_{T3}}{1 + \sqrt{\beta_a}}, \tag{9.36}$$

where we have defined

$$\beta_a = \sqrt{\frac{\beta_2}{\beta_3}} \quad . \tag{9.37}$$

Using the equation for the current I_D through M1 gives the other reference voltage as

$$V_{R1} = \frac{(V_{DD} - |V_{T1}|) + \frac{\sqrt{\beta_b}}{1+\sqrt{\beta_a}}(V_{T2} + V_{T3})}{(1 + \frac{\sqrt{\beta_b}}{1+\sqrt{\beta_a}})} \tag{9.38}$$

where we have defined

$$\beta_b = \frac{\beta_2}{\beta_1}. \tag{9.39}$$

Both V_{R1} and V_{R2} are set by the β-ratios; reversing the analysis gives the design equations for the circuit.

9.5 Current Sources

CMOS current sources can be designed using analogies of well-established bipolar techniques. Consider a saturated nMOS transistor as shown in Figure 9.7. The drain current

$$I_D = \frac{\beta_n}{2}(V_{GSn} - V_{Tn})^2(1 + \lambda V_{DS}) \tag{9.40}$$

is controlled by the voltages V_{GS} and V_{DS} so that the isolated transistor

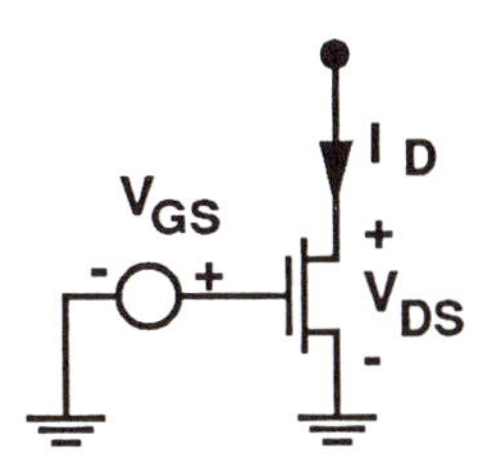

Figure 9.7: Saturated MOSFET Current Source

acts as a simple current source. The output resistance r_o seen looking into the drain is given by

$$r_o = \frac{1}{g_{ds}} \simeq \frac{1}{\lambda I_D}. \tag{9.41}$$

Although a single MOSFET can be used as a current source, it is sensitive to voltage variations and has a small r_o.

Improved performance can obtained using the circuit in Figure 9.8. MOSFET M2 acts as a common-gate amplifier to increase the output resistance. A dual-reference MOSFET stack provides the bias voltages. Analyzing the small-signal network gives

$$\begin{aligned} r_o &= r_{ds1} + r_{ds2} + g_{m2} r_{ds1} r_{ds2}(1+\eta_2) \\ &\simeq (g_{m2} r_{ds2}) r_{ds1} \end{aligned} \tag{9.42}$$

where we have assumed that $(g_{m2} r_{ds2}) >> 1$ in the second step. Qualitatively we see that adding M2 increases r_o by a factor equal to the small-signal gain of the amplifier.

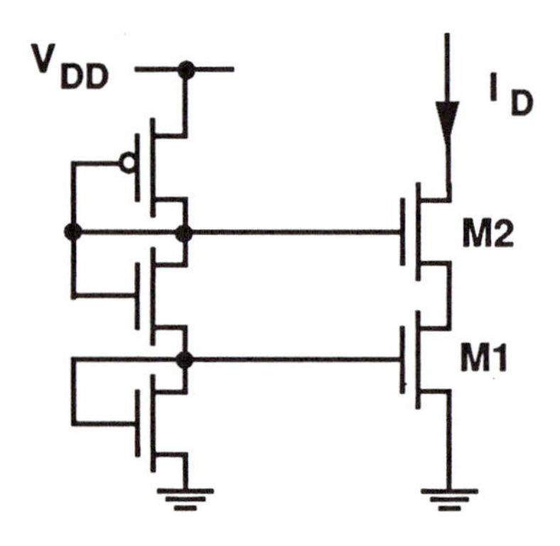

Figure 9.8: Improved Current Source

Current mirrors can be constructed using the basic circuit shown in Figure 9.9. I_R denotes the reference currrent which is controlled by external circuitry, while I_1 is the current used by the network. Note that the gate-source voltages of MR and M1 are equal:

$$V_{GS,R} = V_{GS,1} \equiv V_{GS}. \tag{9.43}$$

Assuming that both transistors are saturated gives

$$\frac{I_1}{I_R} = \frac{\beta_1 (V_{GS} - V_T)^2 (1 + \lambda V_{DS,1})}{\beta_R (V_{GS} - V_T)^2 (1 + \lambda V_{DS,R})}, \tag{9.44}$$

or,

$$\frac{I_1}{I_R} = \frac{(W/L)_1 (1 + \lambda V_{DS,1})}{(W/L)_R (1 + \lambda V_{DS,R})}. \tag{9.45}$$

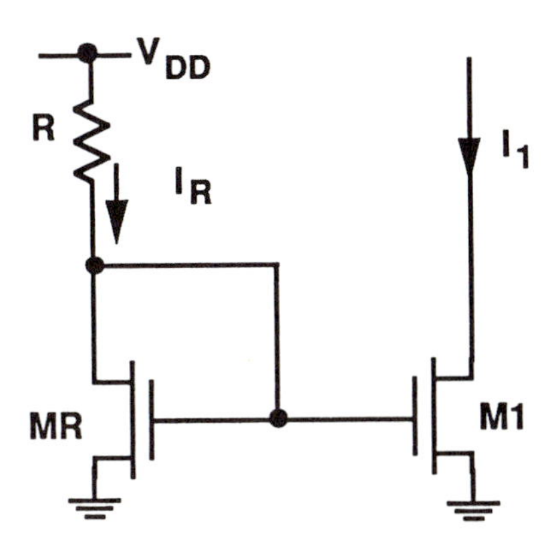

Figure 9.9: Basic Current Mirror

This shows that the current ratio is controlled by the device sizes and the drain-source voltages. In the special case where $V_{DS,1} = V_{DS,R}$, the equation reduces to

$$\frac{I_1}{I_R} = \frac{W_1 L_R}{W_R L_1}; \tag{9.46}$$

this, however, rarely occurs in practice due to normal bias voltages. If M1 and MR have the same aspect ratio, then

$$\frac{I_1}{I_R} = \frac{1 + \lambda V_{DS,1}}{1 + \lambda V_{DS,R}} \tag{9.47}$$

and the current ratio is established by the differences in drain-source voltages. The output resistance of the basic current mirror is given by

$$r_o \simeq \frac{1}{\lambda I_1} \tag{9.48}$$

as expected for a single MOSFET. Larger values of r_o can be achieved using the CMOS version of the Wilson current source shown in Figure 9.10. It can be shown that the output resistance is increased to

$$r_o = r_{ds3} + r_{ds2} A \tag{9.49}$$

where $A > 1$ is a gain factor due to negative current feedback.

Multiple current sources are created by adding MOSFETs as shown in Figure 9.11. Each current I_m ($m = 1, 2, \ldots$) is determined by

$$\frac{I_m}{I_R} = \frac{(W/L)_m (1 + \lambda V_{DS,1})}{(W/L)_R (1 + \lambda V_{DS,R})} \tag{9.50}$$

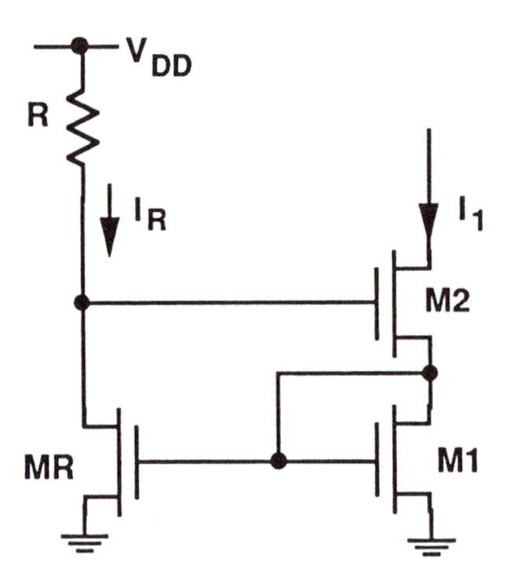

Figure 9.10: Wilson Current Source

which can be adjusted using aspect ratios. It is important to remember that this design equation is valid only if all of the current drivers remain saturated.

9.6 Differential Amplifier

The basic CMOS differential amplifier consists of source-coupled n-channel MOSFETs which are connected as shown in Figure 9.12. In this circuit, nMOSFETS Mn1 and Mn2 provide amplification, while pMOS transistors Mp1 and Mp2 are used in an active load configuration.

To analyze the large-signal properties of the circuit, we first note that KCL gives

$$I_{D1} + I_{D2} = I_{SS} \tag{9.51}$$

at the common source node. Assuming that Mn1 and Mn2 are saturated and ignoring channel-length modulation gives the currents as

$$\begin{aligned} I_{D1} &= \frac{\beta_{n1}}{2}(V_{GS1} - V_{T1})^2, \\ I_{D2} &= \frac{\beta_{n2}}{2}(V_{GS2} - V_{T2})^2. \end{aligned} \tag{9.52}$$

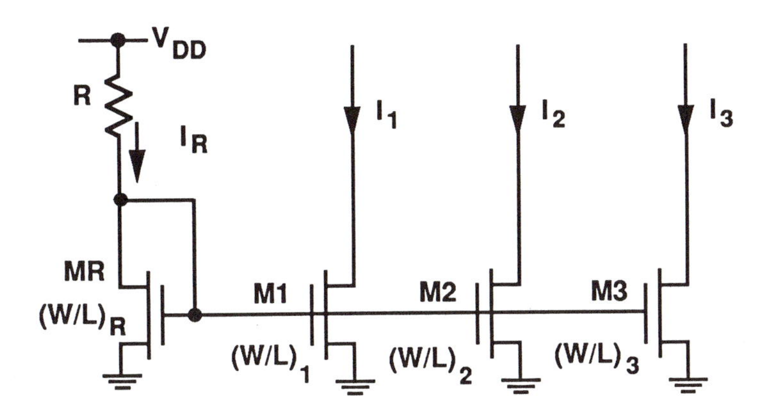

Figure 9.11: Multiple Current Sources

The difference voltage between the inputs is defined by

$$\begin{aligned} V_{ID} &= V_1 - V_2 \\ &= V_{GS1} - V_{GS2}, \end{aligned} \tag{9.53}$$

so that

$$V_{ID} = \sqrt{\frac{2I_{D1}}{\beta_{n1}}} - \sqrt{\frac{2I_{D2}}{\beta_{n2}}}. \tag{9.54}$$

This can be used to derive separate equations for the drain currents. First, we substitute $I_{D2} = I_{SS} - I_{D1}$ to obtain

$$V_{ID} = \sqrt{\frac{2I_{D1}}{\beta_{n1}}} - \sqrt{\frac{2(I_{SS} - I_{D1})}{\beta_{n2}}}. \tag{9.55}$$

This is a quadratic in I_{D1} with solutions of

$$I_{D1} = \frac{1}{2} I_{SS} \left[1 + \sqrt{\frac{\beta_n V_{ID}^2}{I_{SS}} - \frac{\beta_n^2 V_{ID}^4}{4 I_{SS}^2}} \right], \tag{9.56}$$

where we have assumed that the amplifier is symmetrical with $\beta_{n1} = \beta_{n2} = \beta_n$, and that $V_{ID} < \sqrt{2I_{SS}/\beta_n}$. Defining a normalized difference voltage

$$V_d = \sqrt{\frac{\beta_n}{I_{SS}}}\, V_{ID}, \tag{9.57}$$

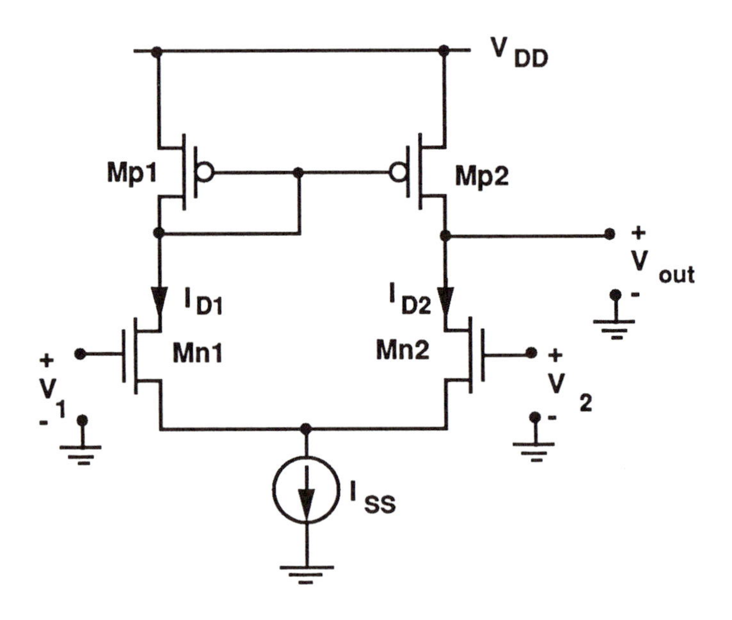

Figure 9.12: CMOS Differential Amplifier

this reduces to

$$I_{D1} = \frac{1}{2} I_{SS} \left[1 + \sqrt{V_d^2 - \frac{V_d^4}{4}} \right]. \quad (9.58)$$

I_{D2} is easily seen to be

$$I_{D2} = \frac{1}{2} I_{SS} \left[1 - \sqrt{V_d^2 - \frac{V_d^4}{4}} \right] \quad (9.59)$$

by direct application of KCL.

The behavior of the currents as functions of V_d is shown in Figure 9.13. Since the two must sum to I_{SS}, changing the difference voltage increases one current while decreasing the other. The minimum/maximum currents occur where

$$V_d = \mp\sqrt{2}, \quad (9.60)$$

corresponding to $V_{ID} = \mp\sqrt{2\beta_n / I_{SS}}$. The balanced behavior provides for the differential amplifying properties.

The small-signal gain of the differential amplifier is given by

$$A_v = g_{m1} r_o, \quad (9.61)$$

where

$$g_{m1} = \sqrt{\beta_n I_{SS}} \quad (9.62)$$

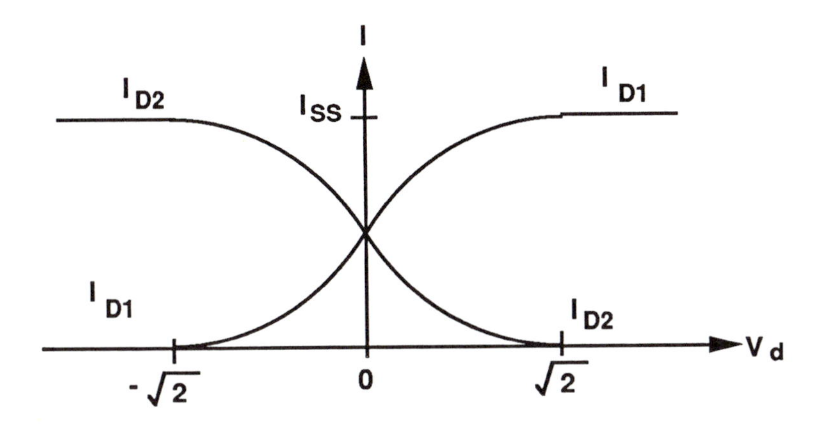

Figure 9.13: Currents in the Differential Amplifier

is the differential-mode transconductance, and

$$r_o = \frac{1}{g_{on2} + g_{op2}} \tag{9.63}$$

is the output resistance.

Figure 9.14 shows a complementary arrangement for the CMOS differential amplifier. The roles of the nMOS and pMOS transistors have been interchanged, so that Mp1 and Mp2 provide amplification. The analysis is identical, and gives the same general results.

9.7 A CMOS Operational Amplifier

A simple 2-stage CMOS operational amplifier is shown in Figure 9.15. The circuit consists of a differential pair (Mp1, Mp2) which is cascaded into a common-source nMOS amplifier (Mn3). MOSFETs Mp3, Mp4, and Mp5 make up the current mirror used to bias the circuit. The open-loop voltage gain is given by

$$a_o = a_1 a_2 \tag{9.64}$$

where

$$\begin{aligned} a_1 &= g_{mp1}(r_{on2}||r_{op2}), \\ a_2 &= -g_{mn3}(r_{on3}||r_{op3}) \end{aligned} \tag{9.65}$$

are the individual gains for stage 1 and stage 2, respectively.

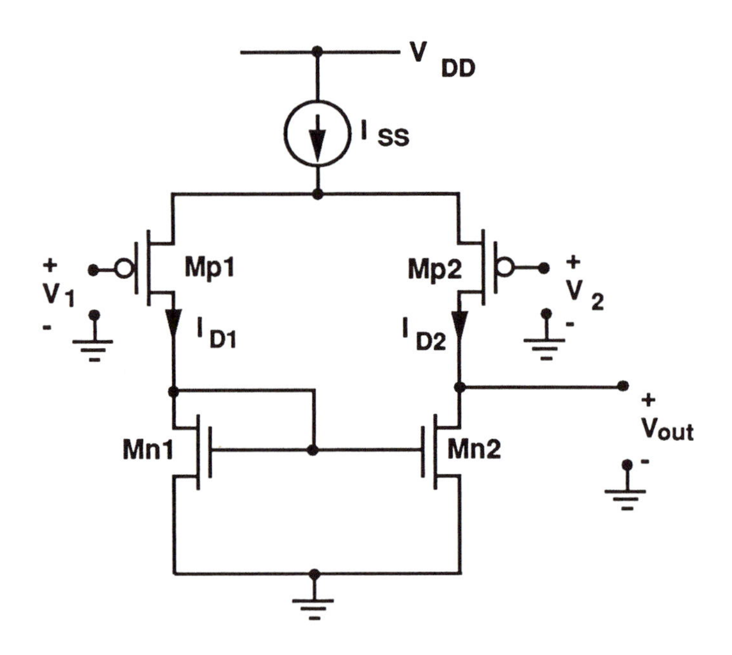

Figure 9.14: Alternate CMOS Differential Amplifier

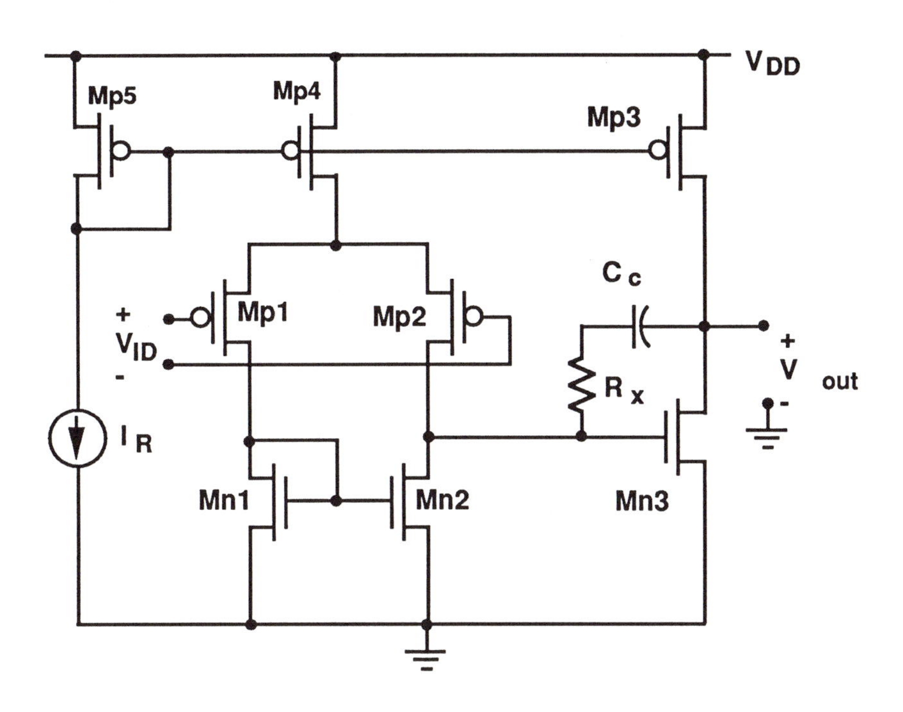

Figure 9.15: 2-Stage CMOS Operational Amplifier

Frequency compensation is achieved using the RC combination of R_x and C_c as shown. We define output resistances

$$\begin{aligned} r_1 &= (r_{on2}||r_{op2}), \\ r_2 &= (r_{on3}||r_{op3}), \end{aligned} \tag{9.66}$$

and examine the small-signal equivalent network shown in Figure 9.16. For

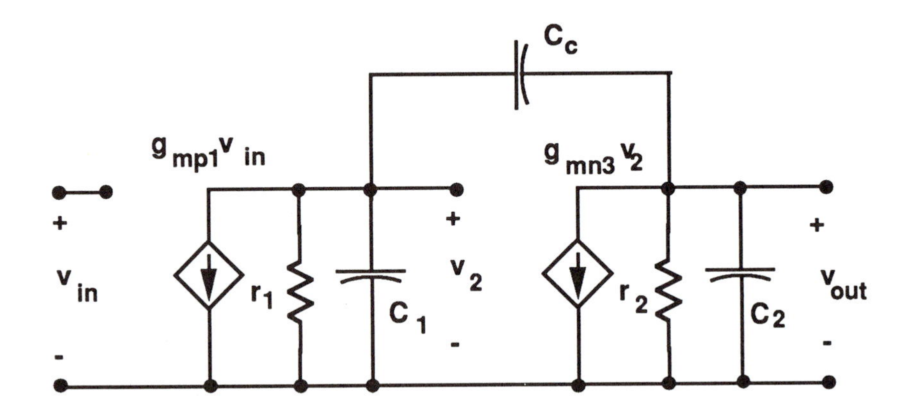

Figure 9.16: Op amp small-signal equivalent circuit

the case where $R_x = 0$, this circuit has two-poles at locations

$$\begin{aligned} p_1 &= \frac{-1}{(1 + g_{mn3}r_2)r_1C_c}, \\ p_2 &= \frac{-g_{mn3}C_c}{C_1C_2 + C_c(C_1 + C_2)}, \end{aligned} \tag{9.67}$$

and a single zero at

$$z_1 = +\frac{g_{mn3}}{C_c}. \tag{9.68}$$

The presence of the RHP zero is significant because g_m in a MOSFET is small enough where it may cause stability problems[3]. Including R_x shifts the zero to a value

$$z_1 \rightarrow \frac{g_{mn3}}{C_c(\frac{1}{g_{mn3}} - R_x)} \tag{9.69}$$

so that choosing $R_x \simeq (1/g_{mn3})$ shifts the zero to a high frequency. A more detailed discussion of compensation may be found in the references.

[3]This problem does not occur in a bipolar op amp due to much larger transconductance values.

9.8 Summary

The discussions in this chapter have been directed towards amplifiers and related circuits. While these provide the basis for analog design in CMOS, the field consists of several major application groups such as

- Switched capacitor networks
- A/D and D/A converters

and others. Earlier techniques developed in nMOS can be directly modified for application in CMOS. In general, analog MOS networks rely on the properties such as

- High input impedances
- Capacitive storage nodes
- Pass transistor switches

to achieve the desired results. The resulting circuits are also attractive to digital designers because they can be directly interfaced without level conversion problems.

9.9 References

All of the books below contain detailed analyses of analog CMOS circuits.

[1] P.E. Allen and D. R. Holberg, **CMOS Analog Circuit Design**, Holt, Rinehart and Winston, New York, 1987.

[2] P.E. Allen and E Sanchez-Sinencio, **Switched Capacitor Networks**, Van Nostrand Reinhold, New York, 1984.

[3] R.L. Geiger, P.E. Allen, and N.R. Strader, **VLSI Design Techniques for Analog and Digital Circuits**, McGraw-Hill, New York, 1990.

[4] R. Gregorian and G. Temes, **Analog MOS Integrated Circuits for Signal Processing**, Wiley, New York, 1986.

[5] P. R. Grey and R.G. Meyer, **Analysis and Design of Analog Integrated Circuits**, 2nd ed., Wiley, New York, 1984.

[6] Y. Tsividis and P. Antognetti (eds.), **Design of MOS VLSI Circuits for Telecommunications**, Prentice-Hall, Englewood-Cliffs, NJ, 1985.

Chapter 10

BiCMOS Circuits

BiCMOS circuits consist of both bipolar junction transistors (BJTs) and MOSFETs on a single substrate. The existence of parasitic bipolar transistors in CMOS structures is well know; for example, latch-up and subthreshold current flow are commonly analyzed using bipolar models as discussed in Chapter 6. BiCMOS technology is different from the classical analysis in that the process flow is specifically designed to allow for both bipolar and MOS transistors. Combining technologies in this manner allows for circuits which have the "best of both worlds": fast switching due to bipolar transistors and low-power/high integration density of CMOS. This evolving field has generated much excitement in recent years.

Many different BiCMOS logic circuits have been discussed in the literature. The most common approach is to use MOSFETs to implement the logic and bipolar transistors to provide a fast, high-current output driver stage. This structuring can be seen in the generalized BiCMOS logic gate shown in Figure 10.1. The inputs are connected to MOSFETs through the logic blocks shown as F and $\overline{F}$ in the drawing. The logic blocks are constructed using transistor switch arrays. Bipolar transistors Q1 and Q2 are connected in a non-inverting stack to drive the output capacitance. When $F = 1$, Q1 is turned on and provides charging current to C_{out}; the circuit specifics determine V_{OH} for this case. If $F = 0$ so that $\overline{F} = 1$, Q2 obtains bias from the output node and conducts to ground. This drains charge off of C_{out} and defines the value of V_{OL}. Two additional impedance devices (Z_1 and Z_2 in the drawing) are included to provide a path to remove base charge when the bipolar transistors are switching off. These are used to increase the switching speed. Either passive resistors or active loads may be used for this purpose.

The intent of this chapter is to introduce the basic properties of BiCMOS

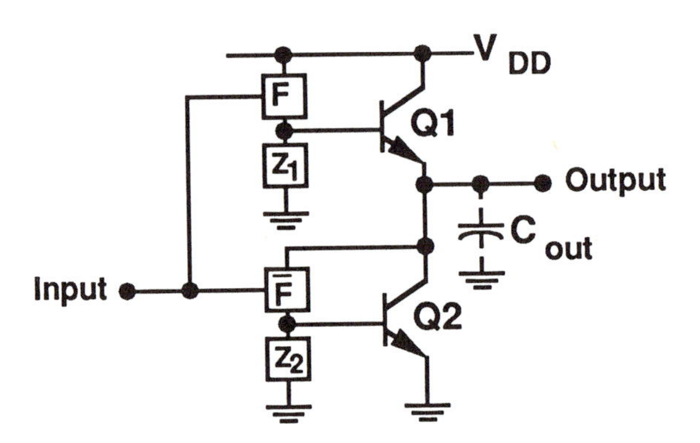

Figure 10.1: Generalized BiCMOS Logic Gate

logic circuits. Bipolar junction transistors are characterized first, and then specific circuits are presented as examples. Alvarez [2] provides thorough discussions of many important topics in BiCMOS.

10.1 Bipolar Junction Transistors

Bipolar junction transistors are three terminal devices formed by layering alternate n-type and p-type regions. This gives two basic BJT types, namely, the **npn** and the **pnp**. Although the current flow is due to both positive and negative charges (hence, **bipolar**), npn transistors are dominated by electron currents, while pnp transistors depend primarily on holes. The npn transistor will be analyzed in this section; the behavior of a pnp is the complement of the npn.

10.1.1 Structure and Operation

Figure 9.1 shows the cross-section for an integrated npn BJT. The terminals are termed the **emitter**, the **base**, and the **collector**, corresponding to the respective n^+-p-n sections. Current flow through the npn transistor relies on the motion of electrons from the emitter n^+-region to the collector n-type layer as controlled by the p-type base.

The three voltages are V_{BE}, V_{BC} and V_{CE}, two of which are independent. For the purposes of analysis, V_{BE} and V_{BC} are chosen as the variables. Similarly, the collector current I_C and the emitter current I_E are assumed to be the independent quantities such that the base current is given by

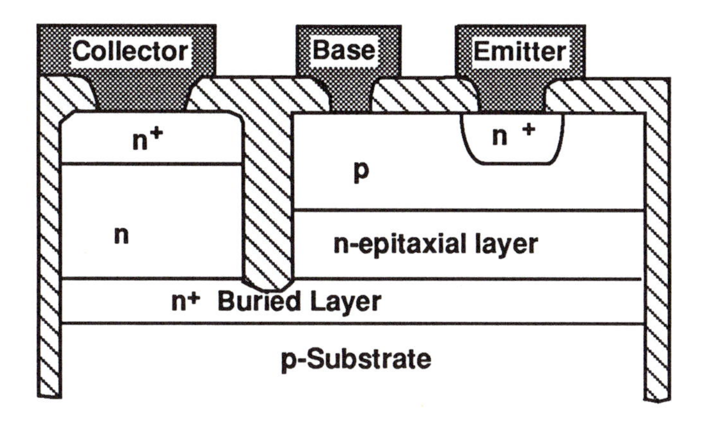

Figure 10.2: Bipolar Junction Transistor

$I_B = I_E - I_C$. Figure 10.2 shows the npn BJT circuit symbol with the definitions of positive voltages and currents.

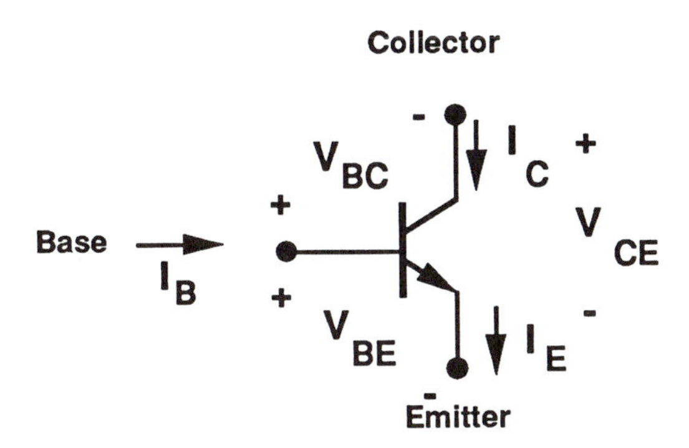

Figure 10.3: Circuit Symbol for the npn Transistor

The current flow through the npn transistor is due to the existence of the two "back-to-back" pn junctions (base-emitter and base-collector) which are in close proximity to one another. The p-type base acts as the intermediate layer connecting the emitter and collector n-type regions.

Consider the case of **forward-active** bias where the base-emitter junction is forward biased and the base-collector junction is reverse-biased. In terms of the usual sign convention, this says that $V_{BE} > 0$ while $V_{BC} < 0$.

Transistor action is understood by tracking the electron flow. Electrons are injected into the base from the emitter. Once in the base, they are minority carriers and diffuse towards the collector. Although some electrons are lost by recombination with holes, most reach the base-collector depletion region and are swept into the collector by the depletion electric field.

The Ebers-Moll equations can be used to model the DC characteristics described above. Applying diode characteristics to each junction gives currents of

$$\begin{aligned} I_{DE} &= I_{ES}(e^{V_{BE}/\phi_T} - 1) \\ I_{DC} &= I_{CS}(e^{V_{BC}/\phi_T} - 1) \end{aligned} \tag{10.1}$$

for the diode current I_D in the emitter and the collector, respectively. In this equation set, $\phi_T = (kT/q)$ is the thermal voltage, and I_{ES} and I_{CS} are the respective saturation currents. Transistor action is included by coupling the diode currents in the form

$$\begin{aligned} I_C &= \alpha_F I_{ES}(e^{V_{BE}/\phi_T} - 1) - I_{CS}(e^{V_{BC}/\phi_T} - 1), \\ I_E &= I_{ES}(e^{V_{BE}/\phi_T} - 1) - \alpha_R I_{CS}(e^{V_{BC}/\phi_T} - 1), \end{aligned} \tag{10.2}$$

where $\alpha_F < 1$ and $\alpha_R < 1$ provide the coupling and are termed the forward and reverse α, respectively. The corresponding Ebers-Moll equivalent circuit is shown in Figure 10.4.

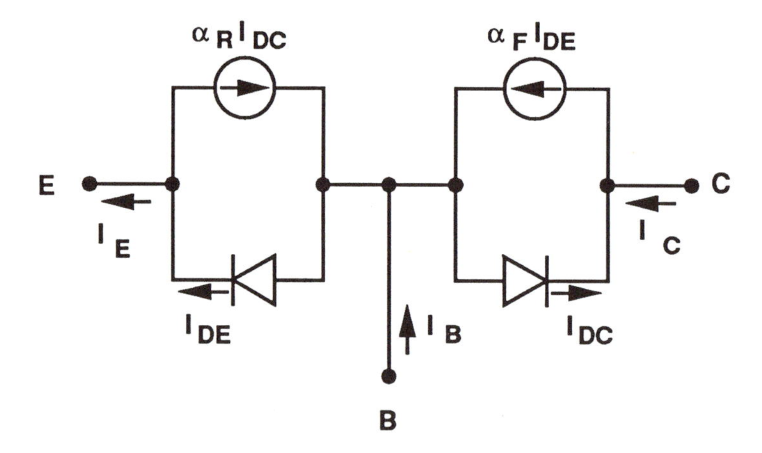

Figure 10.4: Ebers-Moll Equivalent Circuit

The transistor saturation current I_S is defined by

$$I_S = \alpha_F I_{ES} = \alpha_R I_{CS}, \tag{10.3}$$

where the last step follows from reciprocity. I_S is calculated from

$$I_S = \frac{qAD_n n_i^2}{N_B} \tag{10.4}$$

where A is the area of the emitter-to-base junction, D_n is the electron diffusion coefficient in the base, and N_B is the number of base dopants per unit area and is known as the Gummel number [13]. In terms of the base acceptor doping profile $N_a(x)$,

$$N_B = \int_0^{x_B} N_a(x)dx, \tag{10.5}$$

where the base quasi-neutral region extends from $x = 0$ to $x = x_B$.

Forward-Active Bias

Current flow in the BJT can be understood by examining the case of forward-active bias with $V_{BE} > 0$ and $V_{BC} < 0$. The Ebers-Moll equations then reduce to

$$\begin{aligned} I_C &\simeq I_S e^{V_{BE}/\phi_T}, \\ I_E &\simeq \frac{I_S}{\alpha_F} e^{V_{BE}/\phi_T}, \end{aligned} \tag{10.6}$$

where only the dominant terms have been kept. The exponential transfer characteristics contained in $I_C(V_{BE})$ are shown in Figure 10.5. The forward-alpha is given by

$$\alpha_F = \frac{I_C}{I_E} < 1, \tag{10.7}$$

and is the common-base current gain in forward-active bias. Including the neglected terms gives the complete expression

$$I_C = \alpha_F I_E + I_{CO} \tag{10.8}$$

where

$$I_{CO} = I_{CS}(1 - \alpha_F \alpha_R) \tag{10.9}$$

is the leakage current. The common-emitter current gain is specified by

$$\beta_F = \frac{\alpha_F}{1 - \alpha_F} \tag{10.10}$$

such that

$$I_C = \beta_F I_B + (\beta_F + 1)I_{CO}. \tag{10.11}$$

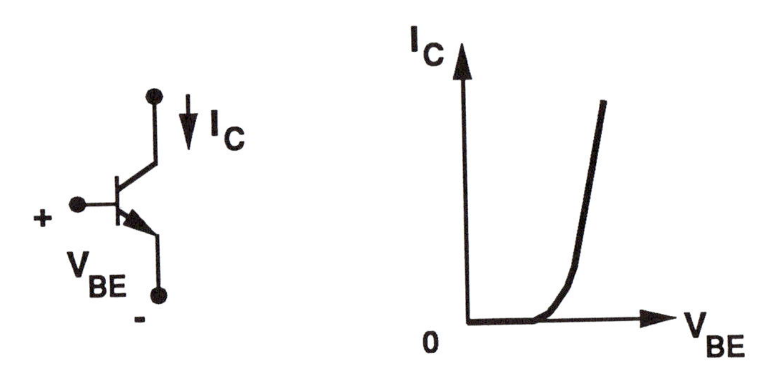

Figure 10.5: BJT Transfer Characteristics

If leakage is ignored, then $\beta_F = (I_C/I_B)$.

The current gain α_F can be written as the product

$$\alpha_F = \alpha_T \gamma, \tag{10.12}$$

where $\alpha_T < 1$ is the **base transport factor** and $\gamma < 1$ is the **emitter injection efficiency**. Both are determined by the doping and geometrical characteristics. In an idealized **prototype transistor** where the doping densities are constants with values N_{dE}, N_a, and N_{dC} for the emitter, base, and collector, respectively, the parameters assume the maximum values [19]

$$\begin{aligned} \alpha_T &\simeq 1 - \frac{x_B^2}{2L_n^2}, \\ \gamma &\simeq \frac{1}{1+\delta}, \end{aligned} \tag{10.13}$$

where L_n is the electron diffusion length in the base. The parameter δ is called the **base defect factor**, and is given by

$$\delta = \frac{N_a x_B D_{pE}}{N_{dE} x_E D_n}, \tag{10.14}$$

with x_E being the length of the neutral emitter region, and D_{pE} the hole diffusion coefficient in the emitter. Including recombination currents from the base-emitter depletion region shows that $\alpha_F(I_E)$ with the behavior illustrated in Figure 10.6.

Constructing a high-gain transistor requires that both α_T and γ be close to unity. This is accomplished by the following rules below as a minimum requirement:

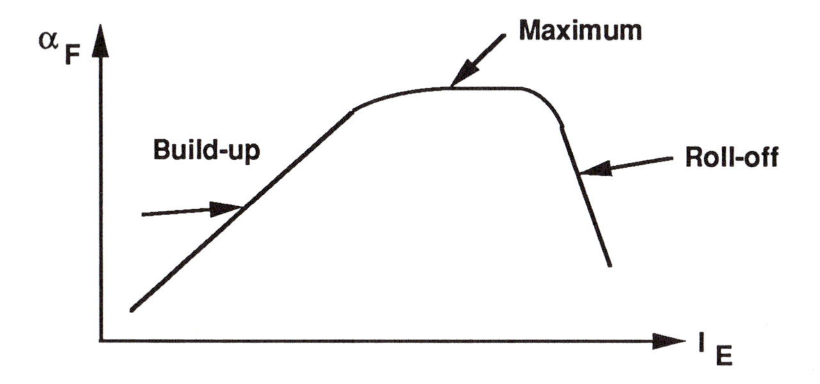

Figure 10.6: Current Gain

- Small base width: $x_B < 1$ [μm]
- High emitter doping: $N_{dE} >> N_a$.

The latter constraint has the effect of simultaneously reducing the reverse current gain α_R. Attaining acceptable transistor performance in BiCMOS requires that the process flow be structured to accommodate both MOS and bipolar devices. The base width value x_B is particularly critical to achieving reasonable values for the bipolar transistor gain.

Reverse-Active Bias

The state of **reverse-active** bias is defined by the bias voltages satisfying $V_{BE} < 0$ and $V_{BC} > 0$. In this case, the emitter and collector exchange roles, and the appropriate current gains are the reverse quantities α_R and β_R. Reverse operation is inefficient since the lightly doped collector cannot supply a large electron flow. Operation in reverse-active bias is generally avoided for this reason.

Saturation

If both junction are forward biased with $V_{BE} > 0$ and $V_{BC} > 0$, the transistor is in a state of **saturation**; this should not be confused with the MOSFET saturation mode of operation, as the two are distinct. The current through a saturated BJT is controlled entirely by the external circuitry; in other words, the device has no control over the current flow.

A saturated bipolar transistor is usually characterized with terminal voltages $V_{CE,sat} \simeq 0.3$ [V] and $V_{BE,sat} \simeq 0.8$ [V] (for room temperature silicon). Saturated bipolar logic can be slow because the transistor stores excess base charge which must be removed to stop conduction. Left to itself, the charge must diffuse through the structure before the device can turn off. There are two basic approaches to solving this problem. The first, and most popular, is to use Schottky-clamped transistors which keep the transistor out of saturation. The second approach is found in BiCMOS and provides additional circuitry to remove the base charge when needed. This will be discussed in the following sections.

Cutoff

The final operational mode which must be examined is that of **cutoff** where both junctions are reverse biased with $V_{BE} < 0$ and $V_{BC} < 0$; only leakage currents flow in this case. Appropriate Ebers-Moll models can be obtained by simply reducing the exponentials. Simplistic approaches use open circuits for both junctions, while more accurate models include the normal leakage contributions.

10.1.2 Bipolar Transistor Capacitances

Bipolar junction transistors inherently possess parasitic depletion capacitances which affect the digital switching characteristics. In large- signal models these are the base-emitter capacitance C_{BE}, the base- collector capacitance C_{BC}, and the collector-substrate capacitance C_{CS}. In general, each capacitance is characterized by a zero-bias value

$$C_0 = \frac{\epsilon_{Si} A}{W_0}, \tag{10.15}$$

with A as the junction area and W_0 as the zero-bias depletion width. Depletion capacitance is nonlinear, and varies with a reverse bias voltage V_R according to

$$C(V_R) = \frac{C_{j0}}{\left(1 + V_R/\phi_o\right)^m} \tag{10.16}$$

with m the grading coefficient. The simplest junction profiles are $m = (1/2)$ (for a step profile) and $m = (1/3)$ (for a linearly graded profile). The capacitances used for circuit design are shown in Figure 10.5.

Bipolar junction transistors also exhibit **diffusion capacitance** which is important in a forward-biased pn junction. This parasitic is due to the fact that diffusing charge carriers of opposite polarity give capacitance by definition. Diffusion capacitance is voltage dependent with the general form

$$C' = C_o e^{V/\phi_T}, \tag{10.17}$$

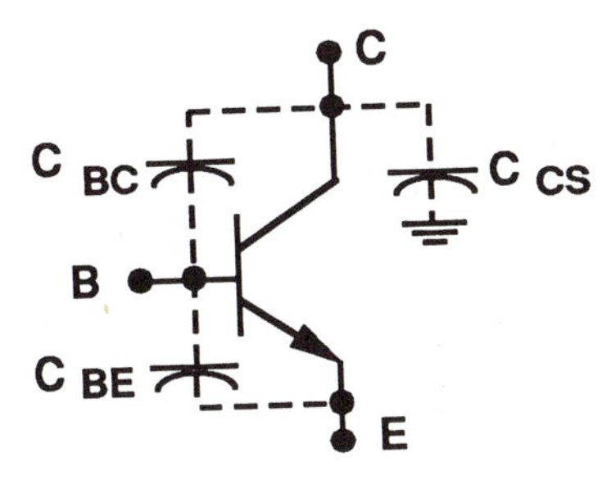

Figure 10.7: Bipolar Transistor Capacitances

where V is the forward voltage on the junction, and the zero-bias value is C_o; C_o is proportional to the junction area. Diffusion capacitance and junction depletion capacitance are in parallel, and bias voltage determines which type dominates. In forward bias, the diffusion term is important and cannot be ignored. When the junction is in reverse-bias, the depletion capacitance is the important contribution.

10.2 BiCMOS Technology

The development of high-performance BiCMOS technology has received much attention in recent years. Figure 10.8 shows a cross-sectional view of a typical oxide-isolated chip with n- and p-channel MOSFETs and npn BJTs; more advanced process flows also provide for high-gain pnp bipolar transistors [14]

Important aspects of a BiCMOS process include

- Device characteristics
- Integration density
- Interconnect levels
- Compatibility with existing CMOS designs

such that both the bipolar and the MOS circuits provide high level performance.

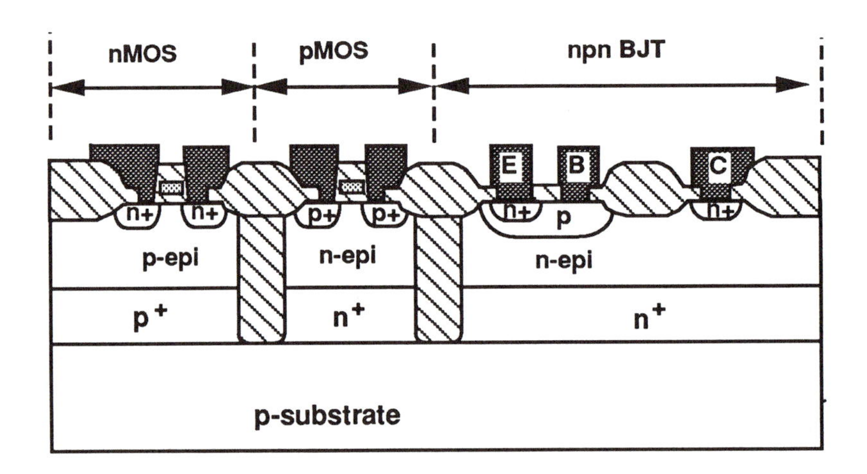

Figure 10.8: BiCMOS Cross-Sectional View

10.3 BiCMOS Inverter

Digital BiCMOS circuits can be designed using standard CMOS logic blocks cascaded into bipolar output stages. This approach combines the area-efficient and low-power characteristics of CMOS layouts with the high- current drive capabilities of bipolar transistors. A BiCMOS inverter forms the basic circuit for the development of a generalized logic family. This section examines typical inverter circuits to illustrate the concepts involved in BiCMOS logic design.

Figure 10.6 shows one way to construct a basic BiCMOS inverter. The MOSFETs are used for logic and control, while the BJTs are connected as line drivers. Transistors Mn and Mp form a split-CMOS inverter sub-circuit which provides the signals to drive output transistors Q1 and Q2. MOSFETs M1 and M2 act as pull-down devices to aid in the switching.

The operation is straightforward. If V_{in} is high, Q1 and M2 are forced into cutoff. Mn is biased into the active region allowing C_{out} to discharge. This supplies base current to Q2, which in turn gives a conducting path to ground. On the other hand, a low input voltage turns on both Mp and M2. The current through Mp acts to bias Q1 into a conducting state; since M2 is conducting, $V_{BE,2}$ is low and Q2 is forced into cutoff. The output node thus has a conducting path to the power supply through Q1, giving a high output voltage V_{out}. From this discussion it is seen that Mn and Mp form the logic, while M1 and M2 act as active pull-down devices (Z_1 and Z_2 in the generic structure shown in Figure 9.1).

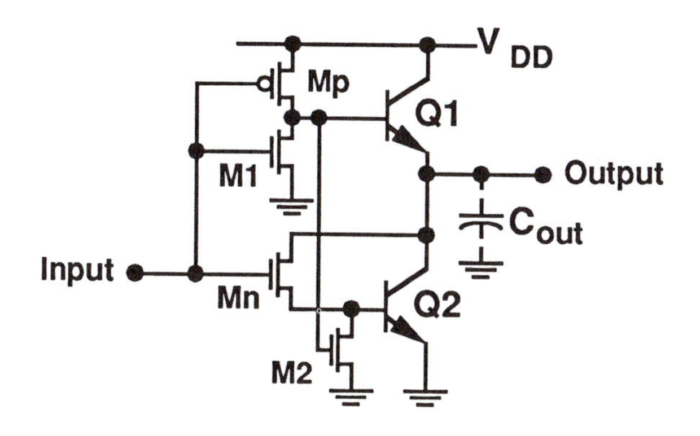

Figure 10.9: BiCMOS Inverter

10.3.1 DC Characteristics

Simple bipolar transistor and MOSFET models may be used to analyze the circuit operation [2]. To perform the DC analysis we examine the bipolar output transistors. We see that both V_{OH} and V_{OL} are controlled entirely by the conducting properties of Q1 and Q2. When Q1 is conducting, Q2 is in cutoff, giving

$$V_{OH} = V_{DD} - V_{BE1} \tag{10.18}$$

as the output high voltage. On the other hand, a high input voltage forces Q1 into cutoff. Q2 is biased on, but V_{out} is clamped to base so that

$$V_{OL} = V_{BE2} \tag{10.19}$$

is the output low voltage. BiCMOS logic of this type does not give a full-rail output swing. Additional level conversion circuits may be necessary to drive CMOS inputs.

10.3.2 Transient Switching Characteristics

The speed advantage of BiCMOS over standard CMOS arises from the large drive currents and fast switching properties which characterize the bipolar output transistors. Milliampere current levels allow fast charging and discharging of C_{out}, while the exponential transfer characteristic $I_C \simeq I_S \exp(V_{BE}/\phi_T)$ indicates a much greater sensitivity than is possible with square-law MOSFETs. A detailed analysis of this circuit has been presented in the literature [6]. For the present discussion, we will use a simplified analysis to illustrate the behavior.

Pull-down MOSFETs M1 and M2 are included to remove charge from the base electrodes of Q1 and Q2 when a switching event is initiated. For example, when V_{in} is switched from a low value to a high value, M1 turns on, allowing stored base charge in Q1 to drain to ground. This aids in rapid shutdown of the output transistor. Similarly, M2 is used to help Q2 change from a conducting to a non-conducting state by providing a discharge path to ground. Because of this type of action, we will assume that only a single output transistor is important in determining the rise and fall times here.

Consider first the case where V_{in} is switched from a high to a low value. This is modeled by the subcircuit shown in Figure 10.10. Capacitor C1

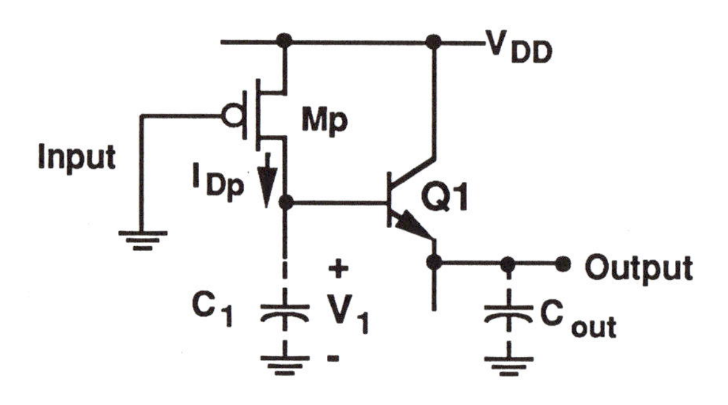

Figure 10.10: Charging Subcircuit

charges through Mp to switch Q1 into a conducting mode. This requires a time

$$t_1 \simeq \frac{C_1}{I_{Dp}} V_{BE(on)}, \tag{10.20}$$

where $V_{BE(on)} \simeq 0.7$ [V] is the turn-on voltage for Q1, and

$$I_{Dp} = \frac{\beta_p}{2}(V_{DD} - V_{OL} - |V_{Tp}|)^2 \tag{10.21}$$

is the pFET current with $V_{in} = V_{OL}$ applied at the gate input. Once Q1 is conducting, the output capacitance C_{out} charges according to

$$I_E = C_{out}\frac{dV_{out}}{dt}$$

$$\simeq \frac{I_S}{\alpha_F} e^{V_{BE}/\phi_T}, \tag{10.22}$$

where

$$V_{BE}(t) = V_1(t) - V_{out}(t) \tag{10.23}$$

with $V_1(t)$ determined by the charging rate of C_1. To approximate the response, assume that I_E is a constant; this is reasonable once Q1 is conducting. Direct integration gives a charge time of

$$t_2 \simeq \frac{C_{out}}{I_E} V_\ell, \tag{10.24}$$

where the logic swing is

$$V_\ell \simeq (V_{DD} - 2V_{BE}). \tag{10.25}$$

The low-to-high time is then estimated to be

$$\begin{aligned} t_{LH} &= t_1 + t_2 \\ &\simeq \frac{C_1}{I_{Dp}} V_{BE(on)} + \frac{C_{out}}{I_E} V_\ell. \end{aligned} \tag{10.26}$$

The initial charge time t_1 is small because C_1 is small and only needs to be charged to $V_{BE(on)}$ to initiate the output response. The bipolar charge time t_2 is the critical quantity. Since the emitter current is generally on the order of milliamperes, this equation shows that large values of C_{out} can be easily driven by the bipolar output.

Large values of I_E indicate the possibility that Q1 will enter saturation. If this occurs, then the output charging transient is limited by the parasitic collector resistance. To account for this situation we write the low-to-high time as

$$t_{LH} = t_1 + t_{sat} + t_3, \tag{10.27}$$

where t_{sat} is the time to saturate Q1, and t_3 the time needed to charge C_{out} when Q1 is conducting in the saturation mode. Denoting the parasitic collector resistance by r_C, KCL gives

$$C_{out}\frac{dV_{out}}{dt} = \frac{V_{DD} - V_{CE(sat)} - V_{out}}{r_C}. \tag{10.28}$$

Solving with the initial condition $V_{out}(0) = V_{sat}$ as the output voltage at the start of saturation gives a charging time of

$$t_3 \simeq \tau_C \ln\left[\frac{V_{DD} - V_{CE(sat)} - V_{sat}}{V_{DD} - V_{CE(sat)} - V_{OH}}\right], \tag{10.29}$$

where we have defined

$$\tau_C = r_C C_{out} \tag{10.30}$$

as the time constant. Collector resistance values on the order of $r_c \sim 100$ [Ω] are typical. These equations are sufficient for first order time estimates; more detailed calculations can be found in the literature.

Discharging can be analyzed by the subcircuit shown in Figure 10.11. In this case we assume that V_{out} is initially at V_{OH}. Switching V_{in} from a low voltage to a high voltage biases Mn into a conducting state with

$$\begin{aligned} V_{GSn} &= V_{in} - V_{BE2}, \\ V_{DSn} &= V_{out} - V_{BE2}. \end{aligned} \tag{10.31}$$

Capacitor C_2 charges through Mn and turns on Q2 in a time

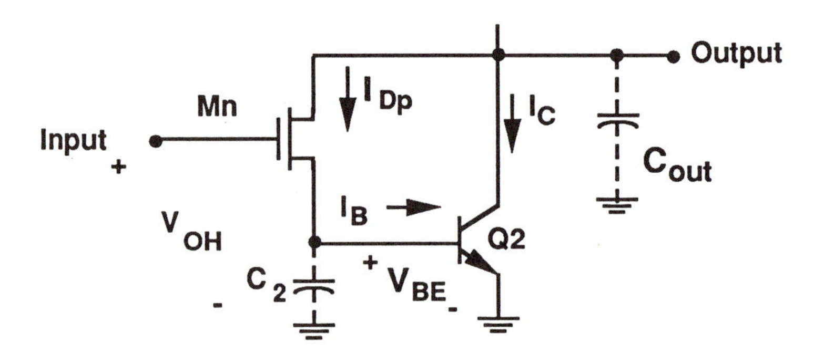

Figure 10.11: Discharging Subcircuit

$$t_4 \simeq \frac{C_2}{I_{Dn}} V_{BE(on)} \tag{10.32}$$

where the nMOS current is given by

$$I_{Dn} = \frac{\beta_n}{2}(V_{OH} - V_{Tn})^2 \tag{10.33}$$

since the MOSFET is initially saturated. Once Q2 is conducting, C_{out} discharges according to

$$I_{Dn} + I_C = -C_{out}\frac{dV_{out}}{dt}. \tag{10.34}$$

Noting that $I_{Dn} = I_B = (I_C/\beta_F)$ and assuming that $I_C \simeq$ constant for simplicity, gives us the time interval

$$t_5 \simeq \frac{C_{out}}{I_C(1 + 1/\beta_F)} V_\ell \tag{10.35}$$

needed to discharge the output. The high-to-low time is then

$$\begin{aligned} t_{HL} &\simeq t_4 + t_5 \\ &\simeq \frac{C_2}{I_{Dn}} V_{BE(on)} + \frac{C_{out}}{I_C(1 + 1/\beta_F)} V_\ell \, . \end{aligned} \tag{10.36}$$

Since $\beta_F >> 1$, $t_{HL} \approx t_{LH}$ as expected from the symmetry of the output circuit. The analysis must be modified if Q2 enters saturation.

10.4 Comparison of CMOS and BiCMOS Performance

BiCMOS logic gates are used to drive large output capacitances at high switching speeds. As we have seen in Chapter 8, CMOS circuits can be designed to accomplish the same objective. It is therefore instructive to compare the performances of CMOS and BiCMOS to place each into proper perspective.

Consider the inverter circuits shown in Figure 10.12. For both circuits, the intrinsic gate capacitance due to the transistors is denoted by C_i, and C_L denotes the external load capacitance. Analyzing the propagation delay time for the CMOS inverter gives the linear approximation [11]

$$t_{P,CMOS} \simeq t_0 + \frac{V_{th}}{I_D}(C_i + C_L) \tag{10.37}$$

where V_{th} is the inverter threshold voltage. The contribution t_0 is due solely to internal circuit capacitances. A similar analysis on the BiCMOS circuit yields an equation of the form

$$t_{P,BiCMOS} \simeq t_1 + \frac{V_{th}}{\beta I_D}(C_i + C_L), \tag{10.38}$$

where $t_1 > t_0$ because the internal circuit capacitances include the bipolar contributions.

Figure 10.13 shows a typical plot of propagation delay t_P versus load capacitance C_L [14] , and illustrates the general trends which are found when comparing the two technologies. The "CMOS" curve starts at a

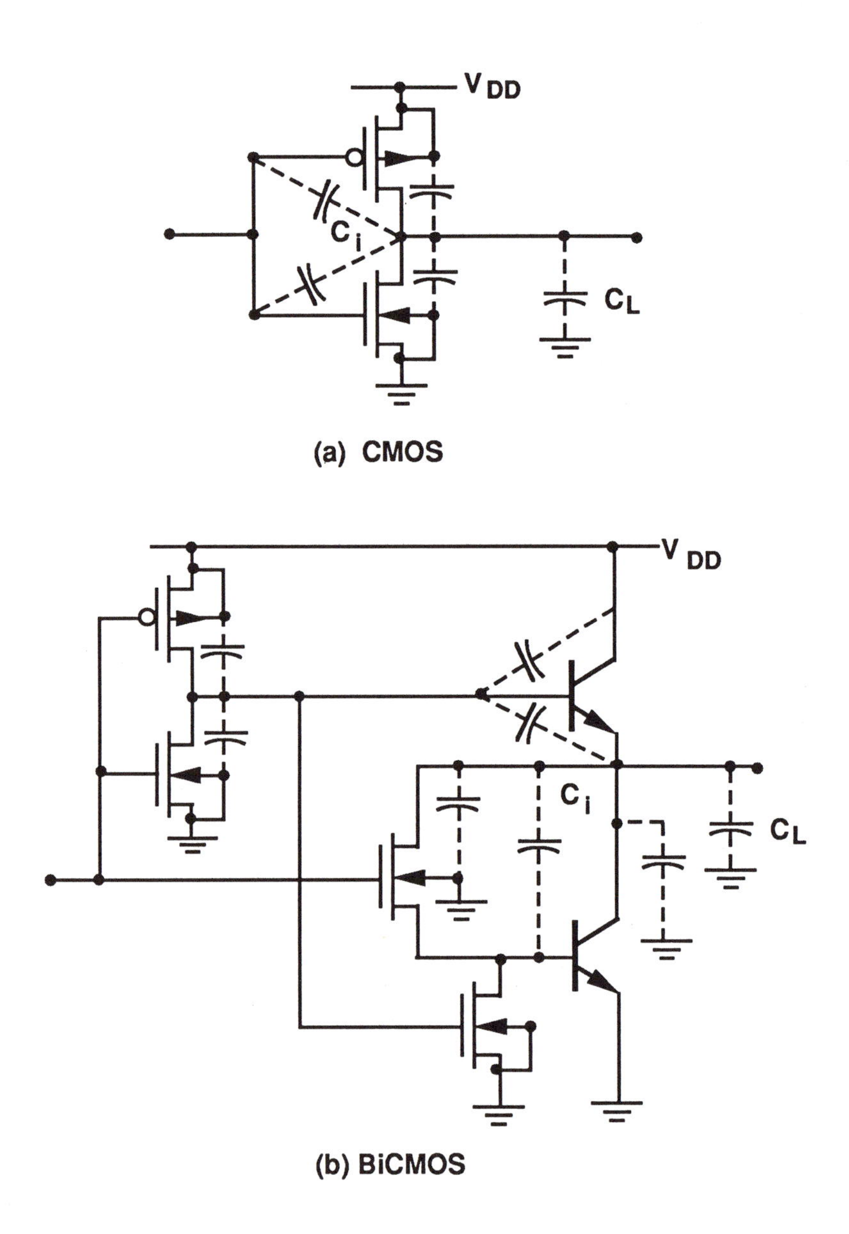

Figure 10.12: Comparison of CMOS and BiCMOS Inverters

lower intrinsic time t_0, and has a slope which is inversely proportional to I_D. The BiCMOS curve, on the other hand, starts at a larger value t_1, but has a reduced slope due to the $(1/\beta)$ factor from the bipolar transistors. For large values of load capacitance (C_L greater than about 0.5 [pF]), the bipolar drivers provide the high currents needed for rapid charging and discharging. However, when C_L is small, the CMOS-only circuit performs better. This is due to the fact that the CMOS circuit has fewer transistors and thus has a smaller level of intrinsic device capacitance. Although the cross-over point is technology-dependent, it typically occurs around a load capacitance value of $C_L \sim 100 - 500$ [fF].

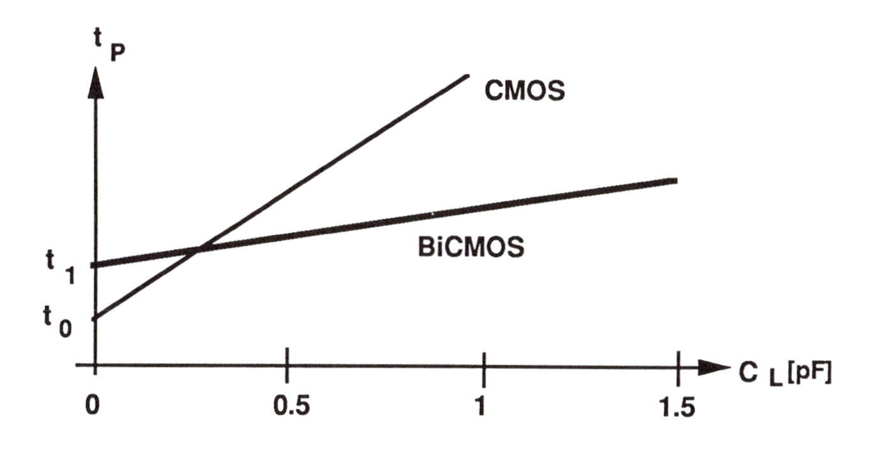

Figure 10.13: Typical CMOS and BiCMOS Switching Times

This comparison must be interpreted with care. In a basic CMOS technology, a single inverter would never be used to drive a large capacitance. Instead, we would design an optimized chain as discussed in Chapter 8. Although the results of the simplified analysis are limited in scope, they do illustrate the point that the placement of BiCMOS stages is important to the overall system performance. A merged combination of standard CMOS and BiCMOS circuits will provide the best characteristics. When BiCMOS is combined with a bipolar technology such as ECL, we again find that strategic placement of the BiCMOS sections is important.

10.5 Circuit Variations

Many variations on the basic BiCMOS inverter have been published in the literature. In general, the differences among the circuits are either in the method used to remove charge from the bipolar base electrodes, or in the value of the output logic swing V_ℓ. In this section we will briefly examine a few alternate circuits which have been studied in the literature. Many more may be found in the references listed at the end of this chapter.

10.5.1 Pull-Down MOSFETs

Figure 10.14 illustrates an alternate approach to removing base charge from Q1. In this case, MOSFET M1 is used to conduct charge from Q1 to the base of Q2. This scheme allows Q2 to turn on faster than for the circuit analyzed above. The circuit shown in Fig. 10.15 illustrates another

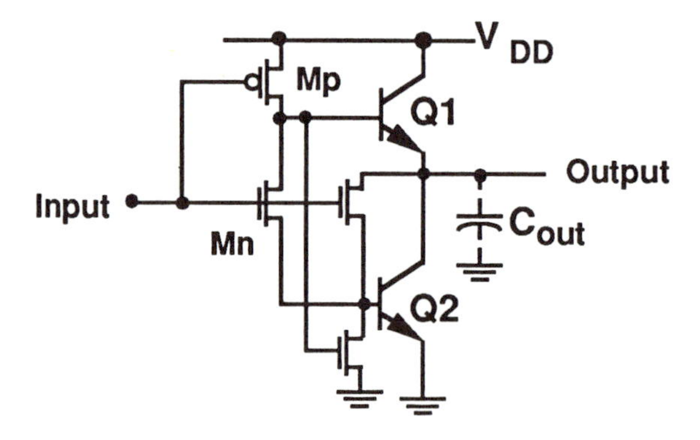

Figure 10.14: Alternate Pull-Down Connections

variation in the placement of the pull-down transistors.

10.5.2 Logic Swing

All of the circuits discussed thus far are characterized by DC output voltage levels of

$$\begin{aligned} V_{OH} &\simeq V_{DD} - V_{BE(on)}, \\ V_{OL} &\simeq V_{BE(on)}. \end{aligned} \tag{10.39}$$

Since $V_{BE(on)} \simeq V_{Tn}, |V_{Tp}|$, the limitations on the output logic swing may cause problem when the circuit is used to drive the inputs to a standard

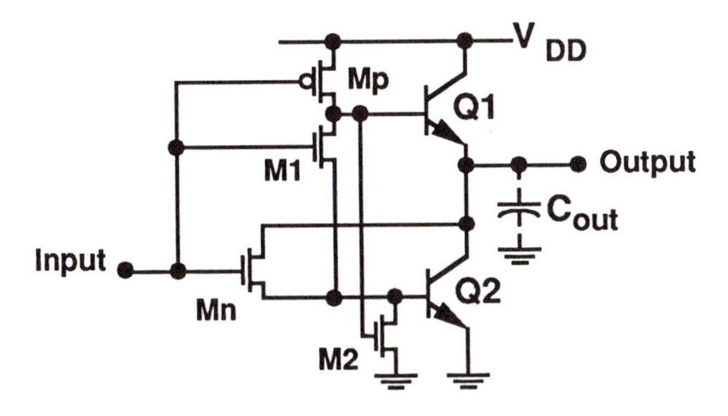

Figure 10.15: Variation of Pull-Down Circuit

CMOS gate. Full-rail output circuits with $V_\ell \simeq V_{DD}$ can be constructed in BiCMOS if necessary. Figure 10.16 illustrates two approaches which have been published to deal with this problem. Note that the circuit in Fig. 10.16(b) uses resistive pull-down devices instead of MOSFETs. More advanced techniques can be found in recent work [4,17].

10.6 Logic Formation

Combinational BiCMOS logic circuits are easily constructed using the generic structure shown previously in Figure 10.1. The logic blocks indicated by F and $\overline{F}$ are simply arrays of pass transistors which provide conduction paths with appropriate inputs. F is constructed using pMOSFETs such that a true state with $F = 1$ connects the logic block to the power supply. Conversely, nMOS array logic is used to construct $\overline{F}$ such that $\overline{F} = 1$ allows the output node to discharge. In this section we will examine some basic logic circuits.

10.6.1 NAND Gate

A 2-input NAND gate is constructed by paralleling 2 pMOS transistors to form F, and series-connecting 2 nMOSFETs for $\overline{F}$ as shown in Figure 10.17. Denoting the inputs by A and B, this circuit gives an output

$$F = \overline{A \cdot B} \tag{10.40}$$

as easily verified by the operation. We may, in principle, build NAND gates with arbitrary numbers of inputs. However, since each input adds

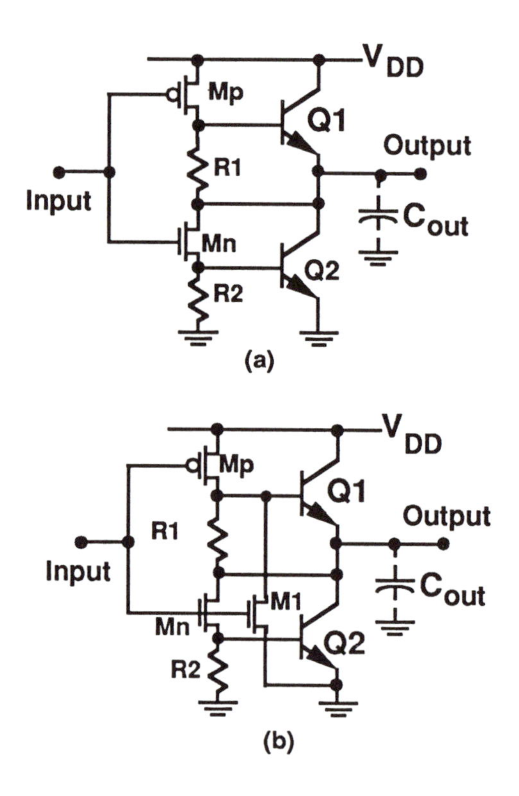

Figure 10.16: BiCMOS With Full-Rail Outputs

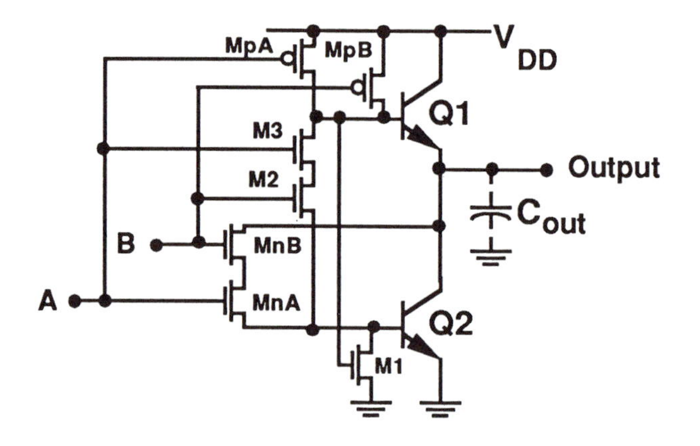

Figure 10.17: NAND2 BiCMOS Gate

a parallel-connected pMOS transistor to F, the node capacitance can get large. This increases the time needed to switch Q1 into a conducting mode and thus increases t_{LH}. Large numbers of inputs are generally avoided for the same reasons as in static CMOS.

10.6.2 NOR Gate

To create a 2-input NOR gate we simply take the dual of the NAND gate discussed above. Logic block F consists of 2 pMOS transistors in series while $\overline{F}$ is implemented using 2 nMOS transistors in parallel. The resulting circuit with an output

$$F = \overline{A + B} \tag{10.41}$$

is drawn in Figure 10.18. This can be extended to an N-input gate. Note,

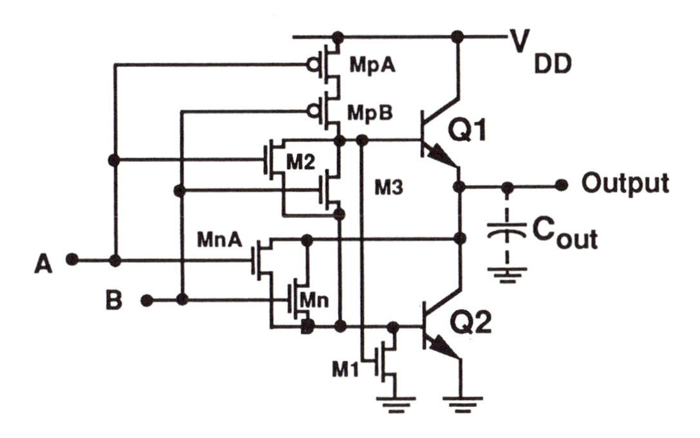

Figure 10.18: NOR2 BiCMOS Gate

however, that the additional parasitic capacitance from the parallel nMOS logic transistors increases the discharge time t_{HL}. Also, an N-input NOR gate requires N series-connected p-channel MOSFETs, which is slower than an equal size chain of nMOS transistors. For this reason, NAND gates tend to be used more than NOR gates in static BiCMOS logic.

10.6.3 AOI and OAI Logic

Generalized logic formation follows directly from standard CMOS theory. The functional block $\overline{F}$ is made using nMOS logic rules listed in Section 4.7; pMOS dual logic is used to build the F segment. Arbitrary logic functions can be implemented by this approach. In particular, both AOI and OAI composite gates are possible; Figure 10.19 provides a simple example for the function $F = \overline{A(B+C)}$. It is seen that the pull-down network connected to

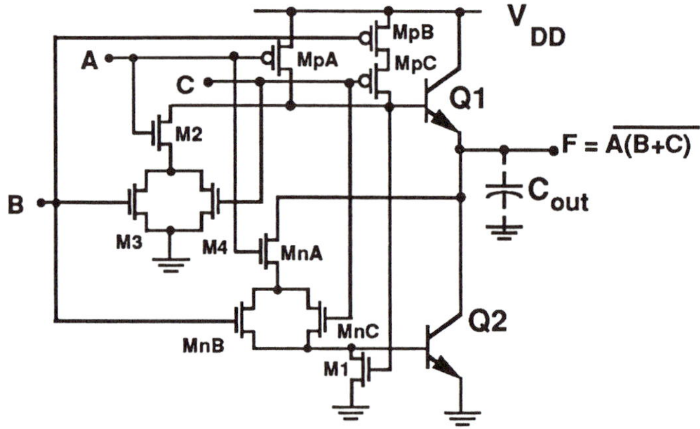

Figure 10.19: AOI Logic Circuit

the base of Q1 is simply a duplicate of the nMOS logic array. This general structure can be used to implement series-parallel logic functions directly.

10.6.4 Pseudo-nMOS Input Circuits

The input capacitance of a conventional BiCMOS gate includes the gate contributions from several MOSFETs. For example, each input to the NAND3 circuit shown in Figure 10.20 drives three transistors. The total gate capacitance seen by A is

$$C_{G,A} = C_{GpA} + C_{Gn4} + C_{G2}, \tag{10.42}$$

with similar expressions for B and C. These transistors must be relatively large to handle base current flow levels to and from the output stage. Since

$$C_{T,A} = C_{ox}[(WL)_{pA} + (WL)_{nA} + (WL)_{n4}], \tag{10.43}$$

the input capacitance will be large compared to that found in standard CMOS.

Reduced-input capacitance BiCMOS gates can be built by employing pseudo-nMOS[1] logic. Consider the modified OR3 circuit in Figure 10.21 where a single constant-bias pMOS transistor MP serves as the load to the nMOS series array [1]. Logic formation is accomplished using a NOR3

[1] See Section 4.11

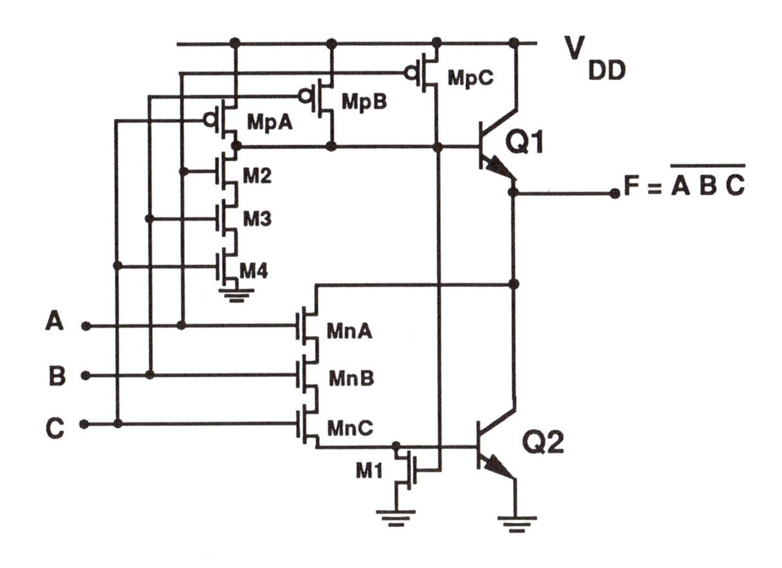

Figure 10.20: Conventional BiCMOS NAND3 Gate

circuit cascaded into an inverter, which in turn drives the bipolar output transistors. The input gate capacitance at A is given by

$$C_A = C_{ox}(WL)_A. \tag{10.44}$$

We assume that the nMOSFETs have the same channel length L in both the original NAND3 and the pseudo-nMOS NOR3 circuit. To reduce the input capacitance we require that ,

$$W_A < W_{pA} + W_{nA} + W_4. \tag{10.45}$$

Since the logic MOSFETs only need to conduct the CMOS inverter currents, this condition can be satisfied while simultaneously increasing the speed.

In the conventional BiCMOS NOR3 gate, the calculation of the aspect ratios is dominated by the switching of the base current. For a single non-saturated MOSFET, the minimum aspect ratio needed to pass the maximum base current I_B is given by

$$\left(\frac{W}{L}\right)_1 = \frac{I_B}{k'[(V_{GS} - V_T)V_{DS} - V_{DS}^2]}. \tag{10.46}$$

With m series-connected logic transistors, each device must be enlarged according to

$$\left(\frac{W}{L}\right)_m = m\left(\frac{W}{L}\right) \tag{10.47}$$

in order to keep the resistance at an acceptable level. Since the base current I_B is large relative to standard MOSFET levels, (W/L) must itself be large to insure fast switching. This, however, increases the input capacitance and loads the driving circuit, thus slowing down the response of the logic cascade. The pseudo-nMOS logic structure buffers the input with the inverter and eliminates this problem.

One complication which arises is that pseudo-nMOS is ratioed logic, so that $(W/L)_A$ must be large enough to insure $\min(V_X) < V_{IL}$, where V_{IL} is the input-low voltage for the inverter made up of Mn1 and Mp1. Analyzing

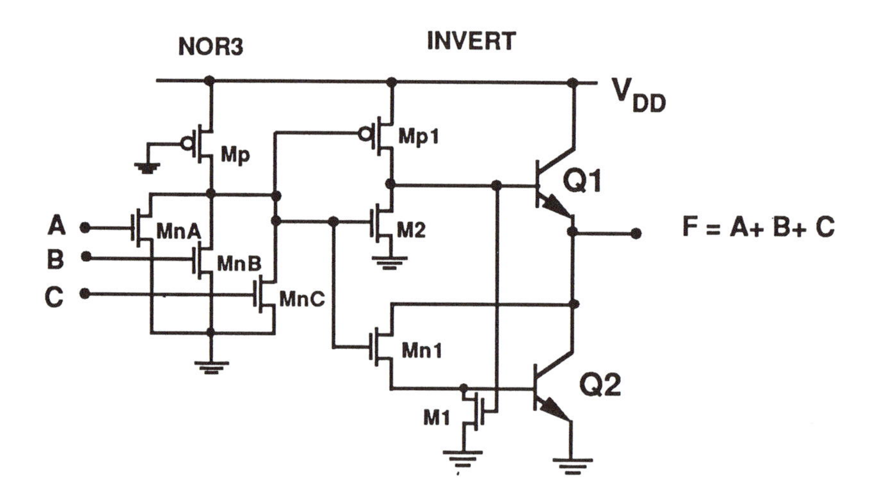

Figure 10.21: Pseudo-nMOS OR3 Gate

the pseudo-nMOS circuit using static logic techniques and ignoring body-bias yields

$$\frac{\beta_{nA}}{\beta_p} \geq \frac{(V_{DD} - |V_{Tp}|)^2}{2(V_{DD} - V_{Tn})V_{IL} - V_{IL}^2} \tag{10.48}$$

as the ratio condition on MOSFET MnA; the equation is also valid for MnB and MnC. The aspect ratio $(W/L)_P$ of Mp is set by the charging time. The critical parameter is the input capacitance C_i of the (Mn1,Mp1) inverter, which is relatively small. Then

$$\left(\frac{W}{L}\right)_P = \frac{C_i}{k'_p \tau_{ch}(V_{DD} - |V_{Tp}|)} \tag{10.49}$$

with τ_{ch} the charging time constant. Once $(W/L)_P$ is found we may write

$$\begin{aligned}\left(\frac{W}{L}\right)_A &= \left(\frac{W}{L}\right)_B = \left(\frac{W}{L}\right)_C \\ &= \frac{k'_p(W/L)_P(V_{DD} - |V_{Tp}|)^2}{k'_n[2(V_{DD} - V_{Tn})V_{IL} - V_{IL}^2]}\end{aligned} \tag{10.50}$$

as the pseudo-nMOS design equations.

10.7 Tri-state Output

Tri-state output drivers are required in many applications. Figure 10.22 shows a BiCMOS implementation which provides the three levels V_{OH}, V_{OL}, and $Z \to \infty$. The output is controlled by the ENABLE signal. When ENABLE=1, the circuit functions as an inverter. This gives the normal

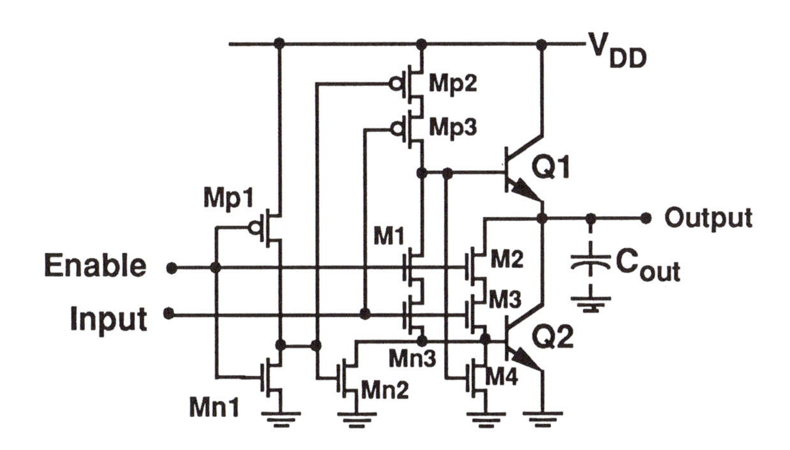

Figure 10.22: Tri-state Inverter

output voltages of V_{OH} and V_{OL} using MOSFETs Mp3 and Mn3 as a standard inverter. Setting ENABLE=0 shuts off M1, M2, and Mp2, and drives Mn2 into a conducting state; both Q1 and Q2 are driven into cutoff, resulting in the desired high-impedance state at the output.

10.8 Level Conversion

When the output of a BiCMOS circuit is used to drive a standard CMOS logic gate, we need to provide a level-conversion stage if the BiCMOS design gives $V_\ell < V_{DD}$. This can be accomplished by modifying the output state as shown in Fig. 10.23. nMOSFET M3 has been added to the output as a pull-down device; this gives

$$V_{OL} \simeq 0 \tag{10.51}$$

for the circuit, which allows direct interfacing into the next stage. Since the output-high voltage is still given by

$$V_{OH} = V_{DD} - V_{BE1}, \tag{10.52}$$

we must design the next stage (CMOS) circuit to have a relatively small value of V_{th}. Alternately, a pMOS pull-up node can be added to the output.

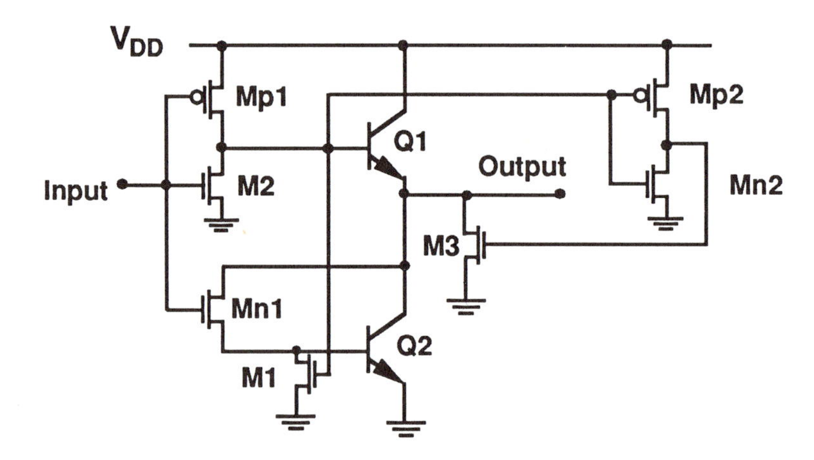

Figure 10.23: BiCMOS to CMOS Interface Circuit

BiCMOS is ideally suited for modified designs in emitter-coupled logic (ECL). It allows one to achieve higher levels of integration and still retain the speed which characterizes ECL switching networks. Interfacing an ECL circuit to a CMOS or a BiCMOS gate is complicated by the fact that the logic swing from an ECL gate is very small with

$$V_{\ell,ECL} \simeq 2V_{BE}. \tag{10.53}$$

Figure 10.24 shows a circuit which accepts an ECL input and provides a CMOS-compatible output. The input voltage is assumed to vary between

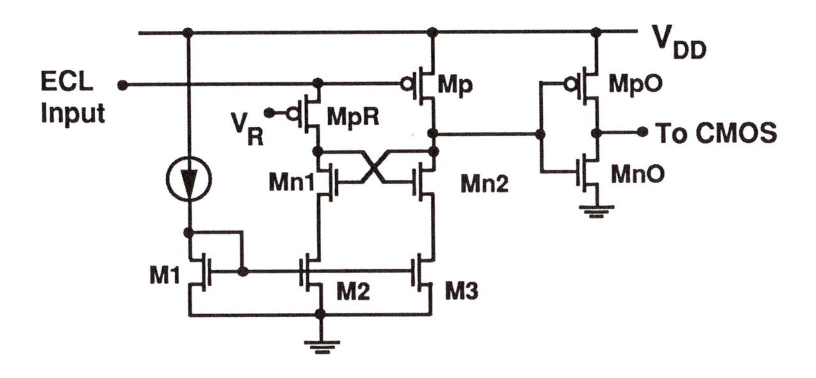

Figure 10.24: ECL-to-CMOS Level Converter

$$\begin{aligned} V_{max} &= V_{DD} - V_{BE}, \\ V_{min} &= V_{DD} - 3V_{BE}, \end{aligned} \tag{10.54}$$

and the reference voltage V_R is set to a value of

$$V_R = V_{DD} - 2V_{BE}. \tag{10.55}$$

Mn1 and Mn2 act as a cross-coupled pair, with a trigger voltage of V_R. The output stage made up of MpO and MnO provides the final buffering. MOSFETs M1, M2, and M3 are used to adjust the bias currents, and is used to optimize the switching speed.

The reverse conversion from CMOS to ECL can be attained using the circuit shown in Fig. 10.28. This consists of an emitter-coupled pair Q1 and Q2 which is driven by a modified CMOS inverter (Mp1 and Mn1). Transistor Q2 is biased so that $V_{B2} = V_{B1} - V_{BE}$ when Mp1 is conducting. If Mn1 is conducting, Q1 is off.

10.9 Summary

Standard CMOS and BiCMOS circuits act together to provide the superior switching performance which characterizes this technology. High integration density is achieved using conventional CMOS techniques, while BiCMOS stages are placed at critical points to attain high-speed switching. In

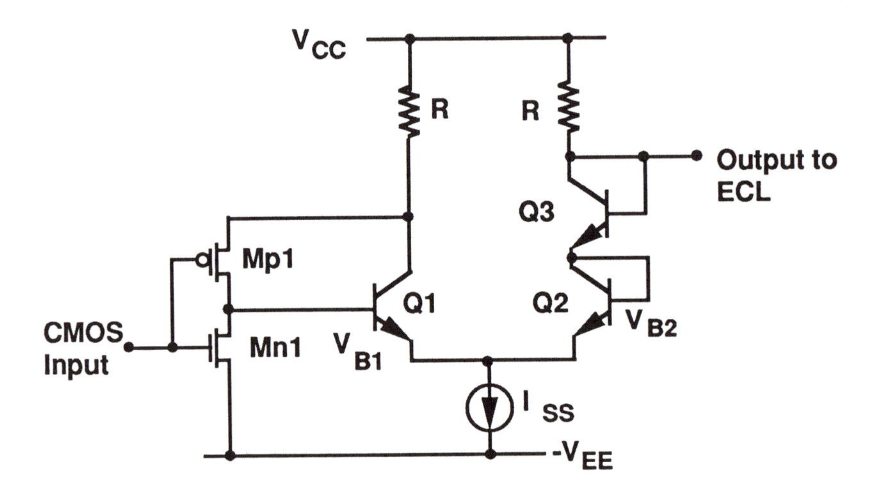

Figure 10.25: CMOS-to-ECL Level Converter

addition, BiCMOS provides a technique to increase the integration density in ECL and other bipolar logic families.

It is important to remember that BiCMOS stages should not be placed indiscriminately in a circuit. The most common usage of BiCMOS stages are for

- Line driver circuits

- Off-chip drivers

or any high-capacitance situation. When applied in this manner, BiCMOS provides a method extending the commercial lifetime of a CMOS process by allowing higher switching speeds.

10.10 References

BiCMOS is an expanding field, with many recent publications associated with it. The short list below provides an introduction into the topics discussed here.

[1] T. Akioka, et. al , "A 6-ns 256-kb BiCMOS TTL SRAM," IEEE J. Solid-State Circuits, vol. 26, no. 3, pp. 439-443, March, 1991.

[2] A.R. Alvarez, **BiMOS Technology and Applications**, Kluwer Academic Publishers, Norwell, MA, 1989.

[3] T. Douseki and Y. Ohmori, "BiCMOS Circuit Technology for a High-Speed RAM," IEEE J. Solid-State Circuits, vol. 2, no. 1, pp. 68-73, Feb., 1988.

[4] S.H.K. Embabi, A. Bellaouar, M.I. Elmasry, and R.A. Hadaway, "New Full-Voltage-Swing BiCMOS Buffers," IEEE J. Solid-State Circuits, vol. 26, no. 2, pp. 150-153, Feb., 1991.

[5] J. Gallia, et al., "High-Performance BiCMOS 100K-Gate Array," IEEE J. Solid-State Circuits, vol. 25, no. 1, pp. 142-149, Feb., 1990.

[6] E. W. Greeneich and K.L. McLaughlin, "Analysis and Characterization of BiCMOS for High-Speed Digital Logic," IEEE J. Solid-State Circuits, vol. 23, no. 2, pp. 558-565, April, 1988.

[7] T. Hanibuchi, et al., "A Bipolar-PMOS Merged Basic Cell for 0.8-μm BiCMOS Sea of Gates," IEEE J. Solid-State Circuits, vol. 26, no. 3, pp. 427-430, March, 1991.

[8] R.A. Kertis, et al., " A 12ns 256K BiCMOS SRAM," ISSCC Dig. Tech. Papers, vol 31, pp. 186-187, Feb., 1988.

[9] G. Kitsukawa, et al., "A 23ns 1-Mb BiCMOS DRAM," IEEE J. Solid-State Circuits, vol. 25, no. 5, pp. 1002-1111, Oct., 1990.

[10] M. Kubo, I. Masuda, K. Miyata, and K. Ogiue, "Perspective on BiCMOS VLSI's," IEEE J. Solid-State Circuits, vol. 23, no. 1, pp. 5-11, February, 1989.

[11] I. Masuda, Y. Nishio, and T. Ikeda, "High-Speed Logic Circuits Combining Bipolar and CMOS Technology," Electronics and Communications in Japan, Part 2, vol. 69, no. 1, pp. 28-36, 1986.

[12] M. Matsui, et al., "An 8-ns 1Mbit ECL BiCMOS SRAM with Double-Latch ECL-to-CMOS-Level Converters," IEEE J. Solid-State Circuits, vol. 24, no. 5, pp. 1226-1232, Oct., 1989.

[13] R.S. Muller and T.I. Kamins, **Device Electronics for Integrated Circuits**, 2nd ed., Wiley, New York, 1986.

[14] Y. Nishio, et al, "0.45ns 7K Hi-BiCMOS Gate Array with Configurable 3-Port 4.6K SRAM," Tech. Dig., IEEE Custom Integrated Circuits Conf., pp. 203-204, 1988,

[15] K. Ogiue, et. al , "13-ns, 500-mW, 64-kbit ECL RAM Using Hi-

BICMOS Technology," IEEE J. Solid-State Circuits, vol. SC-21. no. 5, pp. 681- 684, October, 1986.

[16] G.P. Rossell and R.W. Dutton, "Influence of Device Parameters on the Switching Speed of BiCMOS Buffers," IEEE J. Solid-State Circuits, vol. 24, no.1, pp. 90-99, February, 1989.

[17] H.J. Shin, "Full-Swing BiCMOS Logic Circuits with Complementary Emitter-Follower Driver Configuration," IEEE J. Solid State Circuits, vol 26, no. 4, pp. 578-584, April, 1991.

[18] L. R. Tamura, et al, " A 4-ns BiCMOS Translation-Lookaside Buffer," IEEE J. Solid-State Circuits, vol. 25, no. 5, pp. 1093-1101, October, 1990.

[19] E.S. Yang, **Microelectronic Devices**, McGraw-Hill, New York, 1988.

Index